A C S   S Y M P O S I U M   S E R I E S   **507**

Phenolic Compounds in Food and Their Effects on Health II
Antioxidants and Cancer Prevention

Mou-Tuan Huang, EDITOR
Rutgers, The State University of New Jersey

Chi-Tang Ho, EDITOR
Rutgers, The State University of New Jersey

Chang Y. Lee, EDITOR
Cornell University

Developed from a symposium sponsored
by the Division of Agricultural and Food Chemistry
of the American Chemical Society
at the Fourth Chemical Congress of North America
(202nd National Meeting of the American Chemical Society),
New York, New York,
August 25–30, 1991

D0146862

American Chemical Society, Washington, DC 1992

QP
801
P4
P45
1992
v. 2

Library of Congress Cataloging-in-Publication Data

Phenolic compounds in food and their effects on health II: antioxidants and cancer prevention.

(ACS symposium series, ISSN 0097–6156; 507).

"Developed from a symposium sponsored by the Division of Agricultural and Food Chemistry of the American Chemical Society at the Fourth Chemical Congress of North America (202nd National Meeting of the American Chemical Society), New York, N.Y., August 25–30, 1991."

Includes bibliographical references and indexes.

Contents: [1] Analysis, occurrence, and chemistry—[2] Antioxidants and cancer.

1. Phenols in the body—Congresses. 2. Phenols—Health aspects—Congresses.

I. Ho, Chi-Tang, 1944– . II. Lee, Chang Y., 1935– . III. Huang, Mou-Tuan, 1935– . IV. American Chemical Society. Division of Agricultural and Food Chemistry. V. Chemical Congress of North America (4th: 1991: New York, N.Y.) VI. American Chemical Society. Meeting (202nd: 1991: New York, N.Y.)

QP801.P4P45 1992   612'.0157632   92–23283
                                                      CIP

ISBN 0–8412–2475–7 (v. 1)
ISBN 0–8412–2476–5 (v. 2)

The paper used in this publication meets the minimum requirements of American National Standard for Information Sciences—Permanence of Paper for Printed Library Materials, ANSI Z39.48–1984. ∞

PRINTED IN THE UNITED STATES OF AMERICA

Foreword

THE ACS SYMPOSIUM SERIES was first published in 1974 to provide a mechanism for publishing symposia quickly in book form. The purpose of this series is to publish comprehensive books developed from symposia, which are usually "snapshots in time" of the current research being done on a topic, plus some review material on the topic. For this reason, it is necessary that the papers be published as quickly as possible.

Before a symposium-based book is put under contract, the proposed table of contents is reviewed for appropriateness to the topic and for comprehensiveness of the collection. Some papers are excluded at this point, and others are added to round out the scope of the volume. In addition, a draft of each paper is peer-reviewed prior to final acceptance or rejection. This anonymous review process is supervised by the organizer(s) of the symposium, who become the editor(s) of the book. The authors then revise their papers according the the recommendations of both the reviewers and the editors, prepare camera-ready copy, and submit the final papers to the editors, who check that all necessary revisions have been made.

As a rule, only original research papers and original review papers are included in the volumes. Verbatim reproductions of previously published papers are not accepted.

M. Joan Comstock
Series Editor

Contents

CHEMICAL AND BIOLOGICAL ACTIVITIES
OF PHENOLIC ANTIOXIDANTS

FLAVONOIDS AND CANCER PREVENTION

Contents

Phenolic Compounds in Food and Their Effects on Health I

Analysis, Occurrence, and Chemistry

x

Preface

SCIENTIFIC AND COMMERCIAL INTEREST in phenolic compounds in food has been extremely active in recent years. Apart from the purely academic study of their natural occurrence, distribution, and quality attributes in food, phenolic compounds are becoming increasingly important in applied science. Phenolic compounds aid in the maintenance of food, fresh flavor, taste, color, and prevention of oxidation deterioration. In particular, many phenolic compounds are attracting the attention of food and medical scientists because of their antioxidative, antiinflammatory, antimutagenic, and anticarcinogenic properties and their capacity to modulate some key cellular enzyme functions.

Dietary factors play an important role in human health and in the development of certain diseases, especially cancer. The frequent consumption of fresh fruits and vegetables is associated with a low cancer incidence. It is not known with certainty which components in fruits and vegetables contribute to inhibiting tumor development in humans. Part of the activity may be due to the presence of vitamins A, C, and β-carotene in fruits and vegetables. However, almost all fresh fruits and vegetables contain rich amounts of naturally occurring phenolic compounds such as flavonoids. The amount of ingestion of plant phenolics is proportional to the consumption of fruits and vegetables. In addition, the synthetic phenolic antioxidants, such as butylated hydroxyanisole and butylated hydroxytoluene, are widely used as food antioxidants. Some beverages also contain high amounts of naturally occurring phenolic compounds, for example catechin polyphenols in teas and red wine and chlorogenic acid in coffee. Other phenolic compounds in food include curcumin, a food coloring agent present in turmeric and curry, and carnosol and carnosic acid in rosemary leaves used as a spice.

Humans daily ingest a large amount of phenolic compounds. Do these phenolic compounds contribute many benefits to human health or do they have any adverse effects? In order to probe these questions, the two volumes of *Phenolic Compounds in Food and Their Effects on Health I* and *II* present the recent research data and review lectures by numerous prestigious experts from the international symposium on phenolic compounds in food and health. Contributors from academic institutions, government, and industry were carefully chosen to provide different insights and areas of expertise in these fields.

This subject is presented in two volumes. Volume I consists of occurrence, analytical methodology, and polyphenol complexation in food and the effect of phenolic compounds on the flavor, taste, color, texture, and nutritional quality of food, as well as the utilization of phenolic antioxidants in foods. Volume II includes the source of phenolic compounds, the chemical and biological properties of phenolic compounds in food, and their health effects. Special emphasis is placed on the biological influence of phenolic compounds on the modulation of tumor development in experimental animal models, and possibly in humans.

We are indebted to the contributing authors for their creativity, promptness, and cooperation in the development of this book. We also sincerely appreciate the patience and understanding given to us by our wives, Mary Ho, Ocksoo Kim Lee, and Chiu Hwa Huang. Without their support, this piece of work would not have materialized. We thank Harold Newmark, James Shaw, and Thomas Ferraro for their help in review, suggestions, and editing.

We acknowledge the financial support of the following sponsors: Campbell Soup Company; CPC International, Inc.; Givaudan Corporation; Glaxo, Inc.; Hoffmann–La Roche, Inc.; Johnson & Johnson Consumer Products, Inc.; Kalsec, Inc.; Lipton Foundation; Merck Sharp & Dohme Research Laboratories; Takasago USA; Tea Council of the USA; The Procter & Gamble Company; The Quaker Oats Company; Warner–Lambert Foundation; and the Division of Agricultural and Food Chemistry of the American Chemical Society.

MOU-TUAN HUANG
Rutgers, The State University of New Jersey
Piscataway, NJ 08854

CHI-TANG HO
Rutgers, The State University of New Jersey
New Brunswick, NJ 08903

CHANG Y. LEE
Cornell University
Geneva, NY 14456

June 30, 1992

PERSPECTIVES

Chapter 1

Phenolic Compounds in Food

An Overview

Chi-Tang Ho

Department of Food Science, Cook College, Rutgers, The State University
of New Jersey, New Brunswick, NJ 08903

Phenolic compounds including simple phenols and phenolic acids,
hydroxycinnamic acid derivatives and flavonoids are bioactive
substances occurring widely in food plants. Phenolic compounds are
closely associated with the sensory and nutritional quality of fresh and
processed plant foods. The enzymatic browning reaction of phenolic
compounds, catalyzed by polyphenoloxidase, could cause the
formation of undesirable color and flavor and the loss of nutrient in
fruits and vegetables. Many phenolic compounds in plants are good
sources of natural antioxidants. It is a great interest in recent years that
many phenolic compounds in foods have inhibitory effects on
mutagenesis and carcinogenesis.

The term 'phenolic' or 'polyphenol' can be defined chemically as a substance which
possesses an aromatic ring bearing one or more hydroxy substituents, including
functional derivatives (esters, methyl ethers, glycosides etc.) (*1*). Most phenolics
have two or more hydroxyl groups and are bioactive substances occurring widely in
food plants that are eaten regularly by substantial numbers of people.

Occurrence of Phenolic Compounds

The phenolic compounds which occur commonly in food material may be classified
into three groups, namely, simple phenols and phenolic acids, hydroxycinnamic acid
derivatives and flavonoids.

The Simple Phenols and Phenolic Acids. The simple phenols include
monophenols such as *p*-cresol isolated from several fruits (e.g. raspberry,
blackberry) (*2*), 3-ethylphenol and 3,4-dimethylphenol found to be responsible for
the smoky taste of certain cocoa beans (*3*) and diphenols such as hydroquinone
which is probably the most widespread simple phenol (*4*).
 A typical hydroquinone derivative, sesamol, is found in sesame oil (*4*).
Several derivatives of sesamol, such as sesaminol, found in sesame oil have been
evaluated to have strong antioxidant activity (Osawa, Chapter 10, Vol. II).

0097–6156/92/0507–0002$06.00/0

Vanillin (4-hydroxy-3-methyoxybenzaldehyde) is the most popular flavor. The determination of vanillin in vanilla beans is discussed by Hartman et al. (Chapter 4, Vol. I).

Gallic acid, a triphenol, is present in an esterified form in tea catechins (Balentine, Chapter 8, Vol. I). Gallic acid may occur in plants in soluble form either as quinic acid esters (*5*) or hydrolyzable tannins (Okuda et al., Chapter 12, Vol. II).

The Hydroxycinnamic Acid Derivatives. Hydroxycinnamic acids and their derivatives are almost exclusively derived from *p*-coumaric, caffeic, and ferulic acid, whereas sinapic acid is comparatively rare. Their occurrence in food has recently been reviewed by Herrmann (*6*). Hydroxycinnamic acids usually occur in various conjugated forms, more frequently as esters than glycosides.

The most important member of this group in food material is chlorogenic acid, which is the key substrate for enzymatic browning, particularly in apples and pears (*7*).

A number of chapters in this book discuss the occurrence of hydroxycinnamic acids in foods which include:

Umbelliferous vegetables	Roshdy et al. (Chapter 6, Vol. I)
Citrus fruits	Naim et al. (Chapter 14, Vol. I)
Brassica oilseed	Shahidi (Chapter 10, Vol. I)
Corn flour	Gibson and Strauss (Chapter 20, Vol. I)
Raspberry	Rommel et al. (Chapter 21, Vol. I)
Plums	Fu et al. (Chapter 22, Vol. I)

The Flavonoids. The most important single group of phenolics in food are flavonoids which consist mainly of catechins, proanthocyanins, anthocyanidins and flavons, flavonols and their glycosides.

Although catechins seem to be widely distributed in plants, they are only rich in tea leaves where catechins may constitute up to 30% of dry leaf weight. A number of chapters in Volume II of this book discuss current research on antioxidative and cancer chemopreventive properties of tea and its catechin components. Lunder (Chapter 8, Vol. II) has shown that the antioxidative activity of green tea extract could be related to the content of epigallocatechin. Osawa (Chapter 10, Vol. II) was able to demonstrate that epicatechin gallate and epigallocatechin gallate not only inhibit the free radical chain reaction of cell membrane lipids, but also inhibit mutagenicity and DNA damaging activity. Laboratory studies conducted by Ito et al. (Chapter 19, Vol. II), Conney et al. (Chapter 20, Vol. II), Wang et al. (Chapter 21, Vol. II), Chung et al. (Chapter 22, Vol. II), Laskin et al. (Chapter 23, Vol. II) and Yoshizawa et al. (Chapter 24, Vol. II) are presented. They have shown that tea and tea catechin components can inhibit tumorigenesis and tumor growth in animals.

Proanthocyanidins, or condensed tannins, are polyflavonoid in nature, consisting of chains of flavan-3-ol units. They are widely distributed in food such as apple, grape, strawberry, plum, sorghum and barley (*8*). Proanthocyanidins have relatively high molecular weights and have the ability to complex strongly with carbohydrates and proteins. Comprehensive treatment of polyphenol complexation is given by Haslam et al. in this book (Chapter 2, Vol. I). Hagerman (Chapter 19, Vol. I) and Butler and Rogler (Chapter 23, Vol. I) also present overviews on these topics.

Anthocyanins are almost universal plant colorants and are largely responsible for the brilliant orange, pink, scarlet, red, mauve, violet and blue colors of flower petals and fruits of higher plants (9). Anthocyanins as food colorants have recently been reviewed (10).

Flavones, flavonols and their glycosides also occur widely in the plant kingdom. Their structural variations and distribution have been the subjects of several comprehensive reviews in recent years (11-13). It has been estimated that humans consuming high fruit and vegetable diets ingest up to 1 g of these compounds daily (Leighton et al., Chapter 15, Vol. II). The most common and biologically active dietary flavonol is quercetin. A number of presentations in Volume II of this book discuss current research on the effect of quercetin on mutagenesis and carcinogenesis. Quercetin was found by Verma (Chapter 17, Vol. II) to inhibit both initiation with 7,12-dimethylbenz[a]anthracene (DMBA) and tumor promotion with 12-O-tetradecanoylphorbol-13-acetate (TPA) of mouse skin tumor formation. Starvic et al. (Chapter 16, Vol. II) suggest that quercetin and other polyphenols such as ellagic acid and chlorogenic acid may play a dual protective role in carcinogenesis by reducing bioavailability of carcinogens, and by interfering with their bio-transformation in the liver. By using an experimental model of colon cancer, Deschner (Chapter 18, Vol. II) was able to demonstrate that under conditions of low dietary fat intake, quercetin and rutin have displayed considerable activity in suppressing the hyperproliferation of colonic epithelial cells, thereby reducing focal areas of dysplasia and ultimately colon tumor incidence.

Effect of Phenolic Compounds on Food Quality

Phenolic compounds are closely associated with the sensory and nutritional quality of fresh and processed plant foods (14). The enzymatic browning reaction of phenolic compounds, catalyzed by polyphenoloxidase, is of vital importance to fruit and vegetable processing due to the formation of undesirable color and flavor and the loss of nutrients. For examples, polyphenoloxidase was found to be responsible for the browning of grapes (15), and catechin, polyphenoloxidase and oxygen were reported to be required for the browning of yams (16). The enzymatic browning reaction in fruits often has been considered to be a linear function of the phenolic content and polyphenoloxidase activity. In this book, however, Lee (Chapter 24, Vol. I) show that the rate of browning in fruit products is not a linear function of the total phenolic content and that browning of fruits depends on the concentration and nature of polyphenol compounds that are co-present.

A common approach for the prevention of the browning of food and beverages has been the use of antibrowning agents. The most widespread antibrowning agents used in the food and beverage industries are sulfites. Due to the health risks of sulfiting agents (17), the Food and Drug Administration has banned or limited the use of sulfites in certain foods (18). McEvily et al. (Chapter 25, Vol. I) discuss the isolation and characterization of several 4-substituted resorcinols from figs. These novel 4-substituted resorcinols were shown to be potent polyphenol-oxidase inhibitors.

Oxidative changes of polyphenols during processing are important for the development of color and flavor in certain foods. Browning of polyphenols is a natural process of cocoa fermentation (19). For the manufacture of black tea, the tea leaves are crushed, causing polyphenoloxidase-dependent oxidative polymerization and leading to the formation of theaflavins and thearubigins, the orange and red pigments of black tea. This subject is reviewed in the chapter by Balentine.

Phenolic compounds may contribute directly to desirable and undesirable aromas and tastes of food. Recently, Ha and Lindsay (*20*) reported that highly characterized sheep-mutton aromas in ovine fats were contributed by *p*-cresol, 2-isopropylphenol, 3,4-dimethylphenol, thymol, carvacrol, 3-isopropylphenol and 4-isopropylphenol. They also observed that cresols, especially *m*-cresol, appeared to contribute to beef flavors. Several discussions of flavor characteristics of phenolic compounds are included in this book. Maga (Chapter 13, Vol. I) reviews the roles of hemicellulose, cellulose and lignin thermal degradation on the formation of phenolic compounds which are the major contributors to wood smoke aroma. Fisher (Chapter 9, Vol. I) and Omar (Chapter 12, Vol. I) present overviews on the contribution of phenolic compounds to the aroma and taste of certain spices and plant extracts. Naim et al. (Chapter 14, Vol. I) report that 4-vinylguaiacol is one of the major detrimental off-flavors that form under typical processing and storage of citrus products. Their studies reveal that 4-vinylguaiacol is formed from ferulic acid following the release of ferulic acid from bound forms.

Phenolic Compounds as Natural Antioxidants

Antioxidants are added to fats and oils or foods containing fats to prevent the formation of various off-flavors and other objectionable compounds that result from the oxidation of lipids. BHA and BHT, the most widely used synthetic antioxidants, have unsurpassed efficacy in various food systems besides their high stability, low cost, and other practical advantages. However, their use in food has been falling off due to their suspected action as promoters of carcinogenesis as well as being due to a general rejection of synthetic food additives (*21*).

The most important natural antioxidants which are commercially exploited are tocopherols. Tocopherols have a potent ability to inhibit lipid peroxidation *in vivo* by trapping peroxyl radicals. Their antioxidative mechanism and structure-activity relationship are discussed by Hughes et al. (Chapter 13, Vol. II). Unfortunately, tocopherols are much less effective as food antioxidants. The search and development of other antioxidants of natural origins is highly desirable. Such new antioxidants would also be welcome in combatting carcinogenesis as well as the aging process.

Most natural antioxidants are phenolic in nature. Some of the food materials containing phenolic antioxidants studied and reported herein include:

Chili pepper	Nakatani (Chapter 5, Vol. II)
Ginger	Nakatani (Chapter 5, Vol. II)
Green tea	Lunder (Chapter 8, Vol. II),
	Osawa (Chapter 10, Vol. II)
Pepper	Nakatani (Chapter 5, Vol. II)
Oregano	Nakatani (Chapter 5, Vol. II)
Osbeckia chinensis	Osawa (Chapter 10, Vol. II)
Rice hull	Osawa et al. (Chapter 9, Vol. II)
Rosemary	Nakatani (Chapter 5, Vol. II)
Sesame seeds	Osawa (Chapter 10, Vol. II)
Soybean	Fleury et al. (Chapter 7, Vol. II)
Thyme	Nakatani (Chapter 5, Vol. II)

Phenolic antioxidants not only inhibit the autoxidation of lipids, but sometimes, they also have the ability to retard lipid oxidation by inhibiting lipoxygenase activity. It is believed that the metabolism of arachidonic acid to lipid peroxides and various other oxidative products is significant in carcinogenesis (22). It appears to play an important role in tumor promotion because inhibitors of arachidonic acid metabolism have been observed to inhibit this promotion (23). Four green tea catechin components having strong antioxidant activity also exhibited various degrees of lipoxygenase-inhibitory activities (Ho, C.-T.; Shi, H., unpublished data). (-)-Epigallocatechin gallate, (-)-epicatechin gallate and epigallocatechin displayed IC_{50} values toward soybean 15-lipoxygenase enzyme ranging from 10-21 µM (Table I). (-)-Epicatechin is, on the other hand, relatively inactive. It is also interesting to note that two of the oxidative dimers of tea catechins, the theaflavin monogallate B and theaflavin digallate, which are important polyphenols of black tea, have even stronger lipoxygenase-inhibitory activities than the catechin monomers. The other two structurally closely-related theaflavins, theaflavin and theaflavin monogallate A, have no activity at all (Table I). The detailed chemical structures of these tea polyphenols can be found in the chapter by Balentine (Chapter 8, Vol. I).

Table I. Inhibition of Soybean Lipoxygenase by Tea Polyphenols

Compound	IC_{50} (µM)
(-)-Epicatechin (EC)	140
(-)-Epicatechin gallate (ECG)	18
(-)-Epigallocatechin (EGC)	21
(-)-Epigallocatechin gallate (EGCG)	10
Theaflavin	3604
Theaflavin monogallate A	366
Theaflavin monogallate B	0.62
Theaflavin digallate	0.25

Conclusion

Phenolic compounds are ubiquitous in plant foods, and therefore, a significant quantity is consumed in our daily diet. They are closely associated with the sensory and nutritional quality of fresh and processed plant foods. The antioxidant activities of phenolic compounds have been recognized for decades, and research and development on the use of natural substances or food ingredients containing phenolic antioxidants will continue to be of great interest to the food industry.

Biological activities of phenolic compounds have become well known in recent years. The most important biological activity of phenolic compounds is probably their many observed inhibitory effects on mutagenesis and carcinogenesis. This topic is covered in great depth in the overview chapter by Huang and Ferraro (Chapter 1, Vol. II) and many chapters in Volume II of this book.

Acknowledgements

This publication, New Jersey Agricultural Experiment Station Publication No. D-10205-1-92, has been supported by State Funds. We thank Mrs. Joan Shumsky for her secretarial aid.

Literature Cited

1. Harborne, J. B. In *Methods in Plant Biochemistry, Vol. 1: Plant Phenolics*; Harborne, J. B., Ed.; Academic Press: London, UK, 1989, pp. 1-28.
2. Van Straten, S. *Volatile Compounds in Food*; Central Institute for Nutrition and Food Research: Zeist, The Netherlands, 1977.
3. Guoyt, B.; Gueule, D.; Morcrette, I.; Vincent, J. C. *Coffee, Cocoa, Tea* **1986**, *30*, 113-120.
4. Van Sumere, C. F. In *Methods in Plant Biochemistry, Vol. 1: Plant Phenolics*; Harborne, J. B., Ed.; Academic Press: London, UK, 1989, pp. 29-73.
5. Nishimura, H.; Nonaka, G. I.; Nishioka, I. *Phytochem.* **1984**, *23*, 2621-2623.
6. Herrmann, K. *CRC Crit. Rev. Food Sci. Nutri.* **1989**, *28*, 315-347.
7. Eskin, N. A. M. *Biochemistry of Foods*; Academic Press: San Diego, CA, 1990, pp. 401-432.
8. Haslam, E. *Plant Polyphenols*; Cambridge University Press: Cambridge, UK, 1989.
9. Harborne, J. B. *Comparative Biochemistry of the Flavonoids*; Academic Press: London, 1967.
10. Francis, F. J. *CRC Crit. Rev. Food Sci. Nutri.* **1989**, *28*, 273-314.
11. Harborne, J. B.; Mabry, T. J. *The Flavonoid—Advances in Research*; Chapman and Hall: London, UK, 1982.
12. Harborne, J. B.; Mabry, T. J. *The Flavonoid—Advances in Research*; Chapman and Hall: London, UK, 1988, Vol. 2.
13. Markham, K. R. In *Methods in Plant Biochemistry, Vol. 1: Plant Phenolics*; Harborne, J. B., Ed.; Academic Press: London, UK, 1989, pp. 197-235.
14. Macheix, J. J.; Fleuriet, A.; Billot, J. *Fruit Phenolics*; CRC Press: Boca Raton, FL, 1990.
15. Sapis, J. C.; Macheix, J. J.; Cordonnier, R. E. *J. Agric. Food Chem.* **1983**, *31*, 342-345.
16. Ozo, O. N.; Caygill, J. C. *J. Sci. Food Agric.* **1986**, *37*, 283.
17. Taylor, S. L.; Higley, N. A.; Bush, R. K. *Adv. Food Res.* **1986**, *30*, 1-76.
18. Anonymous. *Food Institute Report* **1990**, *63*, 9.
19. Quesnel, V. C.; Jugmohunsingh, K. *J. Sci. Food Agric.* **1970**, *21*, 537-541.
20. Ha, J. K.; Lindsay, R. C. *J. Food Sci.* **1991**, *56*, 1197-1202.
21. Namiki, M. *CRC Crit. Rev. Food Sci. Nutri.* **1990**, *29*, 273-300.
22. Powles, T. J; Bockman, R. S; Honn, K. V; Ramwell, P. *First International Conference on Prostaglandins and Cancer, Prostaglandins and Related Lipids*; New York: Liss, 1982, Vol. 2.
23. Belman, S; Solomon, J; Segal, A; Block, E; Barany, G, *J. Biochem. Toxicol.* **1989**, *4*, 151-160.

RECEIVED February 11, 1992

Chapter 2

Phenolic Compounds in Food and Cancer Prevention

Mou-Tuan Huang and Thomas Ferraro

Laboratory for Cancer Research, Department of Chemical Biology
and Pharmacognosy, College of Pharmacy, Rutgers, The State University
of New Jersey, Piscataway, NJ 08855-0789

A general overview of the phenolic compounds in food and health is
presented, with emphasis on the actual amounts eaten by humans and
possible effects on cancer. Because of the widespread occurrence of
phenolic compounds in our food, humans ingest a large amount of
phenolic compounds. Most phenolic compounds in food are plant
flavonoids, but others include synthetic antioxidants such as the food
additives butylated hydroxyanisole (BHA) and butylated
hydroxytoluene (BHT), chlorogenic acid in coffee, caffeic acid and
ferulic acid in vegetables and fruits, α-tocopherol and related
compounds in oils from vegetables and grains, the polyphenolic
catechins found in tea and red wine, carnosol in rosemary leaves, and
curcumin in turmeric, curry and mustard. Almost all of these
polyphenolic compounds possess several common biological and
chemical properties: (a) antioxidant activity, (b) the ability to scavenge
active oxygen species, (c) the ability to scavenge electrophiles, (d) the
ability to inhibit nitrosation, (e) the ability to chelate metals, (f) the
potential for autoxidation, producing hydrogen peroxide in the
presence of certain metals, and (g) the capability to modulate certain
cellular enzyme activities. These compounds share some of these
biological and chemical properties with vitamins C and E, and many
have been found, or are likely to be able, to inhibit various steps of
tumor development in experimental animals and probably in humans.
The biological activities and functions of phenolic compounds are
reviewed, especially as they relate to their mechanisms of anticarcinogenicity.

Plant Flavonoids

Occurrence and Human Consumption. Flavonoids are ubiquitous in plants
(1, 2); almost all plant tissues are able to synthesize flavonoids. There is also a wide
variety of types — at least 2000 different naturally occurring flavonoids (3). Among
the most widely found are the flavonols quercetin and rutin (Figure 1), which are

0097–6156/92/0507–0008$07.75/0
© 1992 American Chemical Society

present in tea, coffee, cereal grains, and a variety of fruits and vegetables (*1, 2*). In general, the leaves, flowers, fruit and other living tissues of the plant contain glycosides, woody tissues contain aglycones, and seeds may contain either (*4, 5*).

As a result of their ubiquity in plants, flavonoids are an integral part of the human diet. It is estimated that the average American's daily intake of flavonoids is up to 1 gram (*2*).

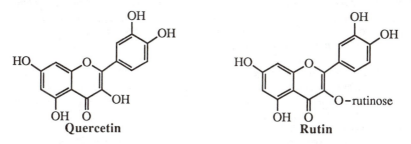

Figure 1. Structures of the flavonoids quercetin and rutin

Some Flavonoids are *In Vitro* Mutagens. Since the Ames *Salmonella typhimurium* short term mutagenesis test system was described in 1975 (*6*), many laboratories have tested the mutagenicity of flavonoids in this test system. Several flavonoids have been found to be mutagens in either the absence or presence of S-9 metabolic activation, most notably quercetin and kaempherol, which are widespread in human dietary vegetables and fruits (*7-12*).

Quercetin has also been found to be genotoxic in a number of *in vitro* short term tests using mammalian cell lines. These include: mutations at the HGPRT locus in V79 hamster fibroblasts (*13*), increased frequency of chromsomal aberrations and sister chromatid exchange in human and Chinese hamster cells (*14*), transformation of Balb/c 3T3 cells, single-strand DNA breaks and mutations at the thymidine kinase locus in L5178Y mouse lymphoma cells (*15*), and induction of chromosomal aberrations in Chinese hamster ovary cells (*16*). Although these studies are limited, being *in vitro*, their results have attracted public attention and concern. Both saliva and intestinal tract bacteria contain glycosidases that catalyze the removal of the sugar moieties from a wide array of flavonoid glycosides, which are not generally found to be mutagenic, yielding their often mutagenic aglycones (*17-20*). Do these widely consumed flavonoids have any *in vivo* toxicity? Do they have any adverse effects in humans?

Several investigators have failed to observe any *in vivo* genetic toxicity of flavonoids (*21, 22*), although increases in the frequency of micronuclei in bone marrow erythrocytes of mice following i.p. injection of quercetin, kaempferol, neohesperidin dihydrochalcone (*23*) or 5,3′,4′-trihydroxy-3,6,7,8-tetramethoxy-flavone (*24*) have been reported. *In vitro* studies have indicated that quercetin induces DNA strand scission in the presence of metal ions (*25, 26*). These results suggest a possible mechanism of mutagenicity of quercetin, kaempferol and other flavonoids found to be mutagenic in *in vitro* assays. Under aerobic conditions in the presence of trace metals, quercetin produces superoxide and other active oxygen species that can cause DNA strand breakage (*26*). Under similar conditions, vitamin C also induces mutations *in vitro* (*27*). In order to accumulate enough hydrogen peroxide in the incubation medium, all of the above *in vitro* mutagenic assays require prolonged incubation. Maruta *et al.* suggested that quercetin is slowly metabolized for activation in their *in vitro* model (*13*). Mammalian cells *in vivo* have sufficient catalase, superoxide dismutase (SOD), and glutathione peroxidase to defend against hydrogen peroxide and superoxide anion radical formation (*28*).

Although addition of quercetin to calf thymus DNA *in vitro* affects the rate of DNA hydrolysis by S_1 nuclease and decreases melting point of DNA after prolonged co-incubation (*29*), quercetin, other flavonoids and their metabolites have not been reported to covalently bind to nucleic acids. Therefore, they may not be genotoxic *in vivo*, even though they are *in vitro* mutagens.

Flavonoids are not carcinogenic in *in vivo* studies. Because several flavonoids are mutagens, they are suspected carcinogens. It is important that compounds that are *in vitro* mutagens and suspected carcinogens be carefully examined in several test systems. There is no evidence that quercetin can form DNA adducts *in vitro* in the absence or presence of S-9 fraction or *in vivo*.

A number of scientists have undertaken systemic *in vivo* studies on flavonoid carcinogenicity. In most of the studies, no significant differences from controls were found in either the number of tumors per animal or the tumor incidence in various tissues after life time adminstration of various levels (0.1 - 10% in the diet) of quercetin or rutin to several strains of mice, rats, and hamsters (Table I). Pamukcu *et al.* reported that 0.1% dietary quercetin induced intestinal and bladder carcinomas in Norwegian rats (*30*) and Ertürk *et al.* found that dietary quercetin significantly increased liver tumors in Sprague-Dawley and Fisher 344 rats (*31, 32*). These latter two results are from the same laboratory and have not been confirmed by other investigors. In addition, quercetin is neither an initiator nor a promoter of mouse skin tumors (Verma, Chapter 17, Volume II).

Anti-carcinogenicity of flavonoids. A variety of dietary plant flavonoids have been found to inhibit tumor development in several experimental animal models and to act through different mechanisms (*43-48*). Plant flavonoids can inhibit several steps of the tumorigenic and/or carcinogenic processes. The hydroxylated flavonoids have been found to: (i) inhibit the metabolic activation of carcinogens by modulation of cytochrome P-450 isozymes, (ii) inactivate ultimate carcinogens, (iii) inhibit generation of active oxygen species and act as scavengers of active oxygen species, (iv) inhibit arachidonic acid metabolism, (v) inhibit protein kinase C and other kinase activity, and (vi) reduce the bioavailibility of carcinogens.

Inhibition of Metabolic Activation. One of the ways flavonoids may inhibit carcinogenesis is by inhibiting the activation of chemical carcinogens to their ultimate carcinogenic metabolites. The varied metabolic capabilities of cytochrome P-450 isozymes, which are present in almost every tissue, play an important role in the metabolic fate, toxicity and carcinogenicity of many drugs and xenobiotics. Many naturally occurring and synthetic flavonoids are able to modulate cytochrome P-450 isozyme activity (*49-54*).

Certain flavonoids which lack hydroxyl groups can induce increased microsomal cytochrome P-450 isozyme activity *in vitro* (*50*) and *in vivo* (*49*). Flavone, methoxyflavonoids and the synthetic flavonoid 7,8-benzoflavone may selectively induce cytochrome P-450 isozymes and/or detoxifying enzymes that accelerate elimination of carcinogens from the body (*53*).

Table I. Experimental carcinogenesis studies of quercetin and rutin

Investigator	Compound Tested	Dietary Level	Species (strain)	Pathology
Wilson *et al.* (*33*)	rutin	1%	rat	normal
Ambrose (*34*)	quercetin	0.25-1%	rat (albino)	normal
Pamukcu *et al.* (*30*)	quercetin	0.1%	rat (Norwegian-derived)	intestinal and bladder tumors
Saito *et al.* (*35*)	quercetin	2%	mouse (ddY)	normal
Hirono *et al.* (*36*)	quercetin rutin	1, 5, 10% 5, 10%	rat (ACI)	normal normal
Hosaka & Hirono (*37*)	quercetin	5%	mouse (A/JJms)	normal
Morino *et al.* (*38*)	quercetin rutin	1, 4, 10% 10%	golden hamster	normal normal
Ertürk *et al.* (*31*)	quercetin	1, 2%	rat (F 344)	hepatomas bile duct tumors
Habs (*39*)	rutin	10-500 mg/kg body weight	rat (SD)	normal
Hirose *et al.* (*40*)	quercetin	5%	rat (F 344)	normal
Takanashi *et al.* (*41*)	quercetin	0.1%	rat (F 344)	normal
Ertürk *et al.* (*32*)	quercetin	1, 2%	rat (F 344)	hepatic tumors biliary adenomas
	quercetin rutin	0.5% 2%	rat (SD)	hepatomas hepatomas
Deschner *et al.* (*42*)	quercetin rutin	2% 4%	mouse (CF-1)	normal normal

Normal.: not different from control group

Flavonoids which possess hydroxyl groups usually inhibit cytochrome P-450 isozyme activity. This has been demonstrated *in vitro* in rat (*52, 54*), rabbit (*51*) and human (*50*) liver microsomes.

Carcinogen Inactivation by Plant Phenolic Flavonoids. Ultimate carcinogens such as the bay-region diol epoxides of polycyclic aromatic hydrocarbons are electrophilic reactants possessing positively charged groups such as carbonium ions that react with electron-rich moieties termed nucleophiles. Cellular macromolecules, including DNA, have many nucleophilic groups capable of being attacked. Plant phenolic flavonoids often contain electron-rich moieties such as phenolic groups which are able to react with ultimate carcinogens, either forming inactive adducts (*55*) or catalyzing the hydrolysis of epoxides to an inactive tetraol form (*55, 56*). A number of flavonoids are very potent antagonists of the mutagenicity *in vitro* (*57, 58*) and carcinogenicity *in vivo* (*45, 59*) of the bay-region diol epoxides of polycyclic aromatic hydrocarbons.

Huang *et al.* investigated the inhibitory effect of a number of flavonoids against the mutagenic activity of the diol epoxide (±)-7β,8α-dihydroxy-9α,10α-epoxy-7,8,9,10-tetrahydrobenzo[*a*]pyrene in *S. typhimurium* TA 100 (*57*). Flavonoids possessing free phenolic groups were the most active, whereas several flavonoids without free hydroxyl groups were weak inhibitors or inactive (Table II). The investigators also found that the antimutagenic flavonoids they tested accelerated the disappearance of the diol epoxide from a cell free solution, suggesting the inhibitory effect on mutagenesis is a result of direct interaction with the diol epoxide.

Several flavonoids, including quercetin and rutin, suppress mutagenesis in *S. typhimurium* strain TA100 NR induced by the direct-acting carcinogen *N*-methyl-*N'*-nitro-*N*-nitrosoguanidine (MNNG) (*60*). These flavonoids were not found to directly interact with MNNG and do not accelerate the inactivation of MNNG. The flavonoids may either inhibit MNNG transport into cells or alter some other cellular processes.

Inhibitory Effects on Active Species of Oxygen by Phenolic Flavonoids. Inflammation and active oxygen species appear to play important roles in tumor promotion (*61*). Quercetin inhibits 12-*O*-tetradecanoyl-phorbol-13-acetate (TPA)-induced mouse skin inflammation (*62*). Flavonoids have been shown to inhibit the production of, and to scavenge, active species of oxygen. Quercetin, for example, inhibits the generation of superoxide anions by neutrophils (*63*). Several flavonoids inhibit myeloperoxidase release while others suppress myeloperoxidase activity — either of which may contribute to the inhibition of cellular production of active oxygen species (*64*). In addition, quercetin, rutin and other flavonoids have been demonstrated to scavenge superoxide anions (*65*).

Inhibition of Arachidonic Acid Metabolism. Certain metabolic products of arachidonic acid from both the lipoxygenase and cyclooxygenase pathways such as hydroxyeicosatetraeonic acids (HETEs), hydroperoxyeicosa-tetraenoic acids (HPETEs), leukotrienes and prostaglandins appear to play important roles in the process of carcinogenesis (*66, 67*). In addition, arachidonic acid metabolism produces lipid peroxides and oxygen free radicals which may also be involved in tumorigenesis (*68*). Certain carcinogens can be activated during the process of arachidonic acid metabolism (*69, 70*).

Table II. Inhibition of the mutagenic activity of
(±)-7β,8α-dihydroxy-9α,10α-epoxy-7,8,9,10-
tetrahydrobenzo[a]pyrene by flavonoids

No.	Flavonoid	Inhibition
1.	Myricetin	++++
2.	Robinetin	++++
3.	Luteolin	++++
4.	Quercetin	++++
5.	7-Methoxyquercetin	++++
6.	Rutin	++++
7.	Quercetrin	++++
8.	Dephinidin chloride	++++
9.	Morin	+++
10.	Myricitrin	+++
11.	Kaempferol	+++
12.	Diosmetin	+++
13.	Fisetin	+++
14.	Apigenin	+++
15.	Dihydroquercetin	++
16.	Naringenin	++
17.	Robinin	++
18.	D-Catechin	++
19.	Genistein	++
20.	Kaempferide	±
21.	Chyrsin	±
22.	Pentamethoxyquercetin	±
23.	Tangeretin	±
24.	Nobiletin	±
25.	7,8-Benzoflavone	±
26.	5,6-Benzoflavone	±
27.	Flavone	±

Each compound was tested at several concentrations for its ability to inhibit the mutagenic activity of 0.05 nmol of (±)-7β,8α-dihydroxy-9α,10α-epoxy-7,8,9,10-tetrahydrobenzo[a]pyrene in *S. typhimurium* TA 100. The amount of flavonoid required to inhibit the mutagenicity by 50% (ID_{50}) is noted. Adapted from ref. 57.

++++: ID_{50} = 2-5 nmol
+++: ID_{50} = 10-40 nmol
++: ID_{50} = 50-100 nmol
±: ID_{50} = >100 nmol or no inhibition

Several compounds that inhibit arachidonic acid metabolism can inhibit carcinogenesis process (*66*) and plant phenolic flavonoids can modify arachidonic acid metabolism (*43, 71*). Kato *et al.* reported that quercetin inhibits epidermal lipoxygenase activity and TPA-induced ornithine decarboxylase activity and tumor promotion in mouse epidermis, but not DNA synthesis (*43*). Thus, inhibition of arachidonic acid metabolism by flavonoids may be one of the mechanisms by which flavonoids inhibit carcinogenesis.

Inhibition of Protein Kinase C Activity. Protein kinase C is a receptor for TPA and is believed to be involved in the regulation of cellular proliferation (*72, 73*). Activated protein kinase C has also been reported to stimulate human neutrophil NADPH-oxidase *in vitro* and increase release of superoxide anion radical (*74*).

Some plant phenolic flavonoids modulate protein kinase C activity *in vitro* and *in vivo*. Several phenolic flavonoids — including quercetin — inhibit the activity of rat brain protein kinase C mediated by TPA (*75, 76*), teleocidin (*77*) and diacylglycerol (*76*). Quercetin inhibits *in vivo* tumor promotion by TPA (*43*) and teleocidin (*78*) in the two-stage skin tumorigenesis model. Although the mechanism of inhibition is unknown, it is possible that this inhibitory effect, and that of other flavonoids, may be mediated through inhibition of protein kinase C (*79*). In addition, the ubiquitous distribution of protein kinase C in mammalian cells suggests that plant flavonoids could exert effects on biochemical reactions and cellular proliferation in many cell types through inhibition of the enzyme.

Reduction of Carcinogen Bioavailabity. Certain plant flavonoids, including quercetin, are capable of interaction with carcinogens in the gastrointestinal tract, thereby reducing their bioavailability (Stavric, *et al.*, Chapter 16, Volume II).

Epidemiological studies. No epidemiological data on the effects of plant flavonoid intake in relation to human cancer incidence are available. Almost all fresh fruits and vegetables contain large amounts of naturally occurring flavonoids so intake of plant flavonoids is proportional to consumption of fruits and vegetables. The frequent consumption of vegetables and fruits is associated with low cancer incidence (*80*), and while the possibility cannot be excluded that the association in these studies is mainly due to the presence of vitamins A, C and β-carotene, the intake of plant flavonoids may contribute to the lower cancer incidence.

Synthetic Phenolic Antioxidants—Butylated Hydroxyanisole and Butylated Hydroxytoluene

Butylated hydroxyanisole (BHA, a mixture of 2- and 3-*tert*-butyl-4-methoxyphenol) and butylated hydroxytoluene (BHT) are synthetic phenolic compounds (Figure 2) that are widely used as antioxidant food additives, mainly in oils and food-coating materials, in the United States, Canada and the European Common Market as well as other countries.

Carcinogenicity of BHA. In 1983 Ito *et al.* first reported that 2% BHA administered in the diet to male and female F344 rats for 104 weeks induced squamous cell carcinoma of the forestomach (*81*). BHA became a suspected carcinogen, but further studies suggested that the carcinogenicity of BHA occurred only in the forestomach of mice, rats, and hamsters. BHA is a not carcinogen in the dog in several studies (Table III) (*82-84*). In addition, BHA-induced forestomach lesions in rats are reversible. Feeding rats 2% BHA in the diet for 15 months induced forestomach lesions which completely regressed within 7 months of the removal of BHA from the diet (*85*). BHT has been shown to promote tumors in the lungs of mice that were initiated with urethan, 3-methylcholanthrene, benzo[*a*]pyrene (BP) and *N*-nitrosodimethylamine (*86*).

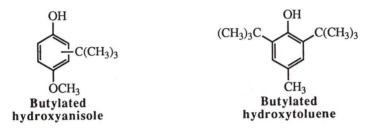

Butylated hydroxyanisole   **Butylated hydroxytoluene**

Figure 2. Structures of the synthetic antioxidants BHA and BHT

Inhibition of Tumorigenesis by BHA and BHT.   Many experiments have indicated BHA and BHT have an inhibitory effect on tumorigenesis in several animal models. Wattenberg has been at the forefront of these studies. He has reported that administration of 0.2 - 1% BHA or 0.5% BHT in the diet inhibited polycyclic aromatic hydrocarbon-induced forestomach tumorigenesis in A/HeJ and Ha/ICR mice and mammary tumorigenesis in SD rats (*92*). Wattenberg also noted that BHA or BHT can reduce 7,12-dimethylbenz[*a*]anthracene (DMBA) toxicity. Another study found that administration of 1% BHA in a diet containing any of several polycyclic aromatic hydrocarbons inhibits carcinogen-induced pulmonary adenoma formation in A/HeJ mice (*93*). Dietary BHA also protected against pulmonary neoplasia resulting from acute exposure to DMBA, BP, urethan, or uracil mustard (*93*).

Other investigators have also found an inhibitory effect of BHA and BHT on tumorigenesis. Oral administration of either compound at 0.25 or 0.5% in the diet, before and one week after an intragastric dose of DMBA, or BHA in the diet administered after the DMBA dose until the end of the experiment, significantly reduced the incidence and delayed the appearance of DMBA-induced mammary gland tumors in Sprague-Dawley rats (*94*). Feeding F344 rats BHA or BHT, 0.1 or 0.6% in diet, one week before, during, and one week after aflatoxin B_1 administration significantly reduced both the average number of tumors per rat and the tumor incidence induced in the liver (*95*).

Inhibition of chemically induced carcinogenesis by BHA or BHT appears, however, to be sex, organ and carcinogen specific. For example, Chung *et al.* have reported that administration of 0.5% BHA in the diet one week before and during the ten weeks of *N*-nitrosodimethylamine or *N*-nitrosopyrrolidine administration in the drinking water significantly inhibits *N*-nitrosodimethylamine-induced lung adenomas, but significantly stimulates *N*-nitrosopyrrolidine-induced lung tumors (*96*). Feeding 0.05 or 0.5% BHT to male or female C3H mice for 10 months markedly increased the incidence of spontaneous liver tumors in male mice, but did not affect the incidence of liver tumors in female mice; the incidence of spontaneous lung tumors was not affected in either sex (*97*). In addition, administration of 0.6% BHT in the diet to male F344 rats given concurrently with the carcinogen

Table III. Test of carcinogenicity of BHA in experimental animals

Investigator	Species	% BHA in the diet	Duration	Pathology
Wilder & Kraybill (*87*)	rat	0.003 - 6	22 mo.	normal
Wilder & Kraybill (*87*)	rat	0.003 - 0.12	21 mo.	normal
Graham, *et al.* (*88*)	rat	0.00135, 0.00675	1 yr.	normal
Wilder, *et al.* (*82*)	Cocker Spaniel dog	0.0176, 0.176 0.88	15 mo.	normal liver damage
Hodge, *et al.* (*89*)	beagle dog	0.001 - 0.3	1 yr.	normal
Allen, *et al.* (*90*)	rhesus monkey	100 mg/kg body wt. BHA/BHT (1:1)	2 years	normal
Ito (*81*)	F344 rat	2	2 years	forestomach carcinoma
Iverson, *et al.* (*91*)	cynomolgus monkey	125, 500 mg/kg body weight	85 days	liver weight increase
Ikeda, *et al.* (*83*)	beagle dog	1.0, 1.3	6 mo.	liver weight increase
Tobe et al. (*84*)	beagle dog	0.25 - 1.0	6 mo.	liver weight increase

N-2-fluorenylacetamide markedly inhibited the carcinogen-induced incidence of liver neoplasia, but resulted in the production of bladder papillomas; BHT or N-2-fluorenylacetamide alone did not induce bladder papillomas (*98*).

BHA and BHT are capable of inhibiting tumor development in various organs of rodents induced by more than three dozen different carcinogens; this review does not include the detailed results of each investigator. The carcinogenic and chemopreventive properties of BHA and BHT have been extensively reviewed by Ito (*99*), Hocman (*100*), Kahl (*101*), and Wattenberg (*46*). The actual effect of BHA and BHT on human cancer risk is, however, unknown. Based on the amount of BHA and BHT in human diets (*102*), it can be estimated that the amounts that have been used in chemoprevention studies in rodents are about a thousand times higher than human intake of BHA and/or BHT. This indicates that further study the effect of very low doses of BHA and BHT on chemically induced tumorigenesis in various organs of rodent is needed.

Possible Mechanisms of Inhibition of Carcinogenesis by BHA and BHT. Wattenberg has suggested that inhibitors of carcinogenesis can be divided into three categories: (i) inhibitors that prevent the formation of carcinogens from precursor substances, (ii) inhibitors that prevent carcinogenic compounds from reaching or reacting with critical target sites in the tissues ("blocking agents"), (iii) inhibitors that act by suppressing the expression of neoplasia in cells previously exposed to a dose of carcinogen that can cause cancer ("suppressing agents") (*44, 46*).

BHA and BHT have been shown to inhibit the hepatotoxicity that occurs after feeding sodium nitrite and dimethylamine in the rat (*103*) suggesting that BHA and BHT are capable of preventing the formation of carcinogens from precursor substances.

Active oxygen species and oxygen free radicals can play important roles in the process of carcinogenesis (*61*). As antioxidants, BHA and BHT can act as scavengers of oxygen free radicals thus inhibiting the carcinogenic process. For example, BHA and BHT have been shown to inhibit TPA- and benzoyl peroxide-induced tumor promotion in mouse skin (*104*). BHA and BHT can also inhibit the binding of the carcinogenic metabolites to cellular DNA *in vivo* (*105-107*).

Induction of detoxification enzymes is thought to play an important role in the anti-carcinogenic activity of BHA and BHT. BHA induces glutathione S-transferase, UDP-glucuronyltransferase and epoxide hydrolase activity in liver, forestomach and other organs (*108, 109*). Quinone reductase has been shown to be induced by BHA *in vitro* and *in vivo* (*110, 111* and Prochaska and Talalay, Chapter 11, Volume II). BHT induced the activity of epoxide hydratase in lung and liver (*112*) and BHA induced cytochrome P-450 activity in liver (*113*). Induction of metabolic activation and detoxification enzymes is species, strain, sex, tissue and inducer specific. This could explain the specificity of carcinogenesis inhibition by BHA and BHT.

Oral Administration of BHT Prolongs the Life Span of Mice. Clapp *et al* have reported that administration of 0.75% BHT in diet to BALB/c mice beginning at 11 weeks of age for the rest of life increase the survival time from 684 days to 890 days in males, and 701 days to 875 days in females (*114*).

The Safety of BHA and BHT in Food. Both BHA and BHT are generally thought to be safe additives to food products. The amount of BHA and BHT that is used in food products ranges from 10 to 200 ppm (0.001-0.02%), as a percent of the total diet it is much less; the estimated intake of BHA in Canada is 7 mg per person per day (*102*). Although the administration of 2% BHA in the diet to rats, hamsters and mice for long time periods induced forestomach carcinoma, this carcinogenic activity is strictly limited to the forestomach. BHA did not cause epithelial cell proliferation in the stomach of dogs, guinea pigs or monkeys which do not have forestomachs, and it is not carcinogenic in rat esophagus or oral cavity (Table III). It is neither a tumor initiator nor promoter in mouse skin (*115-117*). It is unlikely, therefore, that BHA in the amounts used in food is carcinogenic in the gastric epithelium of humans.

The Joint FAO/WHO Expert Committee on Food's recommended acceptable daily intake for either BHA or BHT or the sum of both is 0.5 mg/kg body weight (*118*). Since that amount is on the order of 1/10,000 of the dose used in animal studies that found induction of carcinogenesis, the levels of BHA and BHT in human food are considered to be safe.

Chlorogenic Acid, Caffeic Acid and Ferulic Acid

Occurrence and Consumption. Chlorogenic acid, caffeic acid and ferulic acid are hydroxycinnamic acids (Figure 3) found in many fruits and vegetables in their free state and in a variety of derivatives (*1, 119*). Chlorogenic acid and its isomers are 4% by weight of coffee beans (*119*) and an average cup of coffee contains 0.19 g of these compounds (*120*). They are also present in high amounts in prunes (0.9%), blueberries (0.2%), apples (0.1%), pears (0.2%), sweet potato peels (1%), and grapes (0.2%) (*119*). Oats and other cereals and soybeans (*121*) are also sources of chlorogenic acid, caffeic acid, and ferulic acid.

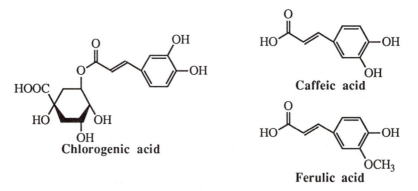

Figure 3. Structures of three common hydroxycinnamic acid derivatives

Anticarcinogenicity of Chlorogenic Acid, Caffeic Acid and Ferulic Acid.

Animal studies. Phenolic compounds can block the nitrosation of amines by reducing nitrite to nitric oxide or by forming C-nitroso compounds. Caffeic acid and ferulic acid react very rapidly with nitrite *in vitro* and effectively block hepatotoxicity and *N*-nitrosodimethylamine formation *in vivo* in rats treated with aminopyrine and sodium nitrite (*122*). Chlorogenic acid also has been reported to block *in vitro* and *in vivo* nitrosation of proline by nitrite (*123*) and to inhibit the *in vitro* formation of nitrosamines from amine and nitrite (*124*).

In addition to blocking the ultimate carcinogenic metabolites of polycyclic aromatic hydrocarbons which attack cellular DNA, caffeic acid, ferulic acid and chlorogenic acid have been shown to inhibit the mutagenicity and to accelerate the disappearance from aqueous solution of bay-region diol epoxides of polycyclic aromatic hydrocarbons which are the ultimate carcinogenic metabolites of polycyclic aromatic hydrocarbons (*56*).

These compounds have also been found to inhibit tumorigenicity in various animal models. Wattenberg *et al.* demonstrated that caffeic and ferulic acid inhibit BP-induced neoplasia of the forestomach in mice (*125*). Dietary chlorogenic acid (0.025%) inhibits methylazoxymethanol-induced colonic tumors in Syrian golden hamsters (*126*) and liver tumors in rats concurrently fed aminopyrene and sodium nitrate (*127*). Topical application of chlorogenic acid, caffeic acid and ferulic acid inhibits TPA-induced inflammation, ornithine decarboxylase activity, DNA synthesis and tumor promotion in CD-1 mouse epidermis (Huang *et al.*, Chapter 26, Volume II).

Epidemological studies. No epidemiological studies relating human consumption of chlorogenic acid, caffeic acid and ferulic acid to cancer incidence have been reported. It is of interest that some epidemiological studies suggest that heavy coffee drinkers may have a decreased risk of colon cancer (*128, 129*). Since chlorogenic acid is a major constituent of coffee, and caffeic acid is readily formed from chlorogenic acid by hydrolysis, it is possible that these compounds contribute to the reported protective effects of coffee.

α-Tocopherol (Vitamin E)

Occurrence. Tocopherol is occurs in many forms, including the dextrorotatory isomer, the racemic mixture, and their acetate and acid succinate esters. The free alcohol forms are easily oxidized, while the esters are more stable. Vitamin E is the α-tocopherol form (Figure 4). Several tocopherol analogs, such as β-, δ-, γ-tocopherols, are used as food preservatives and also present in nature. These analogs have very low vitamin E activity. Tocopherols are widely distributed in plant oils, green vegetables, whole grains, egg yolks, meats, and germ oil (*130*).

Anticarcinogenicity of α-Tocopherol.

In tissue cell culture. Tocopherols have been demonstrated *in vitro* to affect cell growth and inhibit both radiation- and chemical-induced cell tranformation. Addition of 7 μM α-tocopherol succinate to C3H/10T1/2 cells in culture before exposure to x-rays, BP or tryptophane pyrolsate (a carcinogen found in broiled food) markedly inhibited cell transformation (*131*). Various forms of the fat-soluble tocopherols, vitamin E and the esterified form of the vitamin have been reported to inhibit growth in several types of cells *in vitro*, including retrovirus-transformed tumor cells (*132*), human neuroblastoma (*133*), murine neuroblastoma (*134*), rat neuroblastoma and glioma cells (*135*), and murine melanoma (*136*).

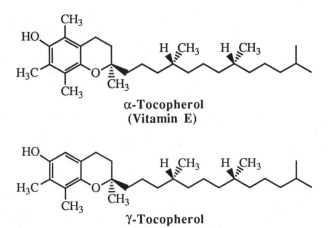

α-Tocopherol
(Vitamin E)

γ-Tocopherol

Figure 4. Structures of α-and γ-tocopherol

Animal Studies. Many animal studies indicate that α-tocopherol inhibits chemically induced carcinogenesis *in vivo*. Topical treatment with vitamin E has been shown to be effective (*137*). Tumorigenesis induced in mice initiated with DMBA and then promoted with croton oil was found to be inhibited by D,L-α-tocopherol acetate administered concomitantly with the croton oil (*138*). Studies by Slaga and Bracken showed that topical application of α -tocopherol inhibits DMBA-induced skin tumorigenesis in mice (*139*). Shklar and colleagues have demonstrated that α-tocopherol inhibits oral mucosal carcinogenesis and induces regression of established epidermoid carcinomas in the buccal pouch of Syrian hamsters treated with DMBA (*140-142*).

α-Tocopherol also effectively inhibits ultraviolet light-induced skin carcinogenesis. Vitamin E administered topically to the backs of C3H/HeN mice inhibits skin carcinogenesis and immunosuppression induced by ultraviolet light (*143*). Topical application of α-tocopherol to hairless mice before and after exposure to ultraviolet light inhibited free radical production and polyamine synthesis in mouse skin (*144*).

Some studies, however, have indicated that high levels of α-tocopherol can enhance tumorigenenesis in the intestines of rodents treated with 1,2-dimethylhydrazine (*145, 146*). In addition, many studies have found no effect of α-tocopherol on tumorigenesis (*147, 148*).

Epidemological Studies. Epidemiological and animal studies have suggested that α-tocopherol decreases the incidence of certain types of tumors (*149, 150*). Higher than average intake of vitamin E has been associated with lower risk of cervical cancer (*151*). Several epidemological studies have indicated that individuals with higher levels of serum vitamin E have lower cancer incidence rates (*150, 152-154*), especially cervical cancer (*154, 155*). Another laboratory, however, has reported no significant differences in the serum levels of vitamin E in patients with cervical cancer vs. controls (*156*). Others suggest there is no vitamin E link with cancer (*157*) or that lower vitamin E levels are a result of, rather than a causal factor for, cancer (*158*).

Possible Mechanism of Anticarcinogenicity. Free radical- and active oxygen species-mediated damage have been implicated in cellular genetic information and cellular proliferation changes that occur during the initiation and promotion stages of carcinogenesis (*61*). α-Tocopherol is a highly efficient antioxidant and is thought to alter tumor development by quenching free radicals, acting as a blocking agent of nitrosating species and scavenging active oxygen species (Mergens, Chapter 27, Volume II). Vitamin E effectively serves as lipid-soluble chain-breaking antioxidant. There is evidence that vitamin E is effectively absorbed, distributed and retained in cellular membrane sites.

Tea Polyphenols

Occurrence and Human Consumption. The tea that is used as a beverage by humans is prepared from the leaves of *Camellia sinensis*, which is further classified into two varieties, *sinensis* and *assamica*. The variety *sinensis* is described in detail by Ballentine, Chapter 8, Volume I. Phenolic catechins, including the flavanols (-)-epigallocatechin gallate, (-)-epicatechin gallate, (-)-epigallocatechin and (-)-epicatechin (Figure 5), are major components of green tea. Fresh leaves of the *assamica* variety contain 9-13% (-)-epigallocatechin gallate, 3-6% (-)epicatechin gallate, 3-6% (-)-epigallocatechin and 1-2% (-)-epicatechin by dry weight as well as other flavonoids and their glycosides (*159*). Thearubigins and theaflavins are found in black tea. Thearubigins is a collective name for large, unidentified, highly colored phenolic catechin oxidation products that are heterogeneous in molecular structure (*159* and Balentine, Chapter 8, Volume I) (Figure 5). Theaflavins and their gallates are of major significance in determining the quality and flavor of tea. Black tea as consumed by humans contains about 36% thearubigins, 3% theaflavins, 5% (-)-epigallocatecin gallate and 1% gallic acid by dry weight. Green tea is consumed mostly in China, Japan, North Africa and the Middle East; tea drinkers in the United States, Canada, Europe and the remainder of the world prefer black tea over green.

Due to the large amounts of these phenolic compounds in tea, they are ingested in significant amounts by humans. Heavy drinkers of green tea in Japan may consume 1 gram of (-)-epigallocatechin gallate per day per person (*160*). Average tea consumption is 7 to 9 g/day/person in Qatar, Ireland, the United Kingdon and Iraq, 4.5 to 7 g/day/person in Hong Kong, New Zealand and Turkey, and about 1 g/day/person in the United States and China (*159*). Since polyphenolic catechins are 20 to 30% of the dry solids in tea extract, total polyphenolic catechin ingestion in humans is in the range of 0.3 to 1 g/day/person.

Anticarcinogenic Activity of Tea Phenols. Recent studies in our laboratory and others have found that green tea and polyphenol compounds isolated from green tea have anticarcinogenic activity in many animal model systems. Oral administration of green tea infusion or (-)-epigallocatechin gallate to rodents significantly inhibits ultraviolet light-induced tumor formation and tumors induced by several carcinogens and tumor promoters in a number of tissues (Conney, *et al.*, Wang *et al.*, Chung, *et al.*, Laskin *et al.* and Yoshizawa *et al.*, Chapters 20-24, Volume II) (Table IV). Oral administration of 0.01 or 0.1% green tea polyphenols in the drinking water has been reported to inhibit the average number of colonic tumors per rat by 60 or 53%, respectively, and tumor incidence by 51% or 38%, respectively, in male Fisher rats previously initiated with azoxymethanol (*161*). Harada, *et al.* reported that administration of green tea polyphenols in the diet (500 mg/kg/day) inhibits the incidence of tumors in the pancreas of golden hamsters promoted by a DL-ethionine, L-methionine, protein deficient diet by 70%. The incidence of severe atypical epithelial hyperplasia was also less frequently observed in the green tea polyphenol treated hamsters (52%) than in the control group (80%). Based on these results the authors suggest that the anti-promotion effect of green tea polyphenols might involve the inhibition of the development of atypical proliferation from hyperplastic pancreatic ductules (162).

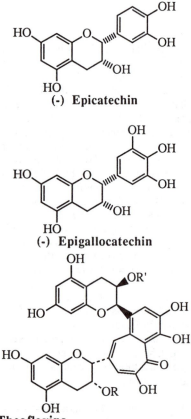

(-) Epicatechin

(-) Epicatechin-3-gallate

(-) Epigallocatechin

(-) Epigallocatechin-3-gallate

Theaflavins

Thearubigins (possible structures)

Theaflavin: R=R'=H
Theaflavin-3-gallate: R=galloyl R'=H
Theaflavin-3'-gallate: R=H R'=galloyl
Theaflavin-3,3'-digallate: R=R'=galloyl

Thearubigin: R=R'=H
Thearubigin-3-gallate: R=galloyl R'=H
Thearubigin-3'-gallate: R=H R'=galloyl
Thearubigin-3,3'-digallate: R=R'=galloyl

Figure 5. Structures of several components of tea

Oral administration of catechins (green tea polyphenols) at 0.1 or 1% in the diet to male F344 rats has been shown to inhibit the incidence of small intestinal tumors formed in both the initiation and promotion stages in a multi-carcinogen,

**Table IV. Some Studies on the Inhibitory Effect of
Green Tea and Its Constituents on Tumorigenesis In Animals**

Organ	Species	Carcinogen treatment	Investigator (reference)
Gastro-intestinal system			
Esophagus	rats	N-Nitrosomethylbenzylamine + sodium nitrite	Xu *et al.* (*164*)
Esophagus	rats	N-Nitrosomethylbenzylamine	Han *et al.* (*165*)
Forestomach	mice	Nitrosamine (Initiation & post-initiation)	Wang *et al.* (*166*)
Duodenum	mice	N-ethyl-N'-nitro-N-nitrosoguanidine (post-initiation)	Fujita *et al.* (*160*)
Small intestine	rats	Multi-carcinogen (Initiation & post-initiation)	Ito *et al.* (Chapter 19, Volume II)
Colon	rats	Azoxymethanol (post-initiation)	Yamane *et al.* (*161*)
Lung	mice	NNK (Initiation & post-initiation)	Wang *et al.* (*166*) Chung *et al.* (Chapter 22, Volume II)
	mice	Nitrosamine	Wang *et al.* (*167*)
Liver	mice	Spontaneous	Muto *et al.* (see Yoshizawa *et al.*, Chapter 24, Volume II)
Mammary gland	rats	Multi-carcinogen	Ito *et al.* (*163*)
Pancreas	golden hamster	BOP (post-initiation)	Harada *et al.* (*162*)
Skin	SKH-1 mice	UV-B light (initiation & post-initiation)	Wang *et al.* (*166, 168*)
	CD-1 mice	DMBA-TPA BP-TPA	Huang *et al.* (*169*)
	BALB/c mice	3-MC (initiation & post-initiation)	Wang *et al.* (*170*)
	CD-1 mice	DMBA-teleocidin (post-initiation)	Yoshizawa *et al.* (*171*)

multi-organ model (Ito, *et al.*, Chapter 19, Volume II). Little or no effect of dietary green tea polyphenols on tumor development in other organs was found. Hirose *et al.* have reported that administration of 1% catechins (green tea polyphenols) significantly reduced the mammary tumor size in virgin female SD rats previously initiated with DMBA (*163*).

Mechanisms of Anticarcinogenicity. Studies on the mechanism of the anticarcinogenicity of green tea and its components have revealed a number of chemical and biological activities that may affect carcinogenesis. Green tea can prevent the formation of carcinogens from precursor substances. Different varieties of Chinese and/or Japanese green tea have been demonstrated to block the formation of nitrosamines *in vitro* and *in vivo* (*164, 172*) and inhibit nitrosamine-induced formation of lesions and papillomas in the esophageal mucosa of rat (*165*). In addition, several green tea components are able to scavenge carcinogenic electrophiles. Addition of green tea polyphenols or (-)-epigallocatechin gallate to (±)-7β,8α-dihydroxy-9α,10α-epoxy-7,8,9,10-tetrahydrobenzo[*a*]pyrene in solution accelerated the disappearance of the diol epoxide (*173*).

Green tea polyphenols possess antioxidant activity (*174, 175*). The polyphenols and (-)-epigallocatechin gallate have been shown to inhibit TPA-induced formation of hydrogen peroxide in human polymorphonuclear cells (*176*) and in mouse epidermis (*169*). Antioxidants enhance immune system function (*177*), and green tea and its components may act in this way to protect against immune function damage due to chemical carcinogens, tumor promoters and ultraviolet light.

Green tea and its components also affect processes and enzymes associated with tumor promotion. Topical application of green tea polyphenols to mice inhibits TPA-induced ornithine decarboxylase activity (*169*) and thus may decrease both the cellular polyamine level and cell proliferation. Topical application of (-)-epigallocatechin gallate has been shown to reduce the number of receptor sites for TPA or teleocidin and to decrease the binding of TPA or teleocidin to the protein kinase C receptors, thereby inhibiting rat brain protein kinase C activity (*171*). This indicates that (-)-epigallocatechin gallate may block signal transduction and inhibit cell proliferation.

Epidemiological Studies. Some epidemiological studies have suggested an inhibitory effect, others an enhancing effect, and still others a lack of effect of tea ingestion on human cancer risk (*129, 159, 178*). None of these studies is considered definitive, however, and more epidemiological research is needed.

Other Dietary Phenols—Carnosol and Curcumin

Rosemary Phenols. The leaves of the plant *Rosmarinus officinalis* L. (rosemary) are commonly used as a spice and flavoring agent; a crude extract of rosemary has been found to inhibit carcinogenesis. Huang *et al.* reported that an extract of rosemary had strong inhibitory effects on TPA-induced inflammation, ornithine decarboxylase activity and tumor promotion as well as on arachidonic acid-induced inflammation (*179*). Topical application of a rosemary extract to the backs of mice inhibited BP-mediated-DNA adduct formation and tumor initiation (*180*). In addition, Singletary *et al.* reported that administration of 1% crude rosemary extract in the diet to female Sprague-Dawley rats for 3 weeks before a single i.g. dose of DMBA reduced the mammary gland tumor incidence by 47% after 16 weeks of DMBA treatment (*181*). They also reported that 0.5% and 1% rosemary extract in the diet inhibited the *in vivo* binding of DMBA to mammary epithelial cell DNA and the formation of two major DNA adducts (*181*).

Carnosol (Figure 6) is a phenolic component of rosemary that has been isolated and has been found to demonstrate the same inhibitory activities as rosemary itself. Topical application of carnosol markedly inhibits TPA-induced inflammation, ornithine decarboxylase activity and tumor promotion in mouse epidermis (*180*).

Curcumin. Curcumin (Figure 6), found in the rhizomes of the plant *Curcuma longa* Linn, is the major yellow pigment in curry, turmeric and mustard and is widely used as a coloring agent in foods, drugs and cosmetics (*182*). For details on the occurrence, analysis and chemistry of curcumin, see Fisher, Chapter 9, Volume I and Tønnesen, Chapter 11, Volume I. Curcumin and turmeric have been used as anti-inflammatory drugs for many years (*182*). More recently, there has been much interest in possible chemopreventive effects of curcumin. Table V summarizes some of the activities of curcumin related to carcinogenesis. Huang *et al.* (Chapter 26, Volume II) describe some of these studies in more detail.

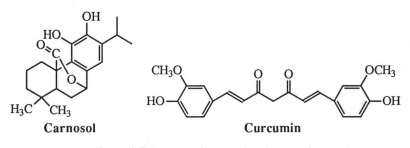

Carnosol   Curcumin

Figure 6. Structures of carnosol and curcumin

Summary and Conclusion

Phenolic compounds are widely consumed in our daily food: plant flavonoids and their glycosides are found in vegetables and fruits; polyphenolic catechins are found in tea and red wine; the synthetic antioxidants BHA and BHT are used as additives in fats, oils and food processing; hydroxycinnamic acids are in many foods, such as chlorogenic acid in coffee, and caffeic acid and ferulic acid in soybeans, cereals, fruits and vegetables; tocopherols are found in vegetable and germ oils and used as food additives; curcumin is in foods containing turmeric and used as coloring agents; carnosol is found in the leaves of the spice rosemary.

These naturally occurring and synthetic phenolic compounds commonly possess the ability to act as antioxidants, to scavenge active oxygen species and electrophiles, to block nitrosation, and to chelate metals. Furthermore, they undergo autoxidation to produce hydrogen peroxide in the presence of metals and are capable of modulating certain cellular enzyme activities.

The phenolic compounds discussed in this chapter are not thought to be genotoxic *in vivo*. The mutagenicity of quercetin, hydroxylated flavonoids and polyphenolic catechins *in vitro* is probably due to mechanisms like the production of hydrogen peroxide by autoxidation in the presence of trace amounts of metals. The amounts of BHA and BHT in our daily food are considered to be safe, but the possibility of several of them acting synergistically at high concentrations to cause harm can not be forgotten.

Table V. Anti-carcinogenic and biological activities of curcumin

Inhibition of the metabolism-mediated mutagenesis of BP and DMBA	*(183, 184)*
Inhibition of the tumor initiating-activities of BP and DMBA	*(183, 185)*
Inhibition of BP-induced forestomach tumorigenesis and DMBA-induced skin tumorigenesis by dietary turmeric	*(186)*
Inhibition of azoxymethanol-induced focal areas of dysplasia in the colon	*(187)*
Inhibition of TPA- and croton oil-induced tumor promotion	*(188, 189)*
Inhibition of the growth of 3-methylcholanthrene-induced sarcomas	*(189)*
Increase in the survival of animals with lymphomas	*(190)*
Inhibition of cell transformation *in vitro*	*(191)*
Inhibition of tumor cell growth *in vitro*	*(190)*
Inhibition of the binding of BP to DNA	*(185)*
Inhibition of lipid peroxidation	*(192, 193)*
Inhibition of lipid peroxide- and smoke-induced DNA damage	*(192, 194)*
Inhibition of epidermal cyclooxygenase and lipoxygenase activities	*(62)*
Inhibition of arachidonic acid-induced inflammation	*(188)*
Inhibition of TPA-induced ornithine decarboxylase activity, DNA synthesis and inflammation	*(188)*
Inhibition of TPA-induced hyperplasia	*(195)*
Inhibition of TPA-induced production of hydrogen peroxide	*(196)*
Hydroxyl and superoxide radical scavenging	*(197, 198)*
Suppression of c-Jun/AP-1 transcription factor activation	*(199)*

Many phenolic compounds have been shown to block different steps of chemical- and/or ultraviolet light-induced carcinogenesis. Eating more fresh vegetables and fruits has many health benefits, including the reduction of cancer risk. On the whole, human daily intake of naturally occurring plant phenolic compounds and synthetic antioxidants in food and beverages is considered to have beneficial rather than adverse effects.

Acknowledgements

The authors wish to thank Professors Allan Conney and Harold Newmark for their review and suggestions during the preparation of this manuscript.

Literature Cited

1. Bate-Smith, E. C. *Adv. Food Res.* **1954**, *5*, pp. 261-300.
2. Kühnau, J. *Wld. Rev. Nutr. Diet.* **1976**, *24*, pp. 117-191.
3. Wollenweber, E.; Dietz, V. H. *Phytochem.* **1981**, *20*(5), pp. 869-932.
4. Harborne, J. B. In *Comparative Biochemistry of the Flavonoids*; Harborne, J. B., Ed.; Academic Press: New York, 1967.
5. Herrmann, K. *J. Fd Technol.* **1976**, *11*, pp. 433-448.

6. Ames, B. N.; McCann, J.; Yamasaki, E. *Mutat. Res.* **1975**, *31*, pp. 347-364.
7. Bjeldanes, L. F.; Chang, G. W. *Science* **1977**, *197*, pp. 577-578.
8. Brown, J. P.; Dietrich, P. S.; Brown, R. J. *Biochem. Soc. Trans.* **1977**, *5*, pp. 1489-1492.
9. Sugimura, T.; Nagao, M.; Matsushima, T.; Yahagi, T.; Seino, Y.; Shirai, A.; Sawamura, M.; Natori, S.; Yoshihira, K.; Fukuoka, M.; Kuroyanagi, M. *Proc. Japan Acad. Ser. B.* **1977**, *53B*(4), pp. 194-197.
10. Hardigree, A. A.; Epler, J. L. *Mutat. Res.* **1978**, *58*, pp. 231-239.
11. MacGregor, J. T.; Jurd, L. *Mutat. Res.* **1978**, *54*, pp. 297-309.
12. Brown, J. P.; Dietrich, P. S. *Mutat. Res.* **1979**, *66*, pp. 223-240.
13. Maruta, A.; Enaka, K.; Umeda, M. *Jpn. J. Cancer Res. (Gann)* **1979**, *70*, pp. 273-276.
14. Yoshida, M. A.; Sasaki, M.; Sugimura, K.; Kawachi, T. *Proc. Japan Acad., Ser. B* **1980**, *56B*, pp. 443-447.
15. Meltz, M. L.; MacGregor, J. T. *Mutat. Res.* **1981**, *88*(3), pp. 317-324.
16. Carver, J. H.; Carrano, A. V.; MacGregor, J. T. *Mutat. Res.* **1983**, *113*(1), pp. 45-60.
17. Tamura, G.; Gold, C.; Ferro-Luzzi, A.; Ames, B. N. *Proc. Natl. Acad. Sci., USA.* **1980**, *77*(8), pp. 4961-4965.
18. Parisis, D. M.; Pritchard, E. T. *Arch. Oral Biol.* **1983**, *28*(7), pp. 583-90.
19. Macdonald, I. A.; Mader, J. A.; Bussard, R. G. *Mutat. Res.* **1983**, *122*(2), pp. 95-102.
20. Macdonald, I. A.; Bussard, R. G.; Hutchinson, D. M.; Holdeman, L. V. *App.l. Environ. Microbiol.* **1984**, *47*(2), pp. 350-355.
21. MacGregor, J. T.; Wehr, C. M.; Manners, G. D.; Jurd, L.; Minkler, J. L.; Carrano, A. V. *Mutat. Res.* **1983**, *124*(3/4), pp. 255-270.
22. Wargovich, M. J.; Newmark, H. L. *Mutat. Res.* **1983**, *121*(1), pp. 77-80.
23. Sahu, R. K.; Basu, R.; Sharma, A. *Mutat. Res.* **1981**, *89*(1), pp. 69-74.
24. Cea, G. F. A.; Etcheberry, K. F. C.; Dulout, F. N. *Mutat Res.* **1983**, *119*(3/4), pp. 339-342.
25. Rahman, A.; Shahabuddin; Hadi, S. M.; Parish, J. H.; Ainley, K. *Carcinogenesis* **1989**, *10*(10), pp. 1833-1839.
26. Fazal, F.; Rahman, A.; Greensill, J.; Ainley, K.; Hadi, S. M.; Parish, J. H. *Carcinogenesis* **1990**, *11*(11), pp. 2005-2008.
27. Stich, H. F.; Karim, J.; Koropatnick, J.; Lo, L. *Nature* **1976**, *260*, pp. 722-724.
28. Stocker, R.; Frei, B. In *Oxidative Stress: Oxidants and Antioxidants*; Sies, H., Ed.; Academic Press: New York, 1991; pp. 213-243.
29. Alvi, N. K.; Rizvi, R. Y.; Hadi, S. M. *Bioscience Rep.* **1986**, *6*(10), pp. 861-868.
30. Pamukcu, A. M.; Yalçiner, S.; Hatcher, J. F.; Bryan, G. T. *Cancer Res.* **1980**, *40*, pp. 3468-3472.
31. Ertürk, E.; Nunoya, T.; Hatcher, J. F.; Pamukçu, A. M.; Bryan, G. T. *Proc. Am. Assoc. Cancer Res.* **1983**, *24*, p. 53.
32. Ertürk, E.; Hatcher, J. F.; Nunoya, T.; Pamukçu, A. M.; Bryan, G. T. *Proc. Am. Assoc. Cancer Res.* **1984**, *25*, p. 95.
33. Wilson, R. H.; Mortarotti, T. C.; Doxtader, E. K. *Proc. Soc. Exp. Bio. Med.* **1947**, *64*, p. 324.
34. Ambrose, A. M.; Robbins, D. J.; DeEds, F. *J. Am. Pharm. Assoc.* **1952**, *41*(3), pp. 119-122.

35. Saito, D.; Shirai, A.; Matsushima, T.; Sugimura, T.; Hirono, I. *Teratogen. Carcinogen. Mutagen.* **1980**, *1*, pp. 213-221.
36. Hirono, I. *CRC Crit Rev Toxicol.* **1981**, *8*, pp. 235-277.
37. Hosaka, S.; Hirono, I. *Jpn. J. Cancer Res. (Gann)* **1981**, *72*, pp. 327-328.
38. Morino, K.; Matsukura, N.; Kawachi, T.; Ohgaki, H.; Sugimura, T.; Hirono, I. *Carcinogenesis* **1982**, *3*(1), pp. 93-97.
39. Habs, M.; Habs, H.; Berger, M. R.; Schmähl, D. *Cancer Lett.* **1984**, *23*, pp. 103-108.
40. Hirose, M.; Fukushima, S.; Sakata, T.; Inui, M.; Ito, N. *Cancer Lett.* **1983**, *21*, pp. 23-27.
41. Takanashi, H.; Aiso, S.; Hirono, I. *J. Food Safety* **1983**, *5*, pp. 55-60.
42. Deschner, E. E.; Ruperto, J.; Wong, G.; Newmark, H. L. *Carcinogenesis* **1991**, *12*(7), pp. 1193-1196.
43. Kato, R.; Nakadate, T.; Yamamoto, S.; Sugimura, T. *Carcinogenesis* **1983**, *4*(10), pp. 1301-1305.
44. Wattenberg, L. W. *Cancer Res. (Suppl.)* **1983**, *43*, pp. 2448s-2453s.
45. Chang, R. L.; Huang, M.-T.; Wood, A. W.; Wong, C.-Q.; Newmark, H. L.; Yagi, H.; Sayer, J. M.; Jerina, D. M.; Conney, A. H. *Carcinogenesis* **1985**, *6*(8), pp. 1127-1133.
46. Wattenberg, L. W. *Cancer Res.* **1985**, *45*(1), pp. 1-8.
47. Verma, A. K.; Johnson, J. A.; Gould, M. N.; Tanner, M. A. *Cancer Res.* **1988**, *48*(20), pp. 5754-5758.
48. Cassady, J. M.; Baird, W. M.; Chang, C.-J. *J. Natural Products* **1990**, *53*, pp. 23-41.
49. Wattenberg, L. W.; Page, M. A.; Leong, J. L. *Cancer Res.* **1968**, *28*(5), pp. 934-937.
50. Buening, M. K.; Chang, R. L.; Huang, M.-T.; Fortner, J. G.; Wood, A. W.; Conney, A. H. *Cancer Res.* **1981**, *41*(1), pp. 67-72.
51. Huang, M.-T.; Johnson, E. F.; Muller-Eberhard, U.; Koop, D. R.; Coon, M. J.; Conney, A. H. *J. Biol. Chem.* **1981**, *256*(21), pp. 10897-10901.
52. Sousa, R. L.; Marletta, M. A. *Arch. Biochem. Biophys.* **1985**, *240*(1), pp. 345-357.
53. Wood, A. W.; Smith, D. S.; Chang, R. L.; Huang, M.-T.; Conney, A. H. In *Plant Flavonoids in Biology and Medicine: Biochemical, Pharmacological, and Structure-Activity Relationships*; Middleton, C. V.; Harborne, J., Eds.; Alan R. Liss, Inc.: New York, 1986; pp. 195-210.
54. Chae, Y.-H.; Marcus, C. B.; Ho, D. K.; Cassady, J. M.; Baird, W. M. *Cancer Lett.* **1991**, *60*(1), pp. 15-24.
55. Sayer, J. M.; Yagi, H.; Wood, A. W.; Conney, A. H.; Jerina, D. M. *J. Am. Chem. Soc.* **1982**, *104*, pp. 5562-5564.
56. Wood, A. W.; Huang, M.-T.; Chang, R. L.; Newmark, H. L.; Lehr, R. E.; Yagi, H.; Sayer, J. M.; Jerina, D. M.; Conney, A. H. *Proc. Natl. Acad. Sci. USA* **1982**, *79*, pp. 5513-5517.
57. Huang, M.-T.; Wood, A. W.; Newmark, H. L.; Sayer, J. M.; Yagi, H.; Jerina, D. M.; Conney, A. H. *Carcinogenesis* **1983**, *4*(12), pp. 1631-1637.
58. Huang, M.-T.; Chang, R. L.; Wood, A. W.; Newmark, H. L.; Sayer, J. M.; Yagi, H.; Jerina, D. M.; Conney, A. H. *Carcinogenesis* **1985**, *6*(2), pp. 237-242.
59. Lesca, P. *Carcinogenesis* **1983**, *4*(12), pp. 1651-1653.

60. Francis, A. R.; Shetty, T. K.; Bhattacharya, R. K. *Carcinogenesis* **1989**, *10*(10), pp. 1953-1955.

61. Trush, M. A.; Kensler, T. W. In *Oxidative stress, Oxidants and Antioxidants*; Sies, H., Ed.; Academic Press: New York, 1991; pp. 277-318.

62. Huang, M.-T.; Lysz, T.; Ferraro, T.; Abidi, T. F.; Laskin, J. D.; Conney, A. H. *Cancer Res.* **1991**, *51*, pp. 813-819.

63. Busse, W. M.; Kopp., D. E.; Middleton, E., Jr. *J. Allergy Clin. Immunol.* **1984**, *73*(6), pp. 801-809.

64. 'T Hart, B. A.; Ip Vai Ching, T. R. A. M.; Van Dijk, H.; Labadie, R. P. *Chem.-Biol. Interactions* **1990**, *73*, pp. 323-335.

65. Robak, J.; Gryglewski, R. J. *Biochem. Pharmacol.* **1988**, *37*(5), pp. 837-841.

66. Fischer, S. M.; Mills, G. D.; Slaga, T. J. *Carcinogenesis* **1982**, *3*, pp. 1243-1245.

67. Nakadate, T.; Yamamoto, S.; Ishii, M.; Kato, R. *Carcinogenesis* **1982**, *3*(12), pp. 1411-1414.

68. Certti, P. A. *Science* **1985**, *227*, pp. 375-381.

69. Zenser, T. V.; Mattamal, M. B.; Armbrecht, H. J.; Davis, B. B. *Cancer Res.* **1980**, *40*(8), pp. 2839-2845.

70. Reed, G. A.; Marnett, L. J. *J. Biol. Chem.* **1982**, *257*, pp. 11368-11376.

71. Baumann, J.; Bruchhausen, F. v.; Wurm, G. *Prostaglandins* **1980**, *20*(4), pp. 627-639.

72. Castagna, M.; Takai, Y.; Kaibachi, K.; Sano, K.; Kikkawa, U.; Nishizuk, Y. *J. Biol. Chem.* **1982**, *257*, pp. 7847-7851.

73. Niedel, J. E.; Kuhn, L. J.; Vandenbank, G. R. *Proc. Natl. Acad. Sci. USA* **1983**, *80*, pp. 36-40.

74. Cox, J. A.; Jeng, A. Y.; Blumberg, P. M.; Tauber, A. I. *J. Immuno* **1987**, *138*, pp. 1884-1888.

75. Gschwendt, M.; Horn, F.; Kittstein, W.; Marks, F. *Biochem. Biophys. Res. Commun.* **1983**, *117*, pp. 444-447.

76. Ferriola, P. C.; Cody, V.; Middleton, E., Jr. *Biochem. Pharmacol* **1989**, *38*(10), pp. 1617-1624.

77. Horiuchi, T.; Fujiki, H.; Hakii, H.; Suganuma, M.; Yamashita, K.; Sugimura, T. *Jpn. J. Cancer Res. (Gann)* **1986**, *77*, pp. 526-531.

78. Nishino, H.; Naito, A.; Iwashima, A.; Tanaka, K.-I.; Matsuura, T.; Fujiki, H.; Sugimura, T. *Jpn. J. Cancer Res. (Gann)* **1984**, *75*, pp. 311-316.

79. Gschwendt, M.; Kittstein, W.; Marks, F. *Cancer Lett.* **1984**, *22*, pp. 219-225.

80. Ziegler, R. G. *Am. J. Clin. Nutr.* **1991**, *53*, pp. 251S-259S.

81. Ito, N.; Fukushima, S.; Hagiwara, A.; Shibata, M.; Ogiso, T. *J. Natl. Cancer Inst.* **1983**, *70*, pp. 343-352.

82. Wilder, O. H. M.; Ostby, P. C.; Gregory, B. R. *Agr. Food Chem.* **1960**, *8*(6), pp. 504-506.

83. Ikeda, G. J.; Stewart, J. E.; Sapienza, P. P.; Peggins, J. O., III; Michel, T. C.; Olivito, V.; Alam, H. Z.; O'Donnell, M. W., Jr. *Fd. Chem. Toxic.* **1986**, *24*(10/11), pp. 1201-1221.

84. Tobe, M.; Furuya, T.; Kawasaki, Y.; Naito, K.; Sekita, K.; Matsumoto, K.; Ochiai, T.; Usui, A. *Fd. Chem. Toxic.* **1986**, *24*(10/11), pp. 1223-1228.

85. Altmann, H. J.; Grunow, W.; Mohr, U.; Richter-Reichhelm, H. B.; Wester, P. W. *Fd. Chem.Toxic.* **1986**, *24*, pp. 1183-1188.

86. Witschi, H.; Morse, C. C. *J. Natl Cancer Inst.* **1983**, *71*, pp. 859-865.

87. Wilder, O. H. M.; Kraybill, H. R. *Am. Meat Inst. Found.* **1948**, *Dec. 1948.*

88.Graham, W. D.; Teed, H.; Grice, H. C. *J. Pharm. Pharmac.* **1954**, *6*, pp. 534-545.
89.Hodge, H. C.; Fassett, D. W.; Maynard, E. A.; Downs, W. L.; Coye, R. D., Jr. *Toxicol. App.l. Pharmacol.* **1964**, *6*, pp. 512-519.
90.Allen, J. R. *Arch. Environ. Health* **1976**, *31*(1), pp. 47-50.
91.Iverson, F.; Truelove, J.; Nera, E.; Wong, J.; Lok, E.; Clayson, D. B. *Cancer Lett.* **1985**, *26*, pp. 43-50.
92.Wattenberg, L. W. *J. Natl Cancer Inst.* **1972**, *48*(5), pp. 1425-1430.
93.Wattenberg, L. W. *J. Natl. Cancer Inst.* **1973**, *50*(6), pp. 1541-1544.
94.McCormick, D. L.; Major, N.; Moon, R. C. *Cancer Res.* **1984**, *44*, pp. 2858-2863.
95.Williams, G. M.; Tanaka, T.; Maeura, Y. *Carcinogenesis* **1986**, *7*(7), pp. 1043-1050.
96.Chung, F.-L.; Wang, M.; Carmella, S. G.; Hecht, S. S. *Cancer Res.* **1986**, *46*(1), pp. 165-168.
97.Lindenschmidt, R. C.; Trylka, A. F.; Goad, M. E.; Witschi, H. P. *Toxicol* **1986**, *38*, pp. 151-160.
98.Williams, G. M.; Maeura, Y.; Weisburger, J. H. *Cancer Lett.* **1983**, *19*, pp. 55-60.
99.Ito, N.; Hirose, M. *Adv. Cancer Res.* **1989**, *53*, pp. 247-302.
100.Hocman, G. *Int. J. Biochem.* **1988**, *20*(7), pp. 639-651.
101.Kahl, R. *J. Envir. Sci. Hlth.* **1986**, *C-4*, pp. 47-92.
102.Kirkpatrick, D. C.; Lauer, B. H. *Fd. Chem. Toxicol.* **1986**, *24*(10/11), pp. 1035-1037.
103.Astill, B. D.; Mulligan, L. T. *Fd. Cosmet. Toxicol.* **1977**, *15*, pp. 167-171.
104.Slaga, T. In *Models, Mechanisms and Etiology of Tumor Promotion*; Morzsomyi, K.; Lapis, N. E.; Day, H.; Yamsslaki, Eds.; IARC Scientific Publications; International Agency For Research on Cancer, World Health Organization: 1984, Vol. 56; pp. 497-506.
105.Wattenberg, L. W.; Sparnins, V. L. *J. Natl Cancer Inst.* **1979**, *63*, pp. 219-222.
106.Anderson, M. W.; Boroujerdi, M.; Wilson, A. G. E. *Cancer Res.* **1981**, *41*(11), pp. 4309-4315.
107.Ioannou, Y. M.; Wilson, A. G. E.; Anderson, M. W. *Cancer Res.* **1982**, *42*(4), pp. 1199-1204.
108.Benson, A. M.; Batzinger, R. P.; Ou, S.-Y. L.; Bueding, E.; Cham, Y.-N.; Talalay, P. *Cancer Res.* **1978**, *38*(12), pp. 4486-4495.
109.Lam, L. K. T.; Sparnins, V. L.; Hochalter, J. B.; Wattenberg, L. W. *Cancer Res.* **1981**, *41*(10), pp. 3940-3943.
110.Perchellet, J.-P.; Owen, M. D.; Posey, T. D.; Orten, D. K.; Schneider, B. A. *Carcinogenesis* **1985**, *6*, pp. 567-573.
111.De Long, M. J.; Prochaska, H. J.; Talalay, P. *Proc. Natl. Acad. USA* **1986**, *83*, pp. 787-791.
112.Kahl, R. *Cancer Lett.* **1980**, *8*, pp. 323-328.
113.Demkowicz-Dobrzanski, K. K.; Henning, E. E.; Juskiewicz-Skiba, M.; Piekarski, L. *Neoplasma* **1984**, *31*, pp. 423-430.
114.Clapp., N. K.; Satterfield, L. C.; Bowles, N. D. *J. Geront* **1979**, *34*(4), pp. 497-501.
115.Boutwell, R. K.; Bosch, D. K. *Cancer Res.* **1959**, *19*(5), pp. 413-427.

116. Berry, D. L.; DiGiovanni, J.; Juchau, M. R.; Bracken, W. M.; Gleason, G. L.; Slaga, T. J. *Res. Commun. Chem. Pathol. Pharmacol.* **1978**, *20*(1), pp. 101-108.
117. Sato, K.; Yoshitake, A.; Asmoto, M.; Hirose, M. *Proc. Jpn. Cancer Assoc.* **1988**, *47*, pp. 117.
118. Wurtzen, G.; Olsen, P. *Fd. Chem. Toxic.* **1986**, *24*, pp. 1229-1233.
119. Sondheimer, E. *Archiv. Biochem. Biophys.* **1958**, *74*, pp. 131-138.
120. Clinton, W. P. In *Coffee and Health (Banbury Report 17)*; MacMahon, B.; Sugimura, T., Ed.; Cold Spring Harbor Laboratory: 1984, Banbury Reports, Vol. 17; pp. 3-10.
121. Pratt, D. E.; Birac, P. M. *J. Food Sci.* **1979**, *44*, pp. 1720-1722.
122. Kuenzig, W.; Chau, J.; Norkus, E.; Holowaschenko, H.; Newmark, H.; Mergens, W.; Conney, A. H. *Carcinogenesis* **1984**, *5*(3), pp. 309-313.
123. Pignatelli, B.; Bereziat, J.-C.; Descotes, G.; Bartsch, H. *Carcinogenesis* **1982**, *3*, pp. 1045-1049.
124. Stich, H. F.; Rosin, M. P.; Bryson, L. *Mutat. Res.* **1982**, *95*, pp. 119-128.
125. Wattenberg, L. W.; Coccia, J. B.; Lam, L. K. T. *Cancer Res.* **1980**, *40*(8), pp. 2820-2823.
126. Mori, H.; Tanaka, T.; Shima, H.; Kuniyasu, T.; Takahashi, M. *Cancer Lett.* **1986**, *30*(1), pp. 49-54.
127. Tanaka, T.; Nishikawa, A.; Shima, H.; Sugie, S.; Shinoda, T.; Yoshimi, N.; Iwata, H.; Mori, H. In *Antimutagenesis and Anticarcinogenesis Mechanisms II*; Kuroda, Y.; Shankel, D. M.; Waters, M. D., Eds.; Plenum Press: New York, 1990, Basic Life Sciences, Vol. 52; pp. 429-440.
128. La Vecchia, C.; Ferraroni, M.; Negri, E.; D'Avanzo, B.; Decarli, A.; Levi, F.; Franceschi, S. *Cancer Res.* **1989**, *49*(4), pp. 1049-1051.
129. Rosenberg, L. *Cancer Lett.* **1990**, *52*(3), pp. 163-171.
130. Tyler, V. E.; Brady, L. R.; Robbers, J. E. In *Vitamins and Vitamin-containing Drugs*; Lea & Febiger: Phildelphia, 1988; pp. 293-294.
131. Borek, C.; Ong, A.; Mason, H.; Donahue, L.; Biaglow, J. E. *Proc. Natl. Acad. Sci., USA* **1986**, *83*(5), pp. 1490-1494.
132. Kline, K.; Cochran, G. S.; Sanders, B. G. *Nutr. Cancer* **1990**, *14*, pp. 27-41.
133. Helson, L.; Verma, M.; Helson, C. In *Modulation and mediation of cancer by vitamins*; Meyskens, F. L.; Prasad, K. N., Eds.; Karger: Basel, 1983; pp. 258-265.
134. Slack, R.; Proulx, P. *Nutr. Cancer* **1989**, *12*, pp. 75-82.
135. Prasad, K. N.; Edwards-Prasad, J.; Ramanujam, S.; Sakamoto, A. *Proc. Soc. Exp. Biol. Med.* **1980**, *164*(1), pp. 158-163.
136. Prasad, K. N.; Edwards-Prasad, J. *Cancer Res.* **1982**, *42*(2), pp. 550-555.
137. Perchellet, J.-P.; Abney, N. L.; Thomas, R. M.; Guislain, Y. L.; Perchellet, E. M. *Cancer Res.* **1987**, *47*(2), pp. 477-485.
138. Shamberger, R. J.; Rudolph, G. *Experientia* **1966**, *22*(2), p. 116.
139. Slaga, T. J.; Bracken, W. M. *Cancer Res.* **1977**, *37*(6), pp. 1631-1635.
140. Shklar, G. *J. Natl. Cancer Inst.* **1982**, *68*, pp. 791-797.
141. Okdukoya, O.; Hawach, F.; Shklar, G. *Nutr. Cancer* **1984**, *6*, pp. 98-104.
142. Shklar, G.; Schwartz, J.; Trickler, D. P.; Niukian, K. *J. Natl Cancer Inst.* **1987**, *78*(5), pp. 987-992.
143. Gensler, H. L.; Magdaleno, M. *Nutr. Cancer* **1991**, *15*, pp. 97-106.
144. Khettab, N.; Amory, M.-C.; Briand, G.; Bousquet, B.; Combre, A.; Forlot, P.; Barey, M. *Biochimie* **1988**, *70*(12), pp. 1709-1713.

145. Toth, B.; Patil, K. *J. Natl Cancer Inst.* **1983**, *70*(6), pp. 1107-1111.
146. McIntosh, G. H. *Nutr. Cancer* **1992**, *17*, pp. 47-55.
147. Newmark, H. L.; Mergens, W. J. In *Inhibition of Tumor Induction and Development*; Zedeck, M. S.; Lipkin, M., Eds.; Plenum: New York, 1980; pp. 127-168.
148. Jenson, H.; Madsen, J. L. *Acta Med. Scand.* **1988**, *223*, pp. 293-304.
149. Watson, R. R.; Leonard, T. K. *J. Am. Diet. Assoc.* **1986**, *86*, pp. 505.
150. Knekt, P.; Aromaa, A.; Maatela, J.; Aaran, R.-K.; Nikkari, T.; Hakama, M.; Hakulinen, T.; Peto, R.; Tepp.o, L. *Am. J. Clin. Nutr.* **1991**, *53*, pp. 283S-286S.
151. Knekt, P. *Int. J. Epidemiol.* **1988**, *17*(2), pp. 281-286.
152. Menkes, M. S.; Comstock, G. W.; Vuilleumier, J. P.; Helsing, K. J.; Rider, A. A.; Brookmeyer, R. *N. Engl. J. Med.* **1986**, *315*(20), pp. 1250-1254.
153. LeGardeur, B. Y.; Lopez-S, A.; Johnson, W. D. *Nutr. Cancer* **1990**, *14*(2), pp. 133-140.
154. Palan, P. R.; Mikhail, M. S.; Basu, J.; Romney, S. L. *Nutr. Cancer* **1991**, *15*(1), pp. 13-20.
155. Verreault, R.; Chu, J.; Mandelson, M.; Shy, K. *Int. J. Cancer* **1989**, *43*, pp. 1050-1054.
156. Heinonen, P. K.; Kuopp.ala, T.; Koskinen, T.; Punnonen, R. *Arch. Gynecol. Obstet* **1987**, *241*, pp. 151-156.
157. Nomura, A. M. Y.; Stemmerman, G. N.; Heilbrun, L. K.; Salkeld, R. M.; Vuilleumier, J. P. *Cancer Res.* **1985**, *45*(5), pp. 2369-2372.
158. Wald, N. J.; Thompson, S. G.; Densem, J. W.; Boreham, J.; Bailey, A. *Br. J. Cancer* **1987**, *56*(1), pp. 69-72.
159. WHO International Agency for Research on Cancer *IARC Monogr. Carcinog. Risks. Hum.* **1991**, *51*, pp. 207-271.
160. Fujita, Y.; Yamane, T.; Tanaka, M.; Kuwata, K.; Okuzumi, J.; Takahashi, T.; Fujiki, H.; Okuda, T. *Jpn. J. Cancer Res.* **1989**, *80*(6), pp. 503-5.
161. Yamane, T.; Hagiwara, N.; Tateishi, M.; Akachi, S.; Kim, M.; Okuzumi, J.; Kitao, Y.; Inagake, M.; Kuwata, K.; Takahashi, T. *Jpn. J. Cancer Res.* **1991**, *82*(12), pp. 1336-1339.
162. Harada, N.; Takabayashi, F.; Oguni, I.; Hara, Y. *The Int. Sympo. on Tea Sci. (Shizuoka, Japan, August 26-29, 1991)* **1991**, p. 45, abstract no. II-A-3-5.
163. Hirose, M.; Hoshiya, T.; Takahashi, S.; Hara, Y.; Ito, N. *The Int. Sympo. on Tea Sci. (Shizuoka, Japan, August 26-29, 1991)* **1991**, p. 47, abstract no. II-A-3-7.
164. Xu, Y.; Han, C. *Biomed. Environ. Sci.* **1990**, *3*(4), pp. 406-412.
165. Han, C.; Xu, Y. *Biomed. Environ. Sci.* **1990**, *3*(1), pp. 35-42.
166. Wang, Z.-Y.; Huang, M.-T.; Ferraro, T.; Wong, C.-Q.; Lou, Y.-R.; Reuhl, K.; Iatropoulos, M.; Yang, C. S.; Conney, A. H. *Cancer Res.* **1992**, *52*(5), pp. 1162-1170.
167. Wang, Z.-Y.; Hong, J.-Y.; Huang, M.-T.; Reuhl, K. R.; Conney, A. H.; Yang, C. S. *Cancer Res.* **1992**, *52*, pp. 1943-1947.
168. Wang, Z. Y.; Agarwal, R.; Bickers, D. R.; Mukhtar, H. *Carcinogenesis* **1991**, *12*, pp. 1527-1530.
169. Huang, M.-T.; Ho, C.-T.; Wang, Z.-Y.; Ferraro, T.; Finnegan-Olive, T.; Lou, Y.-R.; Mitchell, J. M.; Laskin, J. D.; Newmark, H.; Yang, C. S.; Conney, A. H. *Carcinogenesis* **1992**, (in press)

170.Wang, Z. Y.; Khan, W. A.; Bickers, D. R.; Mukhtar, H. *Carcinogenesis* **1989**, *10*(2), pp. 411-415.
171.Yoshizawa, S.; Horiuchi, T.; Fujiki, H.; Yoshida, T.; Okuda, T.; Sugimura, T. *Phytother. Res.* **1987**, *1*, pp. 44-47.
172.Nakamura, M.; Kawabata, T. *J. Food Sci.* **1981**, *46*, pp. 306-307.
173.Wang, Z. Y.; Cheng, S. J.; Zhou, Z. C.; Athar, M.; Khan, W. A.; Bickers, D. R.; Mukhtar, H. *Mutat. Res.* **1989**, *223*, pp. 273-285.
174.Zhao, B. L.; Li, X. J.; Cheng, S. J.; Xin, W. J. *Cell. Biophys.* **1989**, *14*(2), pp. 175-185.
175.Osawa, T.; Namiki, M.; Kawakishi, S. In *Antimutagenesis and Anticarcinogenesis Mechanisms II*; Kuroda, Y., Shankel, D. M.; Waters, M. D., Eds.; Plenum Press: New York, 1990, Basic Life Sciences, Vol. 52; pp. 139-153.
176.Zhong, Z.; Tius, M.; Troll, W.; Fujiki, H.; Frenkel, K. *Proc. Am. Assoc. Cancer Res.* **1991**, *32*, pp. 127.
177.Bendich, A. In *Antioxidant Nutrients and Immune Functions*; Bendich, A.; Phillips, M.; Tengerdy, R., Eds.; Plenum Press: New York, 1989; pp. 35-54.
178.La Vecchia, C.; Negri, E.; Franceschi, S.; D'Avanzo, B.; Boyle, P. *Nutr. Cancer* **1992**, *17*, pp. 27-31.
179.Huang, M.-T.; Ho, C.-H.; Chang, S.-J.; Laskin, J. D.; Stauber, K.; Georgiadis, C.; Conney, A. H. *Proc. Am. Assoc. Cancer Res* **1989**, *30*, p. 189.
180.Huang, M.-T.; Ho, C.-T.; Ferraro, T.; Wang, Z. Y.; Stauber, K.; Georigiadis, C.; Laskin, J. D.; Conney, A. H. *Proc. Am. Assoc. Cancer Res.* **1992**, *33*, p. 165.
181.Singletary, K. W.; Nelshopp.en, J. M. *Cancer Lett.* **1991**, *60*, pp. 169-175.
182.Ammon, H. P. T.; Wahl, M. A. *Planta Med.* **1991**, *57*(1), pp. 1-7.
183.Nagabhushan, M.; Bhide, S. V. *J. Nutr. Growth Cancer* **1987**, *4*, pp. 83-89.
184.Nagabhushan, M.; Amonkar, A. J.; Bhide, S. V. *Food Chem. Toxicol.* **1987**, *25*(7), pp. 545-547.
185.Huang, M.-T.; Wang, Z. Y.; Georgiadis, C. A.; Laskin, J. D.; Conney, A. H. *Carcinogenesis* **1992**, (submitted)
186.Azuine, M. A.; Bhide, S. V. *Nutr. Cancer* **1992**, *17*, pp. 77-83.
187.Huang, M.-T.; Deschner, E. E.; Newmark, H. L.; Wang, Z.-Y.; Ferraro, T. A.; Conney, A. H. *Cancer Lett.* **1992**, (in press)
188.Huang, M.-T.; Smart, R. C.; Wong, C.-Q.; H, C. A. *Cancer Res.* **1988**, *48*, pp. 5941-5946.
189.Soudamini, K. K.; Kuttan, R. *J. Ethnopharmacol.* **1989**, *27*(1-2), pp. 227-233.
190.Kuttan, R.; Bhanumathy, P.; Nirmala, K.; George, M. C. *Cancer Lett.* **1985**, *29*(2), pp. 197-202.
191.Huang, M.-T.; Ma, W.; Lou, Y. R.; Wang, Z. Y.; Lu, Y. P.; Chang, R. L.; Newmark, H.; Manchand, P.; Conney, A. H. *Proc. Am. Assoc. Cancer Res.* **1992**, *33*, p. 167.
192.Shalini, V. K.; Srinivas, L. *Mol. Cell. Biochem.* **1987**, *77*(1), pp. 3-10.
193.Donatus, I. A.; Sardjoko; Vermeulen, N. P. *Biochem. Pharmacol.* **1990**, *39*(12), pp. 1869-1875.
194.Shalini, V. K.; Srinivas, L. *Mol. Cell. Biochem.* **1990**, *95*(1), pp. 21-30.
195.Huang, M.-T.; Lysz, T.; Ferraro, T.; Conney, A. H. In *Cancer Chemoprevention*; Wattenberg, L., Ed.; 1992, (in press).

196.Huang, M.-T.; Ma, W.; Laskin, J. D.; Conney, A. H. (unpublished results)
197.Kunchandy, E.; Rao, M. N. A. *Internatl. J. Pharmaceu.* **1989**, *57*, pp. 173-176.
198.Kunchandy, E.; Rao, M. N. A. *Internatl. J. Pharmaceu.* **1990**, *58*, pp. 237-240.
199.Huang, T.-S.; Lee, S.-C.; Lin, J.-K. *Proc. Natl. Acad. Sci. USA* **1991**, *88*(12), pp. 5292-5296.

RECEIVED April 20, 1992

Chapter 3

Mutagenic, Carcinogenic, and Chemopreventive Effects of Phenols and Catechols

The Underlying Mechanisms

John H. Weisburger

American Health Foundation, One Dana Road, Valhalla, NY 10595

In the context of health effects with emphasis on cancer, phenols are generally not genotoxic, that is, they cannot modify the gene and therefore, they are not considered to be direct cancer risks. Laboratory studies have demonstrated that while not genotoxic, phenols as a class can be co-carcinogens or promoters, increasing the effects of environmental genotoxic carcinogens. The promoting effect is, however, highly dependent on the dosage and chronicity of exposure. These basic mechanistic concepts may account for the fact that critical reviews of occupational health data have revealed no cancer risk in the production and use of phenolic compounds. Recent studies have actually demonstrated that some of the phenols found in fruits and vegetables, as well as synthetic phenolic antioxidants, exert protective effects against cancer. These chemicals are antimutagenic, anti-carcinogenic, including against carcinogens present in tobacco smoke, and can antagonize the effect of promoters. High doses of several of these compounds, however, can cause cancer in animals through mechanisms like cytotoxicity, regenerative cell duplication and hydroxy radical generation. Therefore, the public needs to be much more aware of the importance of dosage and extent of exposure, rather than be told that a chemical is a "carcinogen" with the resulting concern and sometimes expensive, indeed unwarranted, actions. Insight into mechanisms of action is a sound basis for risk evaluation. Naturally occurring and synthetic phenols illustrate the value of judging the benefits and risks as a function of dosage and chronicity of exposure through laboratory and epidemiologic approaches.

Phenols are important industrial products, used as such or as chemical intermediates (1-3). Hydroxylation of benzene, and aromatic compounds in general, is a well known classic metabolic reaction yielding hydroxylated products, i.e. phenols (4). Phenols can be further converted biochemically into polyhydroxybenzenes. Under some conditions phenols, especially the catechols, have displayed adverse effects,

0097–6156/92/0507–0035$06.00/0

including carcinogenicity in model systems, usually at very high dose levels and chronic exposure (1-3). There is no evidence, however, that these chemicals, typified by phenol itself, have caused cancer in humans (5, Table I). The one exception might be the specific instance of appreciable amounts of benzene, chronically inhaled, that can induce leukemia (6). The relevant mechanism is not completely clear, but phenol and the biochemical oxidation products catechol, o-quinone and p-quinone, have been postulated as key reactive agents (7-9).

In the area of nutrition and cancer, many phenols have been found to exhibit protective effects. Polyphenols and their derivatives present in green and black tea have a protective effect against cancer and heart disease. Regular intake of vegetables and fruits and the consumption of tea reduces the risk of diverse cancers in humans. Some of the beneficial effects stem from an adequate intake of vitamins. Also important is the presence of specific antioxidant phenols and polyphenols such as quercetin, ellagic acid, chlorogenic acid and related compounds.

This chapter will deal with the adverse as well as the protective effects of phenols, with emphasis on the underlying mechanisms.

Table I. Phenols and Occupation

1. No clear cancer risk

2. Possibly lower risk for liver cirrhosis, emphysema and arteriosclerotic heart disease

3. Highly toxic to skin upon contact, or by inhalation

Adapted from ref. 5.

Mechanisms Of Carcinogenesis

Basic advances have been realized in the study of the mechanisms of carcinogenesis. It will be useful to outline these concepts as an introduction to the relevant mechanisms applicable to phenols. Reviewed will be the classification of carcinogens according to modes of action, so that the role of specific causative, promoting and inhibiting factors in the action of phenols can be understood. In turn, this knowledge will clarify the possibilities of rational preventive measures based on the properties of each type of carcinogen or modifying agent.

A series of distinct sequential steps in cancer causation, development, and progression are involved (9, Table II).

Genotoxic Events. An early event in carcinogenesis, initiation, is a somatic mutation, an alteration of the genetic material. This critical event can arise through a number of mechanisms that modify cellular DNA (9).

Most phenols do not have the attributes of genotoxicity, as indicated by the negative outcome of *in vitro* tests designed to reveal this property. *In vivo*, the phenolic hydroxy group is a substituent readily conjugated by phase II enzymes. High doses, however, can overload these detoxification systems. Thus, the free phenols can be oxidized to quinones, themselves subject to detoxification by GSH transferases. Overload of these enzymes may yield a DNA-reactive product. The dose-related effects are discussed later in this chapter.

Table II. Types of Carcinogenic Chemicals by Mechanism of Action

Type	Mechanism	Examples
1. Genotoxic (DNA reactive)		
a. biochemical activation independent	In solution, can modify and mutate DNA.	methylnitrosourea, methyl methanesulfonate, nickel, cadmium
b. biochemical activation dependent	Must be converted to DNA-reactive form by host enzymes and cofactors.	dimethylnitrosamine, benzo(a)pyrene, 2-naphthylamine, ethylene dibromide
2. Non-genotoxic (epigenetic)		
a. promoter	Enhances development of transformed cells, can induce ornithine decarboxylase and hydroxy radical and H_2O_2 formation, and increase cell duplication rate.	organochlorine pesticides, Na saccharin, bile acids, phenol
b. co-carcinogen	Enhances the action of a simultaneously present genotoxic carcinogen; may increase conversion of type 1b to 1a.	catechol, ethanol, SO_2, pyrene
c. hormone/endocrine modifying	Causes imbalances in the endocrine system, including gonad/pituitary/thyroid/adrenal.	estrogens, amitrole
d. peroxisome proliferator	Interferes with specific oxidative, often liver-specific, enzymes.	diethylhexylphthalate, clofibrate
e. cytotoxicant	Dose-dependent cell killing, inflammation and cell regulation-duplication.	phenols, catechols, nitrilotriacetate, Na saccharin, carbon tetrachloride
f. immunosuppressant	Inhibits the immune system, especially cell-mediated immunity.	azathioprine, cyclosporin A, TCDD?, corticosteroids, radiation, o-phenylphenol
g. solid state	Complex cellular response, including inflammation, increased cell duplication, generation of hydroxy radicals.	asbestos, plastics (rodents only)

A signal series of events are the DNA synthesis and cell duplication rates that are critical to incorporate the faulty, transformed segment of the genome (9). This also affects the informational message for the specific protein products, like enzymes, and also the chances of successful neoplastic transformation. Thus, any external factor, like phenols or catechols, or internal element increasing replication raises the risk of carcinogenesis. In contrast, inhibition of replication is protective.

Epigenetic, Nongenotoxic Events. The production of abnormal DNA and genetic material, obtained by any of the possible mechanisms, is only the first step in an intricate sequence of molecular and cellular events in the complex neoplastic process. An abnormal cell population needs to achieve a selective growth advantage in the presence of surrounding normal cells that provide growth-controlling factors through intercellular communication via gap junctions. Promoters can function by inhibiting this communication. Cell duplication thus depends on endogenous and exogenous controlling elements operating by epigenetic mechanisms, such as promoters and inhibitors of growth, which either enhance or retard the process. In addition, during the successive generations, early tumor cells can undergo phenotypic changes of expression, perhaps as a result of faulty steps in gene duplication and differentiation. Promotion is complex and stems from a number of operating mechanisms (Table II). The effect is highly dependent on dose, and importantly, it is reversible upon removal of promoters.

Promoters do not produce invasive cancer in the absence of an antecedent cell change. Thus, in exploring the causes of any specific human cancer, a systematic search is needed for agents (genotoxic factors) leading to an abnormal genome and any other agents (epigenetic) involved in the growth and development of the resulting abnormal neoplastic cells and their further progression to malignancy (Table III).

Table III. Identification of Causes of Human Cancer by Mechanisms

1. Types of genotoxic carcinogen; can be chemical, radiation, or virus
2. Types of non-genotoxic promoting agents; can be chemical or virus
3. Amount, duration of exposure, and potency of each type

Phenols and the Mechanisms of Carcinogenesis

In tests for mutagenicity that we deem appropriate to define any genotoxic attributes, specifically mutagenicity in the *Salmonella typhimurium* Ames test, many phenols have failed to display activity (11). Those phenols that were tested were also not active in the DNA repair test of Williams. These are primary tests to reveal genotoxicity (9). Thus, as a class phenols are not directly genotoxic, but are promoters or cocarcinogens (Table IV).

In the special case of catechols, or their oxidation products, *o*-quinones, however, positive results in Ames' mutagenicity test have been found. The mechanism appears to be the generation of oxidized species like hydroxy radicals or hydrogen peroxide that can mutate bacterial species such as the *Salmonella typhimurium* (11, 12). Such chemicals are, nonetheless, usually inactive in the DNA repair test in liver cells of Williams (9), because in this system, there are detoxification reactions blocking the action of hydroxy radicals, or destroying hydrogen peroxide with catalase. Rodent bioassays of typical catechols like quercetin that are mutagenic in the Ames test, but negative in the Williams test, have not yielded any evidence of carcinogenicity (*13*). On the other hand, when catechol itself is administered in the diet of rats at high dose levels (2%), stomach cancer has been observed (*14*). The tissue may be less able to destroy hydroxy radicals, and at the same time high doses yield cytoxicity and increasing cell duplication rates, conditions that favor cancer development.

Table IV. Contrasting Effects Between Phenols, Catechols and Polyphenols

Phenols: Promoter but not co-carcinogen
Mechanisms: selective cytotoxicity and regeneration

Catechols: Co-carcinogen but not promoter
Mechanisms: Increased cell proliferation. Production of •OH radicals?
In vitro, can display mutagenicity in prokaryotes through oxidation-reduction cycles yielding •OH and H_2O_2.
In DNA repair test in liver cells, catechols are negative, since detoxification systems, including phase II conjugation, are present.

Polyphenols: Mainly inhibitory with lower dose levels tested.
Mechanisms: Act as antioxidants, radical and nitrite traps; can act as nucleophiles; may have anti-viral actions, including against HIV; induce enzymes, including cytochrome P-450 systems and phase II conjugation systems.

Low levels of catechol do not have a promoting effect, probably because any hydroxy radicals formed are detoxified and there is no cytoxicity and therefore, no increase in cell duplication. That it takes both generation of hydroxy radicals and increased rates of DNA synthesis and cell duplication is based on the fact that *p*-methoxyphenol yields enhanced cell duplication in the stomach, but fails to promote gastric cancer (*15*). On the other hand, the substituted catechol caffeic acid (3,4-dihydroxycinnamic acid) induced forestomach and renal neoplasms in rats and mice when fed at 2% in the diet for 2 years (*16*).

Chlorophenols and more complex phenols like acidic ethers (2,4-D; 2,4,5-T) used in the agricultural industries have been incriminated in the induction of soft tissue sarcoma and lymphoma in humans (*17-20*). These findings, however, were not universally made but only in select areas. Inasmuch as this class of chemicals is not likely to operate by a genotoxic mechanism, there may be other elements bearing on this type of outcome. For example, chlorophenols and related compounds may depress the immune system and facilitate the occurrence of select neoplasms through this mechanism (*21*).

Phenols As Promoters. Promotion can involve a number of distinct mechanisms (Tables II, V). One that appears relevant to the action of most phenols in distinct target organs is that of a dose-related cytotoxic effect, leading in turn to inflammatory responses, increased DNA synthesis and cell duplication rates. Overall, these phenomena might select cells with altered gene structure, typical of neoplasia.

Several types of markers can be employed to detect promoting potential as a function of dose. One is the induction of ornithine decarboxylase (22). This marker is increased with many types of promoters, including phenolic substances (21). A second marker is increased DNA synthesis, measured through thymidine or bromodeoxyuridine incorporation (9, 10, 22).

Table V. Key Elements Bearing on Adverse and Protective Effects of Phenols

1. •Phenols/catechols are non-genotoxic and are readily detoxified through phase II conjugation reactions at low/moderate exposures.
   •At high doses, conjugation is overloaded and oxidation to reactive quinones, which can be detoxified through GSH-transferase, occurs.
   •At highest doses, overloading all systems, adverse effects like cytotoxicity, increased cell cycling, decreased RNA metabolism and neoplasia in specific organs can occur, especially in tissues with poor detoxification capacity like the stomach, urinary bladder or kidneys.

2. There is no evidence that phenols as a class cause cancer in humans. The active fraction of tobacco smoke, however, has a potent enhancing effect in the events leading to lung cancer. Smoking cessation reverses this action, and the risk of lung cancer is progressively lowered.

3. Phenols, through metabolism, and especially catechols by oxidation-reduction cycles, generate $O_2^{-•}$, $•OH$ and H_2O_2 as a function of dosage. This yields mutagenicity in prokaryotes with low defenses (catalase, superoxide dismutase, etc.), but not in mammalian cells.

4. Phenols act as traps for, and detoxify, nitrite. At high nitrite ratios, however, direct acting mutagens/carcinogens are formed, which are in part associated with gastric cancer risk.

5. Phenols and polyphenols like butylated hydroxyanisole, butylated hydroxytoluene, quercetin, ellagic acid and tea polyphenols: (a) induce detoxifying enzyme systems, (b) can act as traps for electrophiles and (c) can terminate radical reaction chains and thus decrease the effects of genotoxic carcinogens.

6. Polyphenols in green, oolong or black teas are antimutagenic, anticarcinogenic and anti-promoting. Intake of tea as a beverage at moderate temperature may have anticarcinogenic actions.

In general, promoters that operate via cytotoxicity display a sharp dose-response curve. Obviously, dose levels that are not cytotoxic do not display promoting potential. Also important is the chronicity of the application of promoters that eventually yield overt neoplasm.

Phenols have been demonstrated to act as promoters in the classic mouse skin bioassay. Application of a small amount of a genotoxic carcinogen like a polycyclic aromatic hydrocarbon, typically benzo(a)pyrene or 7,12-dimethylbenz(a)anthracene, or N-nitrosomethylurea, followed by repeated application of phenol yields neoplasia at the point of application (1).

Phenols in Tobacco. Tobacco smoke is a complex mixture of genotoxic carcinogens, polycyclic aromatic hydrocarbons, tobacco-specific nitrosamines, and some heterocyclic aromatic amines (*23*). These chemicals are present in relatively small amounts. Tobacco smoke does contain appreciable amounts of phenols in the "acidic" fraction that act as promoters and co-carcinogens (*24, 25*). In part, the relevant mechanism is cytotoxicity by phenols and especially by catechols. This is the reason why the dose-response curve in smokers is sharp. Individuals consuming 30-40 cigarettes a day are at high risk of lung cancer, but with 4 cigarettes a day the effect is difficult to appreciate. This is due in great part to the much lower cytotoxicity, and hence little promoting potential associated with the low smoking rates. Importantly, the reversibility of the effect is demonstrated by the fact that individuals, who have smoked heavily for a number of years and quit, have a progressively lower risk of lung cancer. Interestingly, for the first few years after cessation the risk actually increases because of elimination of the toxic effects of tobacco smoke, including those of the phenolic fraction, releasing tumor cells to develop (*23*).

p-**Hydroxyacetanilid.** The phenolic drug *p*-hydroxyacetanilid (Tylenol) has been used safely for many years. Yet, administration of a very high dose (1.0%) in the diet to rats and mice was hepatotoxic, nephrotoxic and gave a small yield of cancer in the urinary bladder (*26*). Decreasing the dose relatively little from the highest level used abolished the effect. The sharp dose-response curve suggests an indirect mechanism involving in part cytotoxicity due to the presence of the free compound or a cytotoxic metabolite. With lower dose levels, the phenolic hydroxy group is not free but is excreted in urine as an ester conjugate with glucuronic acid or sulfuric acid. At high dose levels there is intracellular conversion of the phenol to a quinone or a quinonimine that may produce the transformed cells (*27-29*). This overall mechanism, involving progressive elimination of available defense mechanisms by the high chronic doses, clearly demonstrates that clinically used dosages will not have such adverse effects. In fact, acetanilide and *p*-hydroxyacetanilide reduced the carcinogenic effect on the liver of 2-acetylaminofluorene, through several reactions affecting the metabolism of the carcinogen, such as *N*-hydroxylation and formation of the reactive sulfate ester of the *N*-hydroxymetabolite (*30*).

BHA and BHT. The useful food antioxidants butylated hydroxyanisole (BHA) and butylated hydroxytoluene (BHT) are synthetic substituted phenols that have displayed a dose-related anticarcinogenic effect through several mechanisms including induction of detoxification enzymes and neutralization of some reactive products of metabolism like hydroxy radicals. Very high doses in the diet, however, cause cancer in animals, through mechanisms like cytotoxicity, regenerative cell duplication and hydroxy radical generation.

These antioxidants display adverse effects in the stomach (BHA), in the liver (BHT), and in the urinary bladder (BHA), when ingested chronically by laboratory rodents at very high dose levels, 2% of the diet (*14*). The effects are sharply lower with a 50% decrease of the dose to 1% of the diet. No neoplasia was seen with yet lower doses but hyperplasia was observed in a dose-related fashion. The mechanism in the stomach and in the urinary bladder with BHA involves cytotoxicity and thence promotion, with a steep, non-linear dose-response. Also involved are specific free, quinone-type metabolites, produced intracellularly, again mainly at the highest doses.

Commercial BHA is a mixture of 3-*tert*-butyl-4-hydroxyanisole, the main component, and the 2-*tert*-butyl isomer, which has lower or negligible adverse effects. Upon metabolism, BHA is not genotoxic by the accepted tests and fails to yield DNA or RNA adducts, but does react with proteins. Nonetheless, a few *in vitro* tests, such as 2-stage initiating-promoting bioassays, displayed moderate positive initiating activity, perhaps due to a more efficient formation of quinones under those conditions (*31*). The toxicity and carcinogenicity of high doses is, thus, indirect. Quinone metabolites have been identified that cause DNA strand breaks through an indirect mechanism (*32-34*). *In vivo*, the effect is distinctly and rapidly reversible, typical of non-genotoxic damage (*35*). In the liver, the action of BHT may involve the same processes involving a reactive quinone (*36*), but this antioxidant is also a powerful inducer of the cytochrome P-450 system and of specific metabolic enzymes (*36-40*). There is a pronounced increase of DNA synthesis and cell duplication, leading to a sharply increased liver weight (*40*).

Hormones That Are Phenols. The naturally occurring estrogen estradiol is a phenol. The effects on the endocrine system are complex (*43, 44*). The specific hormonal effects on specific target organs are elicited by actions on appropriate receptors. Unphysiologically high amounts of estrogen may have adverse effects on the liver, kidney and uterus. The synthetic hormone diethylstilbestrol (DES) is about 10 times as estrogenic as the naturally occurring estradiol. Therefore, administration of unphysiologically large amounts of DES has led to adverse effects including cancer in animals and in humans, especially through transplacental action on the fetus. In part, these effects are mediated through alteration in cell duplication rates favoring carcinogenesis. The effect may be direct on a target organ, or indirect by stimulating the hypothalamic-pituitary-gonadal-adrenal axis to yield abnormal endocrine balances. The underlying mechanisms may involve, in addition, the production of quinone metabolites from DES (*45*).

Phenols As Nitrite Traps. Some phenols act as excellent traps for nitrite, but this reaction is a two-edged sword. Low levels of nitrite are easily detoxified, but in the presence of higher levels of nitrite, such as are present in pickled and smoked foods, phenols can give rise to carcinogens that may account for cancer in the upper gastrointestinal tract. Phenols react readily with nitrite under acidic conditions. With relatively low amounts of nitrite, the product is a nitrosophenol that can be conjugated in the liver with phase II enzymes, and excreted as glucuronide or sulfate esters. However, with high amounts of nitrite, phenols can be converted to genotoxic, reactive diazophenols (*41, 42*). These products have induced neoplasia in the stomach and the oral cavity. Because of their reactive nature they are detoxified in the intestinal tract and therefore do not seem to display adverse effects in other organs.

Protective, Inhibiting Effects Of Phenols

Few studies have been done on simple phenols as inhibitors. The exception might be phenols that can act as traps for limited amounts of nitrite, compete for phase II conjugation reactions, and thus affect activation and detoxification of carcinogens, or act as nuclophilic traps, as discussed above (*1, 14, 31, 41, 42*).

BHA and BHT As Inhibitors. The antioxidants BHA and BHT have been observed to be excellent inhibitors of neoplasia in animal models at specific target organs with the appropriate carcinogens. This effect was originally observed by Wattenberg (*46, 47*) who utilized 5,000 ppm BHA or BHT in the diet and thus discovered the effect now labeled chemoprevention. The underlying mechanism most likely does not directly involve the antioxidant properties of these chemicals. Rather, they may act as powerful inducers of enzymes that increase the detoxification of the carcinogens used (*14, 31, 48, 49*). For example, detoxification of the antioxidants, phenols and catechols is fostered by higher levels of phase II conjugation systems like glucuronyl transferase, sulfotransferase, and GSH-transferase (*4, 50, 51*). These enzyme systems are effective in reacting with electrophilic reactants, the genotoxic carcinogens involved in neoplasia at key organ sites (*30, 31, 49*). A dose-response study with BHA and BHT with several types of carcinogens like 7,12-dimethylbenz(*a*)anthracene affecting the mammary gland in rats, 2-AAF affecting the liver or MAM acetate yielding colon cancer, show that doses used originally by Wattenberg were effective enzyme inducers and also inhibitors (*52, 53*). With lower dose levels, the effect was less pronounced. Doses in the range used as antioxidants in foods, on the order of 60 ppm have no effect as enzyme inducers or inhibitors. Yet, there may be situations like the endogenous generation of hydroxy radicals, where these antioxidants, and also tocopherol, may lower the risk by acting as traps for hydroxy radicals.

In vitro tests, as discussed, may be valuable for studies on relevant mechanisms, but must be interpreted cautiously. Thus, quercetin is mutagenic *in vitro* in cells devoid of detoxification systems (*54*), but not *in vivo* or *in vitro* systems with adequate detoxification (*14*). In fact, quercetin can inhibit mammary gland (*55*) and colon (*56*) carcinogenesis. Catechol blocks the metabolic activation of aromatic amines (*57*), and the metabolism of a tobacco-specific nitrosamines in liver cells (*58*). *In vivo*, catechol and other antioxidants inhibit chemically induced pancreatic carcinogenesis in hamsters (*59*).

BHA and BHT may also serve to modify the promoting potential by acting first on the metabolic detoxification of carcinogens, and also by negating the effect of promoters as inhibitors of intracellular communication (*9*).

Polyphenols In Tea As Inhibitors. Green tea, frequently consumed in the Far East and North Africa, has demonstrated beneficial effects in lowering the risk for heart disease and for specific types of cancer (*60*). Laboratory studies have demonstrated that green tea displays an antimutagenic effect in appropriate test systems. Tea also has lowered the incidence of cancer induced with specific carcinogens and ultraviolet light in laboratory animals at certain target organs like the skin, lung, esophagus or colon (*60-62*). Green tea has also displayed an inhibiting effect in experiments designed to reveal promotion. The active components in green tea are certain polyphenols or polyphenol derivatives, such as the most effective chemical in the series, epigallocatechin gallate.

There have been few studies with black tea, consumed in the Western world, in Southeast Asia, India and in the Near East. The components of black tea are also polyphenols, but they are distinct from these found in green tea (*60*). During the oxidative fermentation converting green to black tea, the original polyphenols undergo partial oxidation and form more complex polyphenols. Since the structure of the polyphenols in black tea still display antioxidant potential, it seems likely that they also have inhibiting effects, but this remains to be documented. In one area of the world, around north east Iran, Turkmenistan, Uzbekistan and Kazakstan, where large amounts, of the order of 10 cups of black tea per day or more, are consumed, traditionally as a very hot beverage, there appears to be an ensuing risk of cancer of the esophagus (60). The effect may be complex. The high temperature of the beverage leads to injury and compensatory repair, meaning cell duplication rates are increased, in the esophagus. It is not known whether the tea polyphenols as such, or through biochemical oxidation products, contribute to this effect.

Fruits And Vegetables As Inhibitors. It is established through studies in worldwide epidemiology that individuals who regularly consume higher amounts fruits and vegetables have a lower risk of diverse types of cancer and of heart disease (42, 46, 47, 63). The relevant mechanisms are certainly complex. People with a high intake of such foods may be on dietary regimens that have other factors that lower risk, such as lower intake of diverse types of fat. Also, the vegetables are sources of vitamins and to some extent minerals with protective effects against cancer in several sites (42). They also include chemicals such as glucaric acid with inhibiting effects (64). In the context of the current chapter, fruits and vegetables also contain phenolic compounds including polyphenols, tannins and ellagic acid that may account, in part, for specific protective effects (60, 65-71).

Concluding Comments

Phenols of diverse structures are found in the human environment. They are large scale industrial products. They are the primary metabolites of aromatic chemicals, with benzene as the simplest prototype. Phenols are metabolized by mammalian enzyme systems, yielding conjugates, like glucuronides or sulfate esters. They can be oxidized to reactive quinones that in turn are detoxified readily by glutathione transferase. Phenolic compounds themselves are not genotoxic, but can cause cancer in animals at high dosages. They can be promoters, co-carcinogens, and also anticarcinogens, depending on chemical structure, host, tissue, dosage and chronicity of exposure (Table VI). Catechols, by generating H_2O_2 or hydroxy radicals, can display evidence of mutagenicity in prokaryotic organisms that have low levels of detoxification systems for these reactive products. In mammalian cell systems, on the other hand, phenols usually have no adverse effects because these systems have adequate defense mechanisms and can detoxify them, mainly by conjugation with type 2 enzymes. It is only with excessive dose levels that the detoxification potential is exceeded, and the oxidized products, quinones that react readily with proteins, yield evidence of toxicity. In turn, there maybe increased cell turnover, which when occurring chronically, can foster the development of neoplasia under conditions where hydroxy radicals are generated, and not neutralized.

Table VI. Diverse Effects of Phenols

"Carcinogenic" — Catechol, sesamol, caffeic acid, hydroquinone, BHA

"Co-carcinogenic" — Catechol, caffeic acid, hydroquinone, BHA

"Promoting" — Phenols, BHT, BHA

"Anticarcinogenic" — Catechol, quercetin, ellagic acid, chlorogenic acid, BHT, BHA, caffeic acid, tannins, flavanols, other polyphenols

More complex phenols such as the synthetic chemicals BHA and BHT or those occurring in vegetables and plants like quercetin, nordihydroguaiaretic acid, ellagic acid, or the tea polyphenols display inhibiting effects in cancers at certain specific sites. Thus, in the context of human health, phenols as a group and in particular those used as antioxidants by humans in foods and beverages have displayed beneficial effects. Optimal nutrition for chronic disease prevention means dietary traditions low in fats and salt, and with adequate amounts of fibers of diverse sources and structure, and of micronutrients and antioxidants from plants and fruits.

Acknowlegements

I am indebted to Ms. Beth-Alayne McKinney for providing excellent editorial service.
Research in my laboratory is supported by USPHS grants CA-42381 and CA-45720 from the National Cancer Institute and grant CN-29 from the American Cancer Society.

Literature Cited

1. IARC Monographs, *Phenol*; International Agency for Research on Cancer, Lyon, France, **1989**, *47*, 263-381.
2. IARC Monographs, *Dihydroxybenzenes*; International Agency for Research on Cancer, Lyon, France, **1977**, *15*, 155-171.
3. IARC Monographs, *ortho-Phenylphenol and its Sodium Salt*; International Agency for Research on Cancer, Lyon, France, **1987**, *30*, 329-344.
4. Witmer, M.C.; Snyder, R.; Jollow, D.J.; Kalf, G.F.; Kocsis, J.; Sipes, I.G. *4th International Symposium, Biological Reactive Intermediates*, Plenum Press, New York and London, 1991.
5. Dosemeci, M.; Blair, A.; Stewart, P.A.; Chandler, J.; Trush, M.A. *Epidemiol.*, **1991**, *2*, 188-193.
6. Andrews, L.S.; Snyder, R. In *Casarett and Doull's Toxicology*, Amdur, M.O.; Doull, J.; Klaassen, C.D., Eds.; 4th edition, Pergamon Press, New York, Oxford, 1991, pp. 681-722.
7. Robertson, M.L.; Eastmond, D.A.; Smith, M.T. *Mutat. Res.*, **1991**, *249*, 201-209.
8. Shibata, M.A.; Hirose, M.; Tanaka, H.; Asakawa, E.; Shirai, T.; Ito, N. *Jpn. J. Cancer Res.*, **1991**, *82*, 1211-1219.
9. Williams, G.; Weisburger, J.H.; In *Casarett and Doull's Toxicology*, Amdur, M.O.; Doull, J.; Klaassen, C.D., Eds.; 4th edition, Pergamon Press, New York, Oxford, 1991, pp. 127-200.
10. Cohen, S.M.; Ellwein, L.B. *Cancer Res.*, **1991**, *51*, 6493-6505.
11. Dean, B.J. *Mutat. Res.*, **1985**, *154*(3), 153-81.
12. Witz, G. *Proc. Exper. Soc. Biol. Med.*, **1991**, *198*, 675.
13. Ito, N.; Hagiwara, A.; Tamano, S.; Kagawa, M.; Shibata, M.A.; Kurata, Y.; Fukushima, S. *Jpn. J. Cancer Res.*, **1989**, *80*, 317-325.
14. Ito, N.; Hirose, M. *Adv. Cancer Res.*, **1989**, *53*, 247-302.
15. Wada, S.; Hirose, M.; Takahashi, S.; Okazaki, S. *Carcinogenesis*, **1990**, *11*, 1891-1894.
16. Hagiwara, A.; Hirose, M.; Takahashi, S.; Ogawa, K.; Shirai, T.; Ito, N. *Cancer Res.*, **1991**, *51*, 5655-5660.

17. Johnson, E.S. *Fund. Appl. Toxicol.*, **1990**, *14*, 219-234.
18. Kelly, S.J.; Guidotti, T.L. *Publ. Hlth. Rev.*, **1989-90**, *17*(1), 1-37.
19. IARC Monographs, *Occupational Exposures to Chlorophenols*; International Agency for Research on Cancer, Lyon, France, **1986**, *41*, 319-56.
20. World Health Organization, *Chlorophenols other than Pentachlorophenol*, Environmntl. Hlth. Criteria, **1989**, *93*, 1-208.
21. Holsapple, M.P.; Morris, D.L.; Wood, S.C.; Snyder, N.K. *Annu. Rev. Pharmacol. Toxicol.*, **1991**, *31*, 73-100.
22a. Guyton, K.Z.; Bhan, P.; Kuppusamy, P.; Zweier, J.L.; Trush, M.A.; Kensler, T.W. *Proc. Natl. Acad. Sci. USA*, **1991**, *88*, 946-950.
22b. Gimenez-Conti, I.; Viaje, A.; Chesner, J.; Conti, C.; Slaga, T.J. *Carcinogenesis*, **1991**, *12*, 563-569.
23. Hoffmann, D.; Hecht, S.S. In *Chemical Carcinogenesis and Mutagenesis I*, Cooper, C.S.; Grover, P.L., Eds.; Springer-Verlag: New York, 1990, pp. 61-102.
24. Slaga,T.J.; Klein-Szanto, A.J.P.; Boutwell, R.K.; Stevenson, D.E.; Spitzer, H.L.; D'Motto, B., *Proceedings of Symposium Dermal Carcinogenesis*, Alan R. Liss, New York, NY, 1989, 331-345.
25. Melikian, A.A.; Bagherti, K.; Hoffmann, D. *Cancer Res.*, **1990**, *50*, 1795-1799.
26. Flaks, B.; Flaks, A.; Shaw, A.P.W. *Acta Path. Microbiol. Immunol. Scand.*, **1985**, *93*, 367-377.
27. Corbett, M.D.; Corbett, B.R.; Hannothiaux, M.H.; Quintana, S.J. *Chem. Res. Toxicol.*, **1989**, *2*, 260-266.
28. Lee, C.A.; Thummel, K.E.; Kalhorn, T.F.; Nelson, S.D.; Slattery, J.T. *Drug Metab. Disp.*, **1991**, *19*, 966.
29. Birge, R.B.; Bulera, S.J.; Bartolone, J.B.; Ginsberg, G.L.; Cohen, S.D.; Khairallah, E.A. *Toxicol. Appl. Pharmacol.*, **1991**, *109*, 443-454.
30. Weisburger, J.H.; Weisburger E.K. *Pharm. Rev.*, **1973**, *25*, 1-66.
31. Stich, H.F. *Mutat. Res.*, **1991**, *259*, 307-24.
32. Schilderman, P.A.E.L.; Verhagen, H.; Schutte, B.; Ten Hoor, F.; Kleinjans, J.C.S. *Fd. Chem. Toxicol.*, **1991**, *29*, 79-85.
33. Morimoto, K.; Tsuji, K.; Iio, T.; Miyata, N.; Uchida, A.; Osawa, R.; Kitsutaka, H.; Takahashi, A. *Carcinogenesis*, **1991**, *12*, 703-708.
34. Lam, L.K.T.; Garg, P. *Carcinogenesis*, **1991**, *12*, 1341-1344.
35. Clayson, D.B.; Iverson, F.; Nera, E.A.; Lok, E. *Annu. Rev. Pharmacol. Toxicol.*, **1990**, *30*, 441-63.
36. Bolton, J.L.; Thompson, J.A. *Drug Metab. Disp.*, **1990**, *19*, 467-472.
37. Huang, M.T.; West, S.B.; Lu, A.Y.H. *J. Biol. Chem.*, **1976**, *251*, 4659-4665.
38. Ryan, D.E.; Levin, W. *Pharmacol. Ther.*, **1990**, *45*, 153-239.
39. Guengerich, F.P. *Pharmacol. Ther.*, **1990**, *45*, 299-307.
40. Powell, C.J.; Connolly, A.K. *Toxicol. Appl. Pharmacol.*, **1991**, *108*, 67-77.
41. Wakabayashi, K.; Nagao, M.; Sugimura, T. *Cancer Surv.*, **1989**, *8*, 385-99.
42. Weisburger, J.H. *Am. J. Clin. Nutr.*, **1991**, *53*, 226S-237S.
43. IARC Monographs, *Sex Hormones II*; International Agency for Research on Cancer, Lyon, France, **1979**, 583.
44. IARC Monographs *Perinatal and Multigeneration Carcinogenesis*, International Agency for Research on Cancer, Lyon, France **1989**, *96*, 436.
45. Gladek, A.; Liehr, J.G. *Carcinogenesis*, **1991**, *12*, 773-776.

46. Wattenberg, L.W. *Cancer Res.*, **1985**, *48*, 1-8.
47. Wattenberg, L.W. *Basic Life Sci.*, **1990**, *52*, 155-166.
48. Ernster, L. In *Xenobiotics and Cancer*, Ernster, L. *et al.*, Eds.; Japn. Sci. Soc. Press, Tokyo/Taylor & Francis Ltd., London 1991, 17-29.
49. Talalay, P.; Prochaska, H.J.; Spencer, S.R.; In *Xenobiotics and Cancer*, Ernster, L. *et al.*, Eds.; Japn. Sci. Soc. Press, Tokyo/Taylor & Francis Ltd., 1991, 177-817.
50. Temellini, A.; Franchi, M.; Giuliani, L.; Pacific, G.M. *Xenobiotica*, **1991**, *21*, 171-177.
51. Shali, N.A.; Curtis, C.G.; Powell, G.M.; Roy, A.B. *Xenobiotica*, **1991**, *21*, 881-893.
52. Cohen, L.A.; Choi, K.; Numoto, S.; Reddy, M.; Berke, B.; Weisburger, J.H. *J. Natl. Cancer Inst.*, **1986**, *76*, 721-30.
53. Tanaka, T.; Maruyama, H.; Maeura, Y.; Weisburger, J.H.; Zang, E.; Williams, G. *Cancer Res.*, **1992**, 52, In Press.
54. Suzuki, S.; Takada, T.; Sugawara, Y.; Muto, T.; Kominami, R. *Jpn. J. Cancer Res.*, **1991**, *82*, 1061-1064.
55. Verma, A.K.; Johnson, J.A.; Gould, M.N.; Tanner, M.A. *Cancer Res.*, **1988**, *48*, 5754-5758.
56. Deschner, E.E.; Ruperto, J.; Wong, G.; Newmark, H.L. *Carcinogenesis*, **1991**, *12*, 1193-1196.
57. Steele, C.M.; Lalies, M.; Ioannides, C. *Cancer Res.*, **1985**, *45*, 3573-3577.
58. Liu, L.; Castonguay, A. *Carcinogenesis*, **1991**, *12*, 1203-1208.
59. Maruyama, H.; Amanuma, T.; Nakae, D.; Tsutsumi, M.; Kondo, S.; Tsujiuchi, T.; Denda, A.; Konishi, Y. *Carcinogenesis*, **1991**, *12*, 1331-1334.
60. American Health Foundation, Proceedings International Workshop of the Health Effects of Camillia Sinensis-Tea, *Prev. Med.*, **1992**, *21*, (in press).
61. Yamane, T.; Hagiwara, N.; Tateishi, M.; Akachi, S.; Kim, M.; Okuzumi, J.; Kitao, Y.; Inagake, M.; Kuwata, K.; Takahashi, T. *Jpn. J. Cancer Res.*, **1991**,*82*, 1336-1339.
62. Wang, Z.Y.; Agarwal, R.; Bickers, D.R.; Mukhtar, H., *Carcinogenesis*, **1991**, *12*, 1527-30.
63. Hocman, G. *Comp. Biochem. Physiol.*, **1989**, *93*, 201-12.
64. Walaszek, Z. *Cancer Lett.*, **1990**, *54*, 1-8.
65. Committee on Diet and Health, Food and Nutrition Board, *Diet and Health: Implications for Reducing Chronic Disease Risk*, Washington, D.C., National Academy Press, 1990.
66. Tanaka, T.; Iwata, H.; Niwa, K.; Mori, Y.; Mori, H. *Jpn. J. Cancer Res.*, **1988**, *79*, 1297-1303.
67. Mandal, S.; Stoner, G.D. *Carcinogenesis*, **1990**, *11*, 55-61.
68. Wood, A.W.; Huang, M.T.; Chang, R.L.; Newmark, H.L.; Lehr, R.E.; Yagi, H.; Sayer, J.; Jerina, D.M.; Conney, A.H. *Proc. Natl. Acad. Sci.*, **1982**, *79*, 5513-5517.
69. Smart, R.C.; Huang, M.T.; Chang, R.L.; Sayer, J.M.; Jerina, D.M.; Wood, A.W.; Conney, A.H. *Carcinogenesis*, **1986**, *7*, 1669-1675.
70. Agarwal, R.; Wang, Z.Y.; Bik, D.P.; Mukhtar, H. *Drug Metab. and Disp.*, **1991**, *19*, 620-624.
71. Rossi, M.; Erlebacher, J.; Zacharias, D.E.; Carrell, H.L.; Iannucci, B. *Carcinogenesis*, **1991**, *12*, 2227-2232.

RECEIVED April 20, 1992

Chapter 4

Plant Phenolic Compounds as Inhibitors of Mutagenesis and Carcinogenesis

Harold L. Newmark

Laboratory for Cancer Research, Department of Chemical Biology
and Pharmacognosy, College of Pharmacy, Rutgers, The State University
of New Jersey, Piscataway, NJ 08855–0789
Memorial Sloan–Kettering Cancer Center, 1275 York Avenue,
New York, NY 10021

It has been known for decades that at least a few types of human
cancer are related to substances in our environment, i.e. the chemical
composition of our food, drink, atmosphere, as demonstrated by
defined tests for mutagenicity and carcinogenicity. Attention has
recently focussed on substances in the environment that act as antimutagens (desmutagens), or protective against carcinogenesis. Plant
phenolics, originally hypothesized to inhibit mutagenesis and/or
carcinogenesis by virtue of antioxidant or electrophile trapping
mechanisms, can also act as modulators of arachidonic metabolism
cascade pathways. Certain plant phenols can be effective inhibitors
of chemical mutagens, *in vitro*, and/or carcinogenesis *in vivo*. The
historical origins, hypothesis of actions, current status and potential
adverse effects of the utility of plant phenolics to reduce risk of cancer
are discussed, as well as future possibilities and needs and objectives
for future research.

It has been known for several decades that there are substances in commonly
consumed foods that reduce the incidence of chemically induced carcinogenesis in
laboratory rodents. In early pioneering studies, Wattenberg found that rodents on a
purified or semi-synthetic diet developed more chemically-induced lesions than on a
mixed "natural food" diet. Further studies elucidated many active "chemopreventative" substances in foods, including terpenes, aromatic isothiocyanates,
organosulfur compounds, protease inhibitors, dithiolthiones and indoles (1-3). Of
particular interest as chemopreventative agents were the monophenols, polyphenols,
flavones, flavonoids and tannins in foods, which may be consumed in large quantities
(up to 1-2 grams per day) in some human diets.

The Antioxidant Hypothesis. Wattenberg found several food antioxidants, such
as butylated hydroxyanisole (BHA), reduced the incidence of neoplasia induced by
some carcinogens in laboratory animals, and expanded the studies to show similar
effects for the plant phenolics caffeic and ferulic acids (4). The mechanism for BHA

0097–6156/92/0507–0048$06.00/0

inhibition of neoplasia was ascribed to alteration of carcinogen metabolism towards inactive products, but no explanation was offered specifically for the tumor inhibitory activities of caffeic and ferulic acids.

Inhibition of Nitrosation. In the mid 1970's Dr. W.J. Mergens and I found that caffeic and ferulic acids were highly effective consumers of nitrite ion, particularly in acid pH (5). This results in strong activity of these plant phenols, commonly present in many human foods, in preventing nitrosation of susceptible secondary amines and amides to form highly potent carcinogenic nitrosamines and nitrosamides *in vitro*, in our foods, and *in vivo* (6). Sources of nitrites in foods are almost ubiquitous, particularly in fermented or smoked foods, or added as aids to preservation, as in processed meats. In addition, nitrates naturally in our foods are readily recycled to the saliva after ingestion and absorption, and then reduced to nitrite by buccal flora, resulting in gastric nitrosation of susceptible amines. The function of dietary plant phenolics in blocking these reactions in foods in food processing and cooking and *in vivo* has probably been underrated as a major cancer prevention process.

Electrophile Radical Trap Hypothesis. The current axiom of chemical carcinogenesis is that many, perhaps most, carcinogens are converted by either non-enzymatic (in the case of direct acting carcinogens) or metabolic activation to highly reactive species that can attack cellular components. The best known form of the reaction species is the electrophilic reactant, possessing a positively charged group such as a carbonium ion, and which reacts with electron-rich moieties chemically termed nucleophiles. Many cellular components can be targets for such electrophilic attack, but the (probably minor) attack and resultant chemical and structurally alteration of DNA is believed to be a key step in carcinogenic initiation in the cells. Protection of the DNA in the cells is largely achieved by competitive efficient chemical nucleophiles in the cell such as glutathione, however, this protection can be overwhelmed.

Prodded by Dr. Allan Conney to consider additional methods of increasing protection of cellular DNA from activated carcinogen electrophilic attack, we realized that some plant phenolics, such as caffeic and ferulic acids could act as potent chemical nucleophiles, based on our previous studies of their reaction with nitrite (5). On testing as inhibitors of mutagenesis *in vitro* induced by benzo[a]pyrene diol epoxide, these plant phenolics were indeed found to be potent, particularly the related ellagic acid. The mechanisms of reaction, involving π bond interactions between the planar molecules involved, and capacity to act as electron-rich donors (i.e. electrophilic trap for electron-poor carcinogenic electrophiles) was reported in a series of papers by Wood, Huang, Chang, Sayer, Jerina, Conney, Newmark, and others as reviewed by Newmark (7,8). Thus, plant phenolics may be inhibitors of initiation processes in carcinogenesis.

Arachidonic Metabolism Modulation. It has long been known that several plant phenolics such as salicylic acid, quercetin and others can inhibit the cyclo-oxygenase pathway of arachidonic acid metabolism to prostaglandins. Some plant phenolics also inhibit lipoxygenase pathways to other prostanoids (9,10). Kato *et al.*, in their demonstration of inhibition of phorbol ester promotion of mouse skin tumors by quercetin suggested the possible involvement of lipoxygenase inhibition (11). Indeed, several inhibitors of lipoxygenase pathways of arachidonic acid metabolism in mammalian cells, as well as cyclo-oxygenase inhibitors have demonstrated anti-tumor activity (12). Arachidonic acid metabolism modulation appears to affect promotion rather than initiation processes in carcinogenesis. Plant phenolics as modulator of arachidonic metabolism (e.g. as lipoxygenase inhibitors) can act as inhibitors of carcinogenic promotion processes.

Alteration of Carcinogen Metabolism. The mechanism of inhibition of chemically induced carcinogenesis by BHA, and a few other food phenolic antioxidants has been related to altered metabolism of certain carcinogens, including: diminished microsomal metabolism to DNA-binding metabolites, decreased epoxidation, increased formation of readily conjugated and excreted metabolites, and enhanced activity and level of GSH and glucuronide conjugating enzymes (4). However, the plant food phenolics have not been extensively studied for activity in altering metabolism of carcinogens, except to suggest a wide range of potency (4).

Current Status. The hypotheses discussed above gave rational cause for experimentatial investigation of inhibitory effects of plant phenolics as inhibitors of mutagenesis (anti-mutagens or desmutagens) and carcinogenesis. With the information currently available, some of the original hypotheses seem less plausible and useful for further studies. The electrophile trap hypothesis is less attractive. The originally promising studies of anti-mutagenic activity were later shown to be partly dependent on reactions *in vitro* of the phenolics with the tested mutagen (often benzo[a]pyrene) outside the cell, before cell entry (Chang, R., Rutgers University, personal communication.) Also the phenolics appear highly reactive within the cells in a variety of functional systems and determination of specificity of effective phenolic amounts in reaching a tissue, entering the cells, and performing a useful tumor inhibitory function, with adequate safety to normal cell function, will probably require much further study. Most of the hypotheses above have some current applicability, but a uniform single mechanism is unlikely, since the plant phenolics appear to have a range of biochemical activities in the cells. Although each plant phenolic component will be somewhat specific in actions, tissue localization, and effects on stresses induced by initiation (e.g. genotoxic) as well as promotion (non-genotoxic) carcinogens, it seems more likely that plant phenolics, being multifunctional, can inhibit carcinogenesis by several activities simultaneously.

Many plant phenolics have been shown to be effective as antimutagens, particularly against activated aromatic carcinogens, somewhat less active against non-aromatic carcinogens (7,8). Several have shown moderate to strong activity as inhibitors of neoplasia development by chemical carcinogens in laboratory rodents.

Problems and Adverse Effects. In a series of studies, Ito and co-workers have shown that caffeic acid (2% of diet), sesamol (2% of diet) and catechol (0.8% of diet) could induce stomach cancer in rodents. However, there is a difference in species sensitivity, rats being more sensitive than mice (13). These effects seem to derive from a gastric mucosal hyperplasia stemming from irritation by the chronically ingested dietary phenolics in the studies.

For decades it has been known that phenols, especially ortho dihydroxy phenols (catechols), can readily oxidize in highly aerobic exposure. Trace metal ions such as copper, or iron present act as potent catalysts for oxidation of phenols *in vitro* where rate increases as the pH rises. The oxidation reaction produces hydrogen peroxide via an intermediate superoxide. Thus, when phenolic substances are tested for "mutagenicity" *in vitro* in an Ames-type assay, in highly aerobic conditions, and appreciable trace metals are present in the media, it is no surprise that the phenols appear to test positive as mutagens. This is probably largely the "mutagenicity" of the hydrogen peroxide produced in the media by the conditions of testing. In several instances, such as the apparent mutagenicity of caffeic acid, addition of catalase enzyme to the *in vitro* system virtually eliminated clastogenic activity, emphasizing the role of artifactual generation of hydrogen peroxide in laboratory tests for mutagenicity of plant phenolics (*14*). A re-evaluation of the literature of apparent mutagenicity of plant phenolics to eliminate artifactual errors due to hydrogen peroxide formation would probably find most of these phenolics free of mutagenic activity. Hydrogen peroxide formation in the dietary systems used in testing may also be partly responsible for promotion of stomach cancer in rats (*13*). Plant phenols, particularly high molecular weight polyphenols such as the gallotannins, can precipitate proteins by physico-chemical interactions. In higher concentrations in food products, such as strong black coffee without added milk as a neutralizing protein source, this can be a source of chronic gastric irritation.

Future Possibilities. A newly emerging area is the endogenous production of phenolic lignans (diphenolic compounds). These are produced from plant precursors (probably plant phenolics) through modification by the colon microflora, possibly the clostridia group (*15*). The two most common mammalian lignans are enterolactone and enterdiol. In limited studies, they appear to have tumor inhibitory properties, particularly as anti-estrogens. These lactones were quantified in human urine in subjects with a varied large range of high fiber cereal diets. Linseed (flaxseed) in the diet gave particularly high levels of urinary lignans (*16*). Recent studies seem to confirm these early reports (Thompson, D., University of Toronto, private communication.) This approach may be of practical use in reducing mammary and possibly colon cancer risk in human studies.

Tyrosine kinase and other protein kinases are enzymes involved in cell proliferation. Plant phenolics could be useful dietary inhibitors of these kinases, and act to reduce hyperproliferation of epithelial cells as a means of reducing cancer risk. Quercetin is an inhibitor of protein kinase C, tyrosine protein kinase and a specific protein kinase in rat colonic epithelium (*17*).

Needs and Objectives for Future Research. I wish to emphasize that plant phenols have multifunctional biochemical activities, as illustrated by quercetin in Table I. Most of these involve modulation of one or more processes thought to be involved in carcinogenesis development. While substances may be chosen as candidates for chemoprevention of carcinogenesis based on a single hypothesized mechanism, probably several activities are involved, adding to the total anticarcinogenic potential.

Table I. Multifunctional Activity of Phenolics: Quercetin

Antioxidant:	Lipids
Antimutagen:	PAH electrophiles
Anti-prostanoid:	Lipoxygenase inhibitor
Anti-kinases:	Inhibitor of tyrosine protein and other kinases

Plant phenolics, components of human foods, have shown interesting activities as inhibitors of mutagenic and carcinogenic processes. In order to utilize these properties for chemoprevention for reduction of risk for human cancer, much further work is needed. This includes further extension of anti-cancer studies, but also fundamental studies in allied areas, including:

1. Reliable food composition data of amounts of specific phenolics in fresh foods, and losses in processing, storage, etc. to get realistic estimates of dietary intake.

2. Absorption and metabolism. Little is known about the fate of most plant phenolics after ingestion. Rutin and quercetin are poorly absorbed (18) while caffeic acid appears well absorbed, but only one-fifth identified as urinary metabolites (19) in human studies. However, these studies were performed with pure crystalline substances, while in foods the substances are usually present as glycosides or esters, or in solution in the terpene-lipid components of the foods.

3. Cellular reactions of the plant phenolics with mammalian tissues, including mode and chemical form of delivery to target tissues, effects on cell membranes, attention of cytosolic enzymes and activation systems. Of theoretical interest would be information on the effects of individual plant phenolics on specific cytochrome P450 systems, and resultant effects on detoxification activities towards endogenous and xenobiotic substances.

Literature Cited.

1. Wattenberg, L.W. *Cancer Res.* **1985**, *45*, 1-8.
2. Wattenberg, L.W. *Proc. of the Nutrition Soc.* **1990**, *49*, 173-183.
3. Hartman, P.E.; Shanekl, D.M. *Env. and Mol. Mutagenesis*, **1990**, *15*, 145-182.
4. Wattenberg, L.W.; Coccia, J.B.; Lam, L.K.T. *Cancer Res.***1990**, *40*, 2820-2823.
5. Newmark, H.L.; Mergens, W.J. In: *Inhibition of Tumor Induction and Development*; Zedeck, M.S.; Lipkin, M., Ed.; Plenum Press: New York, NY, **1981**; pp 127-168.
6. Kuenzig, W.; Chan, J.; Norkus, E.; Holowaschenko, H.; Newmark, H., Mergens, W.; Conney, A.H. *Carcinogenesis* **1984**, *5*, 309-313.
7. Newmark, H.L. *Nutr. Cancer* **1984**, *6*, 58-70.
8. Newmark, H.L. *Can. J. Physio. Pharmacol.* **1987**, *65*, 461-466.
9. Dehirst, F.E. *Prostaglandin* **1980**, *20*, 209-214.
10. Baumann, J.; Wuma, G.; Bruchhausen, F. *Arch. Pharm(Weinheim).* **1980**, *313*, 330-337.
11. Kato, R.; Makadate, S.; Yamamoto, S.; Sugimura, T. *Carcinogenesis* **1983**, *5*, 1301-1305.
12. Karmali, R.A. In: *Biochemistry of Arachidonic Acid Metabolism*; Lands, W.E.M., Ed.; Martinas Nijhoff Publishing: Boston, MA; pp 203-212.
13. Hirosi, M.; Fukushima, S.; Shirai, T.; Hasegawa, R.; Kato, T.; Tanaka, M.; Asakawa, E.; Ito, M. *Jpn. J. Cancer Res.* **1990**, *81*, 202-212.
14. Hanham, A.F.; Dunn, B.P.; Stich, H.F. *Mutation Res.* **1982**, *116*, 333-339.
15. Adlercreutz, M. *Scand. J. Lab. Invest.* **1990**, *50, Suppl 201*, 3-23.
16. Horwitz, C.; Walker, A.P.R. *Nutr. Cancer* **1984**, *6*, 73-76.
17. Schwartz, B.; Fraser, G.M.; Levy, J.; Sharoni, Y.; Guberman, R.; Krawiec, J.; Lamprecht, S.A. *Gut* **1988**, *29*, 1213-1221.
18. Gugler, R.; Leschick, M., Dengler *Europ. J. Clin. Pharmacol.* **1975**, *9*, 229-234.
19. Jacobson, A.E.; Newmark, H.; Baptista, J.; Bruce W.R. *Nutri. Reports Int.* **1983**, *28*, 1409-1417.

RECEIVED November 20, 1991

Sources of Phenolic Antioxidants

Chapter 5

Natural Antioxidants from Plant Material

D. E. Pratt

Department of Foods and Nutrition, Purdue University,
West Lafayette, IN 47907

Natural antioxidants in foods may be from (a) endogenous compounds in one or more components of the food; (b) substances formed from reactions during processing; and (c) food additives isolated from natural sources. Most natural antioxidants are from plants. Most plants contain compounds that possess antioxidant activity. They are polyphenolics that occur in all parts of the plant - wood, bark, stems, leaves, fruit, roots, flowers, pollen and seeds. The antioxidant activities in these plants range from extremely slight to very great. Natural antioxidants may function (a) as reducing agents, (b) as free radical scavengers, (c) as complexers of pro-oxidant metals, and (d) as quenchers of the formation of singlet oxygen. The most common natural antioxidants are flavonoids (flavanols, isoflavones, flavones, catchins, flavanones), cinnamic acid derivatives, coumarins, tocopherols, and polyfunctional organic acids. The antioxidants of plants are phenolics. Some form complexes with metals. However, the major value is in their primary antioxidant activity (i.e., as free radical acceptors and as chain breakers.)

Many endogenous plant compounds retard oxidative processes in their natural environment and in products to which they have been added. The utilization of plant components that have been consumed for many centuries is very tempting. There is an implied assumption of safety for such compounds. One might avoid prolonged and expensive safety studies that are associated with synthetic materials. I do not intent to debate the issue of superiority of either natural or synthetic food components as to the safety or functional properties. However, it is desirable to use substances that do not pose problems of proof of safety. Caution should be employed in the use of natural compounds. They have not usually been subjected to scrutiny and scientific evaluation as have the artificial synthetic compounds (BHA, BHT, TBHQ). Their potential as

0097–6156/92/0507–0054$06.00/0
© 1992 American Chemical Society

mutagens, carcinogens, teratogens, or as other pathogens must be investigated. Certain problems other than toxicity, may be associated with the use of natural antioxidants: (a) the amount of active ingredient may vary with the source and (b)method of extraction; (l) some may impart off colors and off odors; (c) the cost of extraction may often be nearly inhibitive; and (d) undesirable reactions may occur with nutrients in the product.

Plant Sources of Natural Antioxidants

In the plant kingdom, the angiosperms account for between 250 and 300 thousand species. Of this number less than 400 species have been gathered or cullivated as human food which include 33 of the 51 orders and 89 of the 279 families. All parts of plants are eaten - roots, stems, leaves, flowers, fruit, and seeds - but in most species the edible portions are restricted to one part. Antioxidants have been isolated from, or detected in about one-half of these edible plants, but not always in the edible portions. the same compound, or group of compounds, are not always present throughout the plant.

Natural antioxidants occur in all higher plants and in all parts of the plant - wood, bark, stems, pods, leaves, fruit, roots, flowers, pollen, and seeds. These are usually phenolic or polyphenolic compounds. Typical compounds that possess antioxidant activity include tocopherols, flavonoids, cinnamic acid derivatives, phosphatides, and polyfunctional organic acids. The flavonoids include flavones, flavonols, isoflavones, catechins, flavonones, and chalcones. The cinnamic acid derivatives include caffeic acid, ferulic acid, chlorogenic acid and several others.

Use of natural compounds to prevent oxidation of lipids and other organic chemicals is not new. Perhaps the first comprehensive study and certainly one which first stimulated interest in the area of antioxidation was by Morreu and Dufraise (1). Nearly a century ago, these researchers investigated over 500 natural and synthetic compounds for antioxidant activity. Most of the studies immediately following were on ingredients to inhibit oxidation of non-food material, such as rubber, gasoline and plastics. However, it was these early studies that initiated our interest in the search for chemicals to regulate oxidation in food systems. In the early 1930's Musher (2) demonstrated that rancidity in lard could be inhibited by soy flour, oat flour and sesame flour. His greatest antioxygenic activity was from suspending a bag of defatted soy flour in melted lard.

Natural antioxidants in foods may be from (a) endogenous compounds in one or more components of the food; (b) substances formed from reactions during processing; and (c) food additives isolated from natural sources.

Some sources of natural antioxidants are shown in Table I. This is a general and abbreviated list. These materials contain several types of compounds that possess antioxidant activity (Table II).

Plant Compounds with Antioxidant Activity

Many of the flavonoids and related compounds (Tables III-V) have strong antioxidant characteristics in lipid-aqueous and lipid food systems. As may be seen certain flavones, flavonols, flavonones, flavanonals, and cinnamic acid derivatives have considerable antioxidant activity. These also occur in all parts of the plant (Table VI). The very low solubility of these compounds in lipids is often considered a disadvantage and has been considered a serious disadvantage if an aqueous phase is also present (3). However, flavonoids suspended in the aqueous phase of a lipid-aqueous system offer appreciable protection to lipid oxidation. Lea and Swoboda (4), nearly forty years ago, found that flavonols were effective antioxidants when suspended in lipid systems. Polyphenolic antioxidants, sparingly soluble in lipid systems, have been converted into readily fat-soluble form by alkylation or esterification with long chain fatty acids or alcohols. Such a procedure offers promising results with flavonoids.

Table I   Some sources of natural antioxidants

Algae	Oil seed
Amla	Olives
Citrus poly and peel	Osage orange
Cocoa powder or shell	Plant (extracts)
Heated products	Protein hydrolysate
Herbs and spices	Resin
Microbial products	Soy Products
Oat flour	Tempeh

Table II   Some components of natural antioxidants

Amino acid	Other organic acids
Ascorbic acid	Reductions
Carotenoids	Peptides
Flavonoids	Tannins
Melanoidin	Tocopherols

Factors Contributing to Antioxidant Activity

The action flavonol antioxidation is multi-functional. Flavonols form complexes with metals. Chelation occurs at the 3-hydroxy, 4-keto grouping and/or at the 5-hydroxy, 4-keto group, when the α ring is hydroxylated in the 5 position. An o-quinol grouping on the β-ring can also demonstrate metal-complexing activity (5, 6). However, the major value of flavonoids and cinnamic acids is in their primary antioxidant activity (i.e., as free radical acceptors and as chain-breakers).

The major evidence that these compounds work mainly as primary antioxidants is their ability to work equally well in metal catalyzed and uncatalyzed systems. They are also efficient antioxidants in systems catalyzed by relatively large molecules, such as heme and other porphyin compounds. They are also effective against lipoxygenase catalyzed reactions. These compounds cannot be envisaged as forming complexes with flavonols. In addition, hesperitin (5, 7, 3' - trihydroxy-4' methoxyflavone) which possesses as active metal-complexing site has demonstrated negligible antioxidant activity.

Table III Antioxidant activity of some flavonoids and related compounds

Compound[a]	Hours to reach a perioxide value of 50
Control (Stripped corn oil)	110
Hesperidin methyl chalcone	135
D-catechin	410
Chlorogenic acid	505
Caffeic acid	495
Quinic acid	105
Propyl gallate	435
p-comaric acid	120
Ferrulic acid	145

[a] 5×10^{-4M} in stripped corn oil.

Table IV Relative concentration of flavonoids and related compounds in plant tissue

Tissue	Relative concentration
Fruit	Cinnamic acids > catechins ≃ leucoanthocyanins (flavan 3,4-diols) > flavonols
Leaf	Flavonols ≃ cinnamic acids > catechins ≃ leucoanthocyanins > flavonols
Wood	Catechins ≃ leucoanthocyanins > flavonols > cinnamic acid
Bark	As wood but greater concentration

Table V   Antioxidant activity of flavonones

Compound[a]	Hrs to reach a peroxide value of 50
Control Stripped Corn Oil)	105
Aglycones:	
Naringenin (5,7,3'-Trihydroxy)	198
Dihydroquercetin (3,5,7,3',4'-Pentahydroxy)	470
Hesperitin (5,7,3'-Trihydroxy-4'-Methoxy)	125
Glycosides:	
Hesperitin (Hesperitin 7-Rhamnoglucoside)	125
Neohesperidin (Hesperitin 7-glucoside)	135

[a]5 x 10⁻⁴M in stripped corn oil.

^{a}5 x 10^{-4}M in stripped corn oil.

Table VI   Antioxidant Activity of Flavones

Compound[a]	Hrs to reach a peroxide value of 50
Control (Stripped Corn Oil)	105
Aglycones:	
Quercetin (3,5,7,3',4'-Pentahydroxy)	475
Fisetin (3,7,3',4'-Tetrahydroxy)	450
Myricetin (3,5,7,3',4',5'-Hexahydroxy)	552
Robientin (3,7,3',4',5'-Pentahydroxy)	750
Rhammnetin (3,5,3',4'-Tetrahydroxy 7-Methoxy)	375
Glycosides:	
Quercitrin (Quercetin 3-Rhamnoside)	475
Rutin (Quercetin 3-Rhamnoglucoside)	195

^{a}5 x 10^{-4}M in stripped corn oil

The position and the degree of hydroxylation is of primary importance in determining antioxidant activity (Figure 1). There is general agreement that ortho-dihydroxylation of the β ring contributes markedly to the antioxidant activity of flavonoids (9-14). The para-quinol structure of the β ring has been shown to impart even greater activity than the ortho-quinol structure; while the meta configuration has no effect on antioxidant activity. However, para and meta hydroxylation of the β ring apparently does not occur commonly in nature.

All flavonoids with the 3', 4'-dihydroxy configuration possess antioxidant activity. Two (robinetin and myricetin) have an additional hydroxyl group at the 5'-hydroxyl group, fisetin and quercetin. Two flavanones (naringenin and hesperitin) having a single hydroxyl group on the β ring possesses only slight antioxidant activity. Hydroxylation of the β ring is a major consideration for antioxidant activity.

Meta 5, 7-hydroxylation of the α ring apparently has little, if any, effect on antioxidant activity. This is evidenced by the findings that quercetin and fisetin have relatively the same activity and myricetin possesses the same activity as robinetin. Heimann and his associates (12,13) reported that meta 5,7-hydroxylation lowered antioxidant activity. To the contrary, Mehta and Seshadri (11) found quercetin to be a more effective antioxidant than 3,3',4'-trihydroxyflavone. Data from our laboratory support the finding of Mehta and Seshadri.

The importance of other sites of hydroxylation were studied by Lea and Swoboda (4); Mehta and Seshadri (11); Simpson and Uri (10). The two former groups found quercetagetin (3,4,5,7,3,4'-hexa-hydroxyflavone) and gossypetin (3,5,7,8,3',4'-hexahydroxyflavone) to be very effective antioxidants. Uri (14) found that the ortho-dihydroxy grouping on one ring and the paradihydroxy grouping on the other (i.e., 3,5,8,3'4'- and 3,7,8,2',5'-pentahydroxy-flavones) produced very potent antioxidants. These four polyhydroxy flavones are the most potent flavonoids, as antioxidants, yet reported in non-aqueous systems. Simpson and Uri (10) found 7-n-butoxy-3,2',5'-trihydroxyflavone to be the most effective antioxidant of 30 flavones studied in aqueous emulsions of methyl linoleate.

The 3 glycosides possess approximately the same antioxidant activity as the corresponding aglycone when the glycosyl substitution is with monosaccharide. In the case of rutin where the substitution is with a disaccharide antioxidant activity is reduced. The antioxidant capacity of a commercial preparation of rutin is considerable lower than the corresponding aglycone, quercetin. Kelley and Watts (6) studied the antioxidant effect of several flavonoids and found rutin somewhat inferior quercetin and quercitin but the differences were not as great as we have found. Chromatographic purification and the use of several commercially available samples (to eliminate the effect of possible contamination) did not alter the finding. Kelley and Watts (6), using a carotene-lard system also found that quercitin had approximately the same protection as quercetin. Crawford et al, (15) found that methylation of the 3-hydroxyl group of quercetin only slightly lowered antioxidant activity.

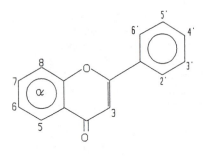

Flavones

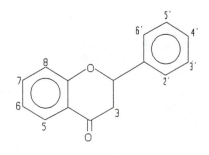

Flavanones

Quercetin 3,5,7,3',4'-penta OH
Fisetin 3,7,3',4'-tetra OH
Luteolin 5,7,3,4-tetra OH
Quercetin 5,7,3,4-tetra OH
3-O rhamnoside

Taxifolin 3,5,6,3',4'-penta OH
Fustin 3,7,3',4'-tetra OH
Eriodictyol 5,7,3,4-tetra OH

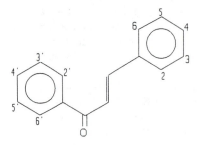

Chalcones

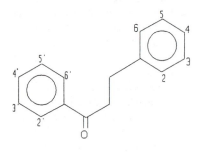

Dihydrochalcones

Butein 2',4',3,4-tetra OH
Okanin 2',3',4',3,4-penta OH

Phleoretin 2',4',6',4-tetra OH

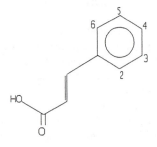

Cinnamic Acids

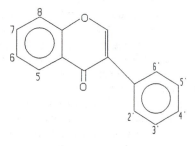

Iso-flavones

Caffeic acid 3,4-di OH
Ferulic acid 4-OH, 3-OMe

Daidzen 7,4'-di OH
Genistein 5,7,4'-tri OH

Figure 1   Some natural occurring flavonoids and cinnamic acids.

However, considerable importance has been attached to the free 3-hydroxyl by others (*10,11,12,16*). Mehta and Seshadri (*11*) postulated that the 3-hydroxyl and the 2,3 double bond allowed the molecule to undergo isomeric changes to diketo forms which would possess a highly reactive-CH group (position 2).

Dihydroquercetin was found to have the same antioxidant activity as quercetin indicating either that the 2,3, double bond is not of major importance to antioxidant activity or that conversion of dihydroquercetin to quercetin took place while the compound was in contract with the oxidizing fat. Mehta and Seshadri (*11*) suggested that conversion might account for the antioxidant activity of dihydroquercetin. However, chromatographic tests demonstrated that dihydroquercetin is not converted to quercetin by the hydrolysis procedure, nor could quercetin be chromatographically detected in the carotene-lard system in which dihydroquercetin was used as antioxidant. Dihydroquercetin was still present after 12 hours in the system.

Perhaps the greatest potential source of flavonoids for food antioxidants is from wood as a by-product of lumber and pulping operations. Whole bark of the douglas fir contains about five percent dihydroquercetin (3,4,7,3',4 pentahydroxyflavonone). The cork fraction, readily separated from the bark, contain up to 22% dihydroquercetin (*17*). Kirth (*18*) reported that approximately 150 million pounds of dihydroquercetin are potentially available annually in Oregon and Washington alone. Quercetin (3,5,7,3'4' pentohydroxyflavone) has been produced commercially as an antioxidant from wood sources (*19*). Quercetin is present in much lower amounts in wood and bark than is dihydroquercetin but quercetin can be obtained in quantity by oxidation of dihydroquercetin.

Pratt (*20*) and Pratt and Watts (*21*) identified a variety of antioxidant substances in hot water extracts from various vegetables (Table VII - X): green onions contained quercetin and green onion tops had quercetin, myristin, three glucosides of quercetin and one of myristin. Quercetin was also found in green pepper and three glucosides of quercetin and caffeic acid in green pepper seeds (Table X).

Table VII[a]   Effect of hydrolysis on aglycone fraction

| | Antioxidant index | |
Extract[a]	before hyd.	after hyd.
Potato	0.9	1.2
Green onion	1.9	3.8
Green pepper	1.8	4.0
Potato peel	1.8	4.3
Green onion top	3.8	7.0
Green pepper seed	4.2	>6.5

[a]10g/100 ml (aqueous)
[b]Coupled oxidation of carotene and linoleic acid. Antioxidant index = rate of bleaching of control/rate of bleaching of B-carotene in test solution. Bleaching rate measure at 470 nm.

Table VIII   Antioxidant activity of vegetable extracts

Extracts[a]	TBA number of roast beef slices at 3° C (3days)	Antioxidant index[a] carotene - lard at 45° C
Control	8.6	1.0
Tomato peel	8.4	1.0
Green onion	4.0	6.0
Potato peel	3.6	7.5
Celery	3.2	7.5
Green pepper	2.4	9.0
Green pepper seed	1.7	> 12.5
Green onion tops	1.7	> 12.6
50% Green onion tops	0.1	-

[a]10g/100 ml (aqueous)
[b]Footnote [b]Table VII

Table IX   Effect of isolated quercetin as antioxidant on roast beef slices

Source	Quercetin concentration x 10^{-5} M	TBA number (3 days)
Control	-	8.2
Potato peel	2.1	4.6
Green onion	2.4	4.1
Green pepper	2.6	3.8
Quercetin	5.0	2.9
Green onion top	6.3	2.5
Green pepper seed	6.8	2.2

Table X   Antioxidant indices of certain flavonoids

Flavonoid 5 x 10^{-5} M	Antioxidant index[a]
Quercetin (3,5,7,3',4' Penta OH)	3.5
Dihydroquercetin 12,3 Dihydro, 3,5,7,3',4' Penta OH	3.7
Quercitrin (5,7,3',4' Tetra OH 3-Rhamnoside)	3.6
Myricetin (3,5,7,3',4',5' Hexa OH)	4.3
Robinetin (3,7,3',4',5' Penta OH	4.3

[a]Footnote b Table VII

As mentioned earlier other plant constituents which might be expected to show antioxidant powers would be primarily phenolic compounds, especially o- and p- dihydroxy phenols such as the hydroxy cinnamic acids, caffeic and ferulic acids. While these acids usually occur in plant tissue as water-soluble esters, commonly chlorogenic acid or caffeoylquinic acid, and sugar esters, they have also been isolated as complex lipophilic esters of glycerol, long-chain diols, and ѡ-hydroxy acids. These lipophilic esters have been revealed as antioxidants in a comprehensive investigation of the antioxidants in oats (*23-26*). The caffeoyl esters have considerably more antioxidant activity than do those of ferrulic acid (*27-29*). Other lipid-soluble esters of ferulic acid with cycloartenol and other triterpenoids have been shown by Ohta et al. (*30*) to occur in rice bran oil, while a ferulate of dihydroxy-B-sitosterol has been isolated from maize by Tamura et al. (*31*). Wheat has also been shown to contain similar steroid esters (*32*).

Tea leaf powder and aqueous extracts of tea have antioxidant activity. Water and alcoholic extracts from black tea are generally less effective than extracts from green tea since phenolic substances in tea are oxidized during fermentation. Montedaro (*22*) has identified cinnamic acid derivatives, benzoic acid derivativies, and caffeic acid in olive oil.

Oilseeds As Source of Antioxidants

The literature is replete with reports of lipid antioxidants and lipid degradation in soybeans and soybean derived products. However, there are considerably fewer reports of antioxidants derived from other oil-seeds. Several phenolic compounds possessing antioxidant activity have been identified and isolated form soybeans. Flavonoids of soybeans are unique in that all identified and isolated flavonoids are isoflavones. The isoflavones occur in soybeans primarily as 7-0-monoglucosides of three iso-flavones (Figure 2). The glycosides are present in concentration of approximately 100 times that of the corresponding aglycone. The 7-0-monoglucosides of 5,7,5'-trihydioxyisoflavone (genistein) and 7,4'-dihydroxyisoflavone (daidzein) accounted for nearly 90% of the flavonoids. The genistein glucoside was present in 3.5 times the concentration of the daidzein glucoside. Only one other isoflavone glycoside has been found in fresh or dried soybeans, e.3., 7,4'-dihydroxy, 6-methoxy-isoflavone-7-0-monoglucoside. This compound was characterized and identified (*38-41*) and the aglycone named glycitein.

Another isoflavone, 6,7,4'-isoflavone has been shown to be present in several fermented soybean products and in extremely "browned" samples of soybean flakes. Several investigators have attempted to identify this compound in fresh and/or dried soybeans without success. TLC, HPLC, and GLC analyses, in our laboratory failed to demonstrate the presence of 6,7,4'-isoflavone in any unfermented soybean product. Antioxidant activities of soybean isoflavones are shown in Figure 2. Quercetin is shown only for comparison. Antioxidant activity shown here was determined by measuring the coupled oxidation of the β-carotene and linoleic acid (*42, 43*).

Genistein R_1 = H, R_2 = OH
Daidzein R_1 = H, R_2 = H
Glycitein   7 - O - R_1 = OCH, R_2 = H

Figure 2   Soybean isoflavones

Phenolic acids, including chlorogenic, isochlorogenic, caffeic, ferulic, p-coumaric, syringic, vanillic, and p-hydroxybenzoic acids are present in soybeans (*34, 35*), cottonseeds (*47*) and peanuts (*35*). The cinnamic acid derivatives, chlorogenic, isochlorogenic, and caffeic acids were found in significant concentrations in the three oil-seeds. These hydroxylated cinnamic acid derivatives possess appreciable antioxidant activity in lipid-aqueous systems. The presence of two isomers of chlorogenic acid, also ferulic acid, and several other phenolic acids, has been confirmed in hexane defatted soy flour. Antioxidant activity of soybean cinnamic acid derivatives are shown in Figure 3.

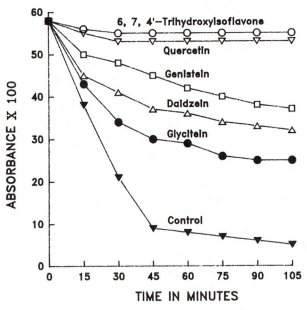

Figure 3   Antioxidant activity of soybean isoflavone (each 5 x 10^{-4}M). Measured by the bleaching time of β-carotene[a].

[a]In Figures 3, 4, and 5, 5 ml of an emulsion containing 0.01 mg β-carotene and 0.2 mg linoleic acid was incubated at 50 °C. Spectrophotometric readings were made at 15-minute intervals at 470 nm.

Soy HVP (hydrolyzed vegetable protein) can serve as both a primary antioxidant and synergist in food products (*44-46*). A number of low molecular weight phenolic compounds have been identified in soy HVP. Pratt et al (*34*) isolated three isoflavone aglycones-genistein, dizdzein and glycitein-from soy HVP (Table XI). They also found caffeic acid present in concentrations of 3.6 x 10^{-3} moles/kg in HVP. Thus it is not surprising that soy HVP is a very potent lipid antixidant. From the data presented by Pratt et al. (34), it is obvious that flavonoids and phenolic acids of soybeans are able to withstand the severe acid and heat treatments used in the preparation of HVP. Phenolic acids of soy HVP are shown in Table XII. These compounds are eseentially the same as those reported earlier (*33*); the exceptions are that chlorogenic and salicylic acids were not found in HVP. Obviously, the chlorogenic acid was hydrolyzed, accounting the for the high caffeic acid concentration. The potent antioxidant activity of soy HVP apparently results from the combined effect of phenolic compounds (isoflavones and phenolic acids) and peptide-amino acid mixtures (Figure 3, 4). Both may serve as primary and synergistic antioxidants.

Table XI   Isoflavones concentration of soy hydrolyzed vegetable protein and fresh soybeans (mole/kg)

	HVP	Fresh Soybeans
Genistein	2.0 x 10^{-3}	3.5 x 10^{-3}
Daidzein	0.6 x 10^{-3}	1.1 x 10^{-3}
Glycitein	0.3 x 10^{-3}	0.6 x 10^{-3}

Table XII   Phenolic acids with antioxidant activity of soy hydrolyzed vegetable protein[a]

Acid	Concentration mole/kg
Caffeic	3.6 x 10^{-3}
Ferulic	1.5 x 10^{-4}
p-coumaric	Trace
Syringic	1.8 x 10^{-4}
Vanillic	1.2 x 10^{-5}
Gentisin	Trace
p-hydroxybenzoic	Trace

[a]Pratt et al. (1982)

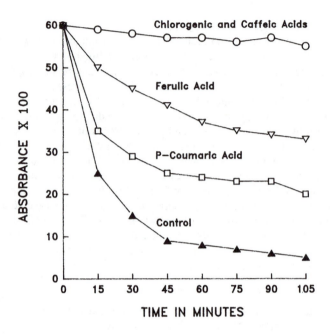

Figure 4   Antioxidant activity of soybeans cinnamic acids (each 5 x 10⁻⁴M). Measured by the bleaching time of β-carotene[a].

There have been considerable fewer reports on flavonoids of cottonseed and peanuts than there have been on soybeans. The work in our laboratory has not been as comprehensive on these oil-seeds as they have been on soybeans. Four flavonol aglycones and one flavanonol aglycone has been identified in cottonseed (47). The flavonols are quercetin, kaempferol, gossypetin and heracetin. The flavanonol is dihydroquercetin (Pratt, 1979). The flavonol glycosides that have been identified are rutin (quercetin 3-rhamnoglucosides) quercetrin (quercetin 3-rhamnoside) and isoquercitrin (quercetin 3-glucosides). The flavonoids of peanuts are apparently in very low concentrations. The only flavonoid that we have found in Spanish peanuts is dihydroquercetin.

The desert plant, chia, has been shown to be an excellent source of antioxidants (40) (*Table XIII*, Figure 5). In fact chia is the most potent source of antioxidants we have studied. It is interesting to note that the fatty acids of chia lipids are extremely unsaturated (42) (Table XIV).

[a]In Figures 3,4, and 5, 5 ml of an emulsion containing 0.01 mg β-carotene and 0.2 mg linoleic acid was incubated at 50 °C. Spectrophotometric readings were made at 15-minute intervals at 470 nm.

Table XIII   Concentration of Phenolic Acids in Chia Seed Extract[a]

Compounds	Concentration moles/Kg of chia seeds
Ferulic Acid	1.1×10^{-2}
Caffeic Acid	1.7×10^{-3}
Vanillic Acid	4.1×10^{-4}
Gentistic Acid	1.4×10^{-4}
Syringic Acid	1.3×10^{-4}
p-Coumaric Acid	1.2×10^{-4}

[a]Determined as TMS derivatives using BSTFA and pyridine. GLC separation on a 304 x 0.16 cm, 10% SE 30 column; column temperature 240°C, carrier gas nitrogen at 40 ml/min.

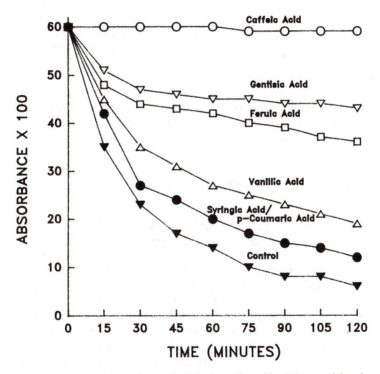

Figure 5   Antioxidant activity of chia phenolic acid. Measured by the bleaching time of β-carotene (concentration as in Table XIII)[a].

[a]In Figures 3, 4, and 5, 5 ml of an emulsion containing 0.01 mg β-carotene and 0.2 mg linoleic acid was incubated at 50 °C. Spectrophotometric readings were made at 15-minute intervals at 470 nm.

Table XIV   Fatty Acid Composition of Chia Seed Oil

Fatty Acid	A[a]	B[b]	C[c]
16:0	7.6	9.9	5.2
16:1	Tr	Tr	Tr
18:0	6.7	16.2	2.9
18:1	8.0	21.3	7.6
18:2	17.4	46.3	15.3
18:3	60.3	6.3	69.0
20:0	Tr	Tr	Tr
%Fat	22.8	26.2	32.5

[a]GLC separation of fatty acid methyl esters on a 180x0.3 cm, 10% DEGS column. Column Temperature, 185°C; carrier gas nitrogen at 40 ml/min.
[b]Determined by Taga et al., 1984.

A major portion of the antioxidant activity of oil-seed and oil-seed flours and concentrates is attributable to flavonoid and hydroxylated cinnamic acids. The combined influence of these isolated phenolic compounds accounts for nearly all antioxidant activity of soybeans, soy flours and concentrates, cottonseed, and cottonseed flour. They account for an appreciable amount of activity of peanuts and peanut flour. Although flavonoids and hydroxylated cinnamic acids are present in soy isolates, their combined antioxidant power by no means accounts for all of the antioxidant activity.

Spices As Source of Antioxidants

Spices and herbs have been used for many centuries to enhance flavour and extend the keeping times of various foods. However, it was not until the work of Chipault et al.(48, 49) that spices were compared as antioxidants in various fat sources. These authors showed the effects of spices in several food systems. In oil-in-water emulsions cloves appeared to be the most impressive as they did in ground pork. Allspice, cloves, sage, oregano, rosemary and thyme possess antioxidant properties in all fats in which they were tested.
    In general flavor and odor have made antioxidants from spices and herbs objectionable in foods. Rosemary and sage have been extensively studied as antioxidants. Chang et al (50) developed a technique for purifying the antioxidants of rosemary. Bracco et al (51) reported an industrial process for obtaining the antioxidant. The antioxidant activity of rosemary depends primarily on the concentration of carnosic acid and rosmaric acid (Figure 6), clearly a derivative of caffeic acid. Flavonoids and related polyphenolic compounds may also have detrimental effect in food systems. Certain flavonoids and related hydroxylated cinnamic acids serve as substrates for

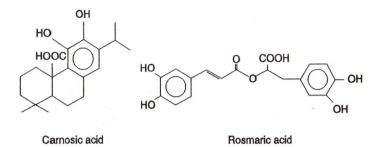

Carnosic acid   Rosmaric acid

Figure 6   Structures of carnosic and rosmaric acids

poly-phenoloxidase catalyzed browning reactions. Flavanonols possessing the o-diphenolic grouping on the β ring (e. g., dihydroquercetin) and o-diphenolic cinnamic acids (e. g., caffeic and chlorogenic acids) are substrates for enzymatic browning. Compounds with these configurations also possess antioxidant activity. When one or both of the o-hydroxyl groups are replaced by methyl esters, theses compounds inhibit polyphenoloxidase catalysis. Another detrimental effect of some p-diphenolic compounds is thiamine inactivation. Caffeic acid and catechol have been shown to destroy thiamine activity presumably by an oxidation-reduction reaction (*52, 53*). More recent work in this laboratory has demonstrated that certain flavonoids including quercetin and dihydroquercetin also had antithiamine activity. The reactions are both pH and temperature dependent.

In light of several recent investigations, flavonoids (particularly flavonols) as mutagens and possible carcinogens must be seriously considered. Other polyphenolic antioxidants have been eliminated from our food as carcinogenic (for example, nordihydroguaiaretic acid, Figure 7).

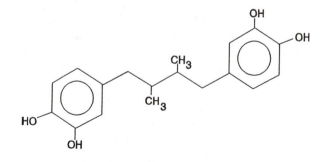

Figure 7   Nordihydroguaiaretic acid

Biological significance of Antioxidants

The biological significance of food antioxidants can be divided into two phases. The first phase would be in the prevention of autoxidation of polyunsaturated fatty acids. There is general consensus among scientists that free radicals formed during the autoxidation processes are toxic and detrimental to health. Although there is abundant literature on the consumption of free radicals, the ramification of such ingestion has not been completely elucidated. Secondly, there is ample evidence that fatty acid in various tissues of the body can peroxidize in vitro and form toxic free radicals. Such perioxidation can be retarded by ingestion of suitable antioxidants. Both of these phases are discussed in detail in other presentations at this symposium.

Literature Cited
1. Blanck, F.C.; *Handbook of Food and Agriculture,* Reinhold, New York, 1955.
2. Musher, S., *Food Ind.* 1935, *7,* 167.
3. Heimann, W.; Heimann, A.; Gremminger, M. & Holland, H.; Fette u. Seifen; 1953, *55,* 394.
4. Lea, C. H.; Swoboda, P. A. T. *Chem. Ind.* 1956, 1426.
5. Ramsey, M. B.; Watts, B. M. *Food Technol.* 1963, *17,* 1056.
6. Kelley, G. G.; Watts, B. M. *Food Research.* 1957, *22,* 308.
7. Pratt, D. E.; Watts, B. M. J. *Food Sci.* 1964, *29,* 27.
8. Pratt, D. E. J. *Food Sci.* 1965, *30,* 737.
9. Hudson, B. J. F.; Lewis, J. I. *Food Chem.* 1965, *10,* 47.
10. Simpson, T. H.; Uri, N. *Chem. Ind.* 1956, 956.
11. Mehta, A. C.; Seshadri, T. R. J. *Sci. Ind. Research* 1959, *18B,* 24.
12. Heimann, W. Heimann, A.; Gremminger, M.; Holman, H. H. Fette u. Soifen. 1953, *55,* 394.
13. Heimann, W.; Reiff, F. Fette u. Soifen. 1953, *55,* 451.
14. Uri, N. 1961. *Autoxidation and Antioxidants.* Mechanism of antioxidation. Chapter 4. (Edited by W. O. Lundbert) Interscience Publishers, New York.
15. Crawford, D. L.; Sinnhuber, R. O.; Aft. H. J. *Food Sci.* 1962, *26,* 139.
16. Lea, C. H.; Swoboda, P. A. T. *Chem. Ind.* 1956, 1426.
17. Hergert, H. L.; Kurth, E. F. Tappi 1952, *35,* 59.
18. Kirth, E. F. Ind. *Eng. Chem.* 1953, *45,* 2096.
19. Anon. *Chem. Eng. News.* 1958, *36,* No.7, 58.
20. Pratt, D. E. J. *Food Sci.* 1965, *30,* 737.
21. Pratt, D. E.; Watts, B.M. J. *Food Sci.* 1964, *29,* 27.
22. Montedaro, G. *Sci. Technol.* degli Alin. 1972, *2,* 177.
23. Daniels, D. G. H.; King, H. G. C.; Martin, H. F. J. *Sci. Food Agr.* 1963, *14,* 385.
24. Daniels, D. G. H., Martin, H. F. *Chem. Ind.* 1964, *2058.*
25. Daniels, D. G. H.; Martin, H. F. J. *Sci. Food Agr.* 1967, *18,* 589.
26. Daniels, D. G. H.; Martin, H. F. J. *Sci. Food Agr.* 1968, *19,* 710.

27. Dziedzic, S. Z. F; Hudson, B. J. F. *Food Chm.* 1984, *14*, 45.
28. Dziedzic, S. Z. F.; Hudson, B. J. F. *Food Chm.* 1983, *12*, 205.
29. Dziedzic, S. Z.; Hudson, B. J. F.; Barnes, G. *J. Agric. Food Chem.* 1985, *33*, 244.
30. Ohta, G.; Shimuzu, M. *Pharm. Bull.* 1957, (Tokyio) *5*, 40.
31. Tamura, T.; Sakaedani, N.; Matsumoto, T. Nippon Kagaku Zasshi. 1958, *29*, 1011.
32. Tamura, T.; Hibino, T.; Yokoyama, D.; Matsumoto, T. Nippon Kagaku Zasshi 1959, *80*, 215.
33. Hammerschmidt, P. A.; Pratt, D. E. *J. Food Sci.* 1978, *43*, 556.
34. Pratt, D. E.; Birac, P. M. *J. Food Sci.* 1979, *44*, 1720.
35. Pratt, D. E. *In Flavor Chemistry of Fats and Oils*, ed. D. B. Min & T. M. Smouse. American Oil Chemical Society, Champaign, IL 1985.
36. Phillip, F., PhD thesis, Purdue University, W. Lafayette, IN, 1974.
37. Pratt, D. E. *In Antioxidation and Antioxidants*, ed. M. G. Simic. Plenum Publishing Corp., New York, 1980.
38. Naim, M.; Gestetner, B.; Bondi, A.; and Dirk, Y. J. Agric. Food Chem. 1976, *24;* 1174.
39. Naim, M.; Gestetner, B.; Zikah, S.; Birk, Y.; Bondi, A. *J. Agric. Food Chem.* 1974, *22*, 806.
40. Pratt, D. E. *J. Food Sci.* 1972, *37*, 322.
41. Hemmerschmidt, P. A.; Pratt, D. E. *J. Food Sci.* 1978, *43*, 556.
42. Taga, M. S.; Miller, E. E.; Pratt, D. E. *JAOCS.* 1984, *65*, 928.
43. Araujo, J. M. A. PhD thesis, Purdue University, W. Lafayette, IN 1981.
44. Bishov, S. J.; Henich, A. S. *Food Technol.* 1972, *37*, 873.
45. Bishov, S. J.; Henich, A. S. *Food Technol.* 1975, *40*, 345.
46. Pratt, D. E.; Di Pietro, C.; Porter, W. L.; Giffee, J. W. *J. Food Sci.* 1981, *47*, 24.
47. Whiltern, C. C.; Miller, E. E.; and Pratt, D. E. *JAOCS.* 1984, *61*, 1075.
48. Chipault, J. R.; Mizuno, G. K.; Hawkins, J. M.; Lundbergh, W. O. Fd. Res. 1952, *17*, 46.
49. Chipault, J. R. 32 spices gauged as antioxidants. *Food Eng.* 1957, *29*, 134.
50. Chang, S. S.; Ostric-Matijasevic, B.; Hsieh, O. A. H.; Huang, C-Li. Natural antioxidants from rosemary and sage. *J. Food Sci.* 1977, *42*, 1102.
51. Bracco, U.; Löliger, J.; Viret, J. L. Production and use of natural antioxidants. *JAOCS.* 1981, *58;* 686-90.
52. Wesuing, P. H.; Freed, A. M.; Hoag, J. B. *J. Biol. Chem.* 1946, *165*, 737.
53. Davis, J. S.; Somogyi, J. C. Inter. *J. Vita Res.* 1969, *39*, 401.
54. Yang, P. F.; Pratt, D. E. *J. Food Sci.* 1984, *49*, 489, 7.

RECEIVED January 13, 1992

Chapter 6

Natural Antioxidants from Spices

Nobuji Nakatani

Department of Food and Nutrition, Osaka City University, Sumiyoshi,
Osaka 558, Japan

There is currently great worldwide interest in finding new
and safe antioxidants from natural sources to prevent oxidative deterioration of foods and to minimize oxidative
damage to living cells. Our study has focused on spice constituents or isolates which are functional antioxidants yet
do not have the overpowering or undesireable organoleptic
properties of the whole spice. New antioxidative compounds
were isolated from various spices and herbs and their structures were determined by chemical and spectroscopic means.
Examples of new antioxidants include: phenolic diterpenes
from rosemary and sage, phenolic carboxylic acids from
oregano, biphenyls and flavonoids from thyme, phenolic
amides from pepper and chili pepper, and diarylheptanoids
from ginger. All the phenolic compounds have antioxidant
activities that are more effective than the natural occurring antioxidant, α-tocopherol, and are comparable to the
synthetic antioxidants, BHA and BHT. In addition to antioxidant activity, this study assessed the stability and
utility of these compounds in food systems.

Much research is concerned with finding ways to prevent or to delay
the deterioration of foods. One of the principle causes of food
deterioration is the oxidation of fats and oils. Even a 1% oxygen
level can cause reactions that produce undesirable flavor, rancid
odor, discoloration and other spoilage. In particular, lipids
containing large amounts of unsaturated fatty acids can be easily

oxidized to lipid peroxide or hydroperoxide, which then rapidly decomposes by radical chain reaction to give lower molecular compounds such as aldehyde, ketone, epoxide, carboxylic acid and others. These compounds contribute to food rancidity. This process is outlined in Figure 1. These lipid peroxides damage not only foods but also living organisms. Various efforts have been made to minimize the oxidation reaction in foods such as by reducing the amount of oxygen in the headspace of food, filling with inert gas or addition of antioxidants.

Antioxidants from Natural Souces

Synthetic antioxidants such as *tert*-butyl-4-hydroxyanisol (BHA) and *tert*-butyl-4-hydroxytoluene (BHT), and to a lesser extent natural antioxidants as α-tocopherol, have been widely used to depress rancidity of fats and oils. The use of synthetic anti-oxidants as food additives is restricted in several countries, because of the possibility they could cause an undesirable effect on the enzymes of human organs. Consequently, many reseachers have searched for antioxidative compounds from natural sources. For example, tocopherols and flavonoids are abundant plant consti-tuents, nordihydroguaiacol (NDGA) is found in creosote bush (*Larrea divaricate*), gossypol in cotton seed and sesamol in sesame oil and orizanol in rice germ.

Antioxidants from Spices

Our attention for the developement of safe and effective antioxi-dants has been focused on edible plants, especially spices, because man has used them not only for flavoring foods but also for anti-septic or medicinal properties, since the prehistoric era (Table I). Researchers such as Chipault *et al.* examined more than 70 spices for antioxidative activity. They reported that rosemary and sage were remarkably effective antioxidants and that oregano, thyme, nutmeg, mace and turmeric also retarded the oxidation of lard. In an oil-in-water emulsion, clove showed extremely high antioxidant activity, and to a lesser degree did turmeric, allspice, mace, rose-mary, nutmeg, ginger, cassia, cinnamon, oregano, savory and sage (*1, 2*). The practical effectiveness of ground spices was examined in different types of food (*3*). Other studies have been concerned with the antioxidative effect of spices and spice extracts (*4–9*). Results of these studies have promoted research to isolate active components. Herrmann reported the activity of labiatic acid (rosmar-inic acid), caffeic acid and other phenolic carboxylic acids from

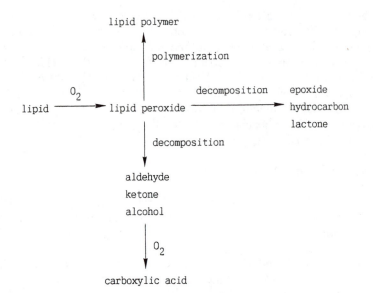

Figure 1. Oxidation process of lipid.

Table I.   Taxonomical classification of spices

Angiospermae	Dicotyledoneae	Sympetalae	Tubiflorae	Labiatae	basil, marjoram, mint, oregano, rosemary, sage, savory, thyme
				Solanaceae	chili, paprika, red pepper
				Pedaliaceae	sesame
			Campanulatae	Compositae	chamomile, chicory, tarragon
		Archichlamydeae	Piperales	Piperaceae	cubeba, long pepper, pepper
			Ranales	Myristicaceae	mace, nutmeg
				Lauraceae	bay leaf, cassia, cinnamon
				Magnoliaceae	star-anise
			Rhoeadales	Cruciferae	mustard, wasabi
			Myrtiflorae	Myrtaceae	allspice, clove
			Umbelliflorae	Umbelliferae	anise, caraway, celery, chervil, coriander, cumin, dill, fennel, parsley
	Monocotyledoneae		Liliiflorae	Liliaceae	garlic, onion
				Iridaceae	saffron
			Scitamineae	Zingiberaceae	cardamom, ginger, turmeric
			Orchidales	Orchidaceae	vanilla

the family Labiatae such as rosemary, sage and pepermint (*10*).

Eugenol, isoeugenol, thymol and other volatile phenolics exhibit remarkable antioxidative activity, but they possess too strong a characteristic odor to be used as food additives. Other spices have characteristics which limit their practical use. Curcumin is isolated from turmeric which has an intense yellow color while capsaicin is contained in pungent red peppers. Consequently, our studies and research interest is aimed at the isolation of components from the nonvolatile fractions of spices. These will have the desirable antioxidant properties without the taste, odor or color limitations of the whole spice or essential oil. The following sections will address a number of spices with known antioxidative activity. Current and previous research findings will be described with emphasis placed on the phenolic compounds responsible for the antioxidative activity.

Rosemary (*Rosmarinus officinalis* L.). Rosemary leaves is one of the most effective spices widely used in food processing. Chang *et al.* patented a process for the extraction of its active components (*11*). Brieskorn *et al.* isolated and described a phenolic diterpene, carnosol (1)(*12*). Houlihan *et al.* found additional two compounds (5,6) (*13,14*).

We examined in detail the extracts of rosemary leaves by fractionation and purification. Diterpenoids were isolated from the weakly acidic fraction of nonvolatile components (*15–17*). Four of these compounds (1–4,Fig. 2) showed remarkable antioxidative effect against the oxidation of lard (Fig. 3) and linoleic acid (*18*).

A new compound named rosmanol (2) was the most active and showed more activity than α−tocopherol or BHT. The molecular structure was determined to be an abietane skeleton bearing a five−membered lactone moiety, which was clarified by IR spectum. [1]H−NMR and [13]C−NMR spectra confirmed the stereo structure, which was further supported by X−ray analysis. Two more related compounds were newly obtained, being accompanied with carnosol. All these diterpenoids are suitable for food additives being colorless, odorless and tasteless.

The stability of antioxidant, carnosol (1) was determined by adding it to a buffered solution of linoleic acid and allowing it to stand at $40^{\circ}C$ in the dark. In the first week, carnosol was rapidly oxidized to orthoquinone, but the oxidation of linoleic acid was still depressed. During the second week, hydrolysis occurred on the lactone ring to give carboxylic acid. After three weeks, more oxidized compounds were formed and the peroxide value of linoleic acid began to increase.

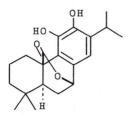

carnosol

1

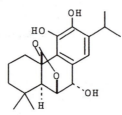

rosmanol

2

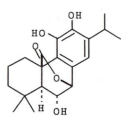

isorosmanol

3

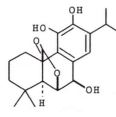

epirosmanol

4

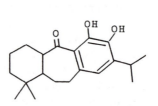

rosmaridiphenol

5

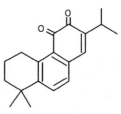

rosmariquinone

6

Figure 2. Structures of antioxidative compounds isolated from rosemary.

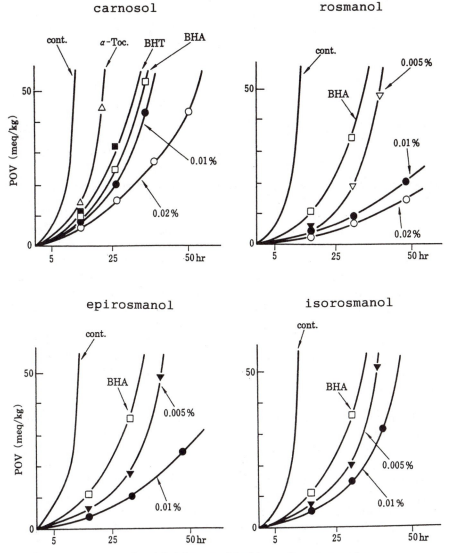

Figure 3. Antioxidative activity of compounds
isolated from rosemary (AOM) (α-Toc., BHA, BHT: 0.02%)
(Reproduced with permission from ref. 17, 18. Copyright 1983, 1984 The Agricultural Chemical Society
of Japan.)

Oregano (*Origanum vulgare* L.). Oregano is one of the species belonging to the family Labiatae. The water soluble fraction of methanol extract was purified with polyamide chromatography to give five polar compounds (7– 11) (Fig. 4). All of them showed higher antioxidative than did α–tocopherol, and in particular the activity of a glucoside 10 was comparable to that of BHA against linoleic acid oxidation (*19, 20*). The structure of 10 was confirmed by synthesis.

Thyme (*Thymus vulgaris* L.). Thymol and carbacrol are the primary components which contribute to the characteristic aroma of essential oil of thyme. These are also known to have antioxidative properties. After removal of essential oil, the nonvolatile portion was frac–tionated by different pHs. Antioxidative activity was found in the weakly acidic fraction, which was repeatedly purified by chromato–graphy to give five new biphenyls, dimers of thymol, and six highly methoxylated flavonoids (Fig. 5). Among them, compounds 12, 14, 18, 20 and 21 showed activity as strong as BHT (*21, 22*). The biphenyls obtained here also possessed significant deodorant properties. (*23*).

Pepper (*Piper nigrum* L.). In Chipault's report (*1*), the antioxi–dative activity of black and white pepper was low. However, when the methylene chloride extract was fractionated, the weakly acidic frac–tion exhibited potential activity. Five phenolic amides (20–24) were isolated and determined as major antioxidatives (Fig. 6) (*24–26*). These compounds showed no pungent taste in spite of their structure which similar to that of piperine, a pungent principle of pepper. This lack of pungency is necessary if the compounds are to find application as a food antioxidant.

Chili pepper (*Capsicum frutescens* L.). Chili pepper and red pepper (*C. annum* L.) are popular spices belonging to the genus *Capcicum.* They contain capsaicin (25), a pungent principle, and shows a significant antioxidative effect. In the course of our study, a new capsaicinoid, named capsaicinol (26), was isolated from chili pepper (Fig. 7) (*27*). This compound exhibited a potent antioxidative effect without the taste or pungency of capsaicin. Capsaicinol was synthesized in 18% overall yield starting from δ–valelolactone and the stereochemistry of the natural component was determined to be R–(–)–capsaicinol (Fig. 8) (*28*).

Ginger (*Zingiber officinale* Roscoe). Rhizomes of ginger are com–monly used as spices and in folk medicine. Pungent components, [6]–gingerol and [6]–shogaol, are known to possess moderate antioxi–

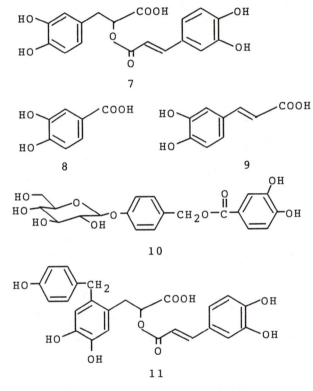

Figure 4. Structures of antioxidative compounds isolated from oregano.

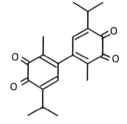

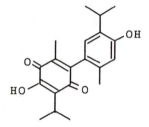

12   13   14

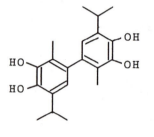
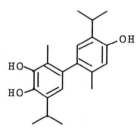

15   16

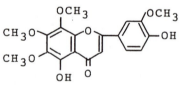

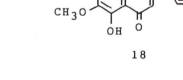

17   18

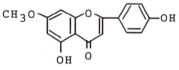

19

Figure 5. Structures of antioxidative compounds
isolated from thyme.

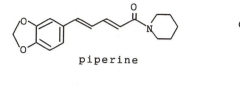

piperine

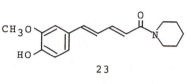

22

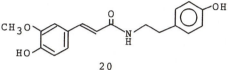

20

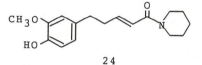

23

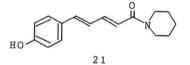

21

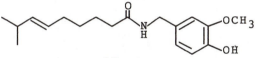

24

Figure 6. Structures of antioxidative compounds isolated from pepper.

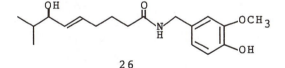

25

26

Figure 7. Antioxidative capsaicinoids isolated from chili pepper.

dative activity. After removing the essential oil by steam distil–lation from the ginger extract, a nonvolatile fraction was purified by chromatography to afford four gingerol related compounds and four diarylheptanoids (*29–30*). All these compounds displayed strong antioxidative activity at a concentration as low as 4μM as shown in Figs. 9 and 10. Diarylheptanoids are also found in turmeric (*Curcuma domestica* L.), a spice which contains curcumin, a yellow pigment.

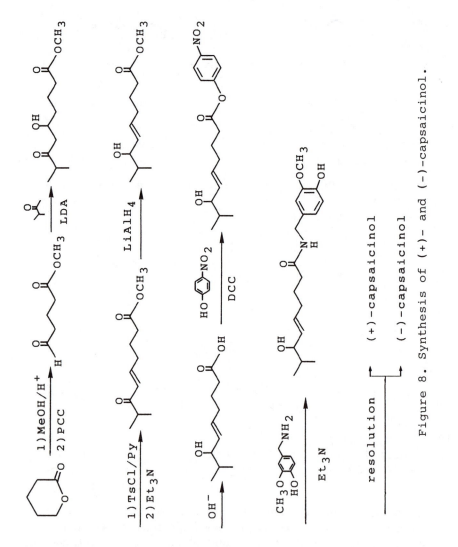

Figure 8. Synthesis of (+)- and (-)-capsaicinol.

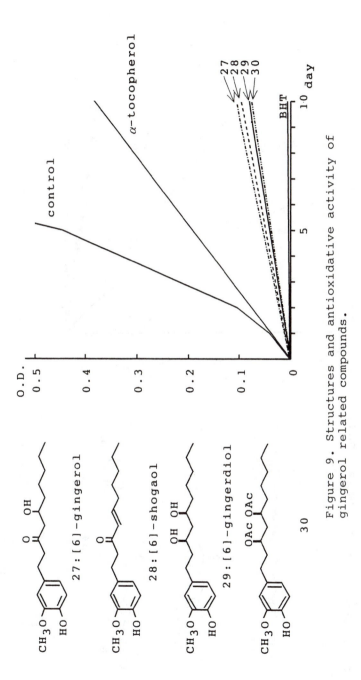

Figure 9. Structures and antioxidative activity of gingerol related compounds.

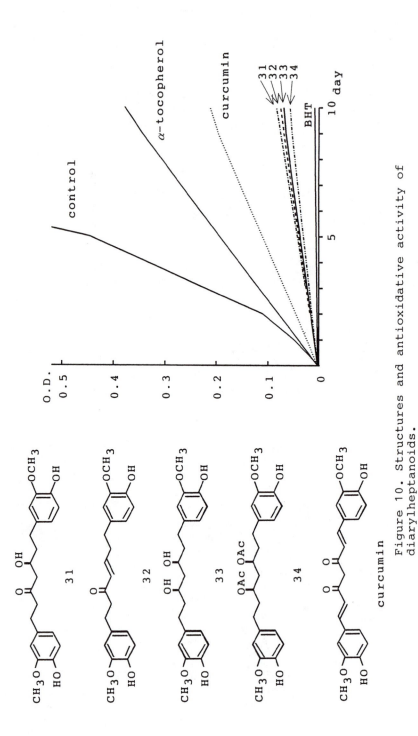

Figure 10. Structures and antioxidative activity of diarylheptanoids.

Antioxidative compounds have been searched and investigated not only to prevent oxidative deterioration in food but also to depress the production of lipid hydroperoxide in living cells. We have focused on the property of spices as one of the natural sources and studied their chemical constituents. Several remarkably active compounds were isolated and their structures were elucidated. Structure–activity relationship is of much interest and should be investigated. Thus it is seen that components of spices offer significant opportunities in the search for new and safe antioxidants.

Literature Cited

1   Chipault, J.R.; Mizuno, G.R.; Hawkins, J.M.; Lundberg, W.O. *Food Res.* 1952, 17, 46.
2   Chipault, J.R.; Mizuno, G.R.; Hawkins, J.M.; Lundberg, W.O. *Food Res.* 1955, 20, 443.
3   Chipault, J.R.; Mizuno, G.R.; Lundberg, W.O. *Food Technol.* 1956, 10, 209.
4   Sethi, S.C.; Aggarwal, J.S. *J. Sci. Ind. Res. Sect. B.* 1956, 15B, 34.
5   Lewis, E.J.; Watts, B.M. *Food Res.* 1959, 23, 274.
6   Cort, W.M. *Food Technol.* 1974, 28, 60.
7   Palitzsch, A.; Schulze, H.; Lotter, G.; Steichele, A. *Fleischwirtschaft* 1974, 54, 63.
8   Bishov, S.J.; Masuoka, Y.; Kapsalis, J.G. *J. Food Processing and Preservation* 1977, 1, 167.
9   Economou, K.D.; Oreopoulou, V.; Thomopoulos, C.D. *J. Amer. Oil. Chem Soc.* 1991, 68, 109.
10 Herrman, K. *Z. Lebensmitt. –Untersuch* 1962, 116, 224.
11 Chang, S.S.; Ostric–Matijevic, B.; Hsieh O.A.L.; Huang, C.L. *J. Food Sci.* 1977, 42, 1102.
12 Brieskorn, C.H.; Doemling, H.J. *Z. Lebensm. Unters. Forsch.* 1969, 141, 10.
13 Houlihan, C.M.; Ho, C–T.; Chang, S.S. *J. Amer. Oil Chem Soc.* 1984, 61, 1036.
14 Houlihan, C.M.; Ho, C–T.; Chang, S.S. *J. Amer. Oil Chem Soc.* 1985, 62, 96.
15 Nakatani, N.; Inatani, R. *Agric. Biol. Chem.* 1981, 45, 2385.
16 Inatani, R.; Nakatani, N.; Fuwa, H.; Seto, H. *Agric. Biol. Chem.* 1982, 46, 1661.
17 Nakatani, N.; Inatani, R. *Agric. Biol. Chem.* 1984, 48, 2081.
18 Inatani, R.; Nakatani, N.; Fuwa, H. *Agric. Biol. Chem.* 1983, 47, 521.
19 Nakatani, N.; Kikuzaki, H. *Agric. Biol. Chem.* 1987, 51, 2727.

20 Kikuzaki, H.; Nakatani, N. *Agric. Biol. Chem.* 1989, 53, 519.
21 Miura, K.; Nakatani, N. *Chem Express* 1989, 2, 237.
22 Miura, K.; Nakatani, N. *Agric. Biol. Chem.* 1989, 53, 3043.
23 Miura, K.; Inagaki, T.; Nakatani, N. *Chem, Pharm. Bull.*
    1989, 37, 1816.
24 Nakatani, N.; Inatani, R.; Fuwa, H. *Agric. Biol. Chem.* 1980,
    44, 2831.
25 Inatani, R.; Nakatani, N.; Fuwa, H. *Agric. Biol. Chem.* 1981,
    45, 667.
26 Nakatani, N.; Inatani, R.; Ohta, H.; Nishioka, A. *Environ*
    Health Perspect 1986, 67, 135.
27 Nakatani, N.; Tachibana, Y.; Kikuzaki, H. In *Medical,*
    Biochemical and Chemical Aspects of Free Radicals, Elsevier:
    Amsterdam, 1989, pp.453–456.
28 Masuda, T.; Nakatani, N. *Agric. Biol. Chem.* 1991, 55, 2337.
29 Kikuzaki, H.; Usuguchi, J.; Nakatani, N. *Chem. Pharm. Bull.*
    1991, 39, 120.
30 Kikuzaki, H.; Kobayashi, M.; Nakatani, N. *Phytochem* 1991,
    in press.

RECEIVED June 11, 1992

Chapter 7

Antioxidant Effects of Tannins and Related Polyphenols

Takuo Okuda, Takashi Yoshida, and Tsutomu Hatano

Faculty of Pharmaceutical Sciences, Okayama University, Tsushima, Okayama 700, Japan

Antioxidant effects were exhibited by polyphenolic compounds isolated from medicinal plants, in various experimental systems: autoxidation of ascorbic acid and methyl linoleate, lipid peroxidation in liver mitochondria and microsomes, lipoxygenase-dependent lipid peroxidation, arachidonic acid metabolism, lipid metabolic injury in rat (oral administration), cytotoxicity in cultured hepatocytes, oxidative damage model of ocular lens, superoxide anion radical generated in the hypoxanthine-xanthine oxidase system. Mechanistic study showed participation of radical-scavenging activity of polyphenols.

A large number of polyphenolic compounds of a variety of structures have been isolated from diverse plants used as food and medicine, and their biological and pharmacological activities have been investigated, as described in the chapter "Polyphenols from Asian Plants —Structural Diversity, and Their Antitumor and Antiviral Activities—" in this book. These polyphenolic compounds can be defined as tannins, based on the comparisons of their structures and properties with those of known polyphenolic compounds regarded as tannins, and also with those of a large number of new compounds isolated in recent years from "tannin-containing plants."

Inhibition of various actions of active-oxygen species, which is one of the most important activities underlying the biological and pharmacological activities of these polyphenols (1,2), is reviewed in this chapter.

Inhibition of Autoxidation of Ascorbic Acid and Methyl Linoleate

Ascorbic acid, an antioxidant and a radical scavenger, is quickly decomposed via a free radical upon aerobic oxidation, particularly in the presence of metallic catalysts. However, this compound is more stable when it is in the infusion of green tea which is rich in "green-tea tannin."

In the experiments of Cu(II)-catalyzed aerobic oxidation of ascorbic acid, geraniin and tannic acid (Figure 1) remarkably inhibited the oxidation, while polyphenols of small molecules such as gallic acid and (+)-catechin did not inhibit it appreciably. The inhibitory activity of geraniin was stronger than that of tannic acid (3).

0097–6156/92/0507–0087$06.00/0

This inhibition of oxidation may be attributable to the chelation of these phenolic compounds with the metallic ion. However, the complexation constant of Cu(II) with geraniin, determined by the method of Scatchard plots, was smaller than that with tannic acid. On the other hand, the electron spin resonance (ESR) signals of several tannins in alkaline dimethylsulfoxide (DMSO) showed that tannin radicals are stable. These results strongly suggested that the inhibition of the oxidation is due to the radical scavenging activity associated with the formation of stable tannin radicals, rather than the chelation of the tannins with Cu(II) (4). Stronger inhibitory effects on the autoxidation of ascorbic acid in the same experimental system were exhibited by ellagic acid and quercetin (4).

Marked inhibitory effect of tannins and polyphenols of related structures was also observed in an experiment of autoxidation of methyl linoleate, initiated by photo-irradiation of 2,2'-azobisisobutyronitrile (AIBN) in the solution, which is a model system of lipid peroxidation (5) (Figure 2). The mechanism of this inhibition will be discussed later.

Inhibition of Lipid Peroxidation

Inhibition of Lipid Peroxidation in Liver Mitochondria and Microsomes. Inhibition of lipid peroxidation, occurring both in food and in living tissues, has been attracting interests from the angle of health effects. Several tannins and related compounds, isolated from various medicinal plants, markedly inhibited lipid peroxidation in rat liver mitochondria and microsomes (6,7).

In this experiment, the lipid peroxide produced by incubating a mitochondrial fraction from rat liver with adenosine 5'-diphosphate (ADP) and ascorbic acid, and that produced in a rat liver microsomal fraction incubated with ADP and nicotinamide adenine dinucleotide phosphate (NADPH), were remarkably lowered by some hydrolyzable tannins [i.e., pedunculagin (Figure 3), penta-O-galloyl-β-D-glucose and isoterchebin) (6).

Dicaffeoylquinic acids (Figure 3), the main components in "tannin-rich" *Artemisia* species, also markedly inhibited lipid peroxidation in these systems. The inhibitory effects of dicaffeoylquinic acids were noticeably stronger than those of chlorogenic acid (monocaffeoylquinic acid) and caffeic acid (7).

Mechanistic Study of Lipid Peroxidation. The inhibitory effects of polyphenols upon the autoxidation of methyl linoleate were studied by kinetic study and *in situ* ESR measurements using 25 compounds. The strength of this effect was dependent on the type of polyphenolic groups and their number in each molecule, and the effect of these polyphenols generally lasted longer than those of the antioxidants on the market. For instance, the duration of inhibition of autoxidation of methyl linoleate by geraniin was several times of that of ascorbic acid or α-tocopherol. The radical-scavenging effect of ellagitannins such as geraniin, having a hexahydroxydiphenoyl (HHDP) group in the molecule, was stronger than that of gallotannins composed of galloyl groups. *In situ* ESR detection of tannin radicals, under the experimental condition inhibiting autoxidation, showed transient ESR signals of tannin radicals (Figure 4), which were identical with the signals obtained in separate measurement of aerial oxidation of tannins (5).

These results indicated that tannins are proton-donors to lipid free radicals in the peroxidation. Stable tannin radicals were formed upon this action, and stopped the chain-reaction of lipid autoxidation.

Upon comparison of the intensity of radical-scavenging effect of these phenolic compounds with their scavenging effect on 1,1-diphenyl-2-picrylhydrazyl (DPPH), which gives violet solution of its free radical, and is decolorized when its radical is scavenged, the orders of intensity of each compound in these two experiments were similar. The radical-scavenging activity of the polyphenols of comparatively large

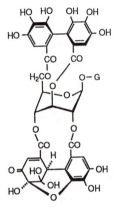

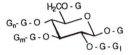

Tannic acid
(l + m + n = 0 ~ 7)

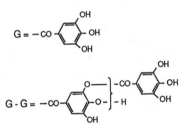

Geraniin

Figure 1. Chemical structures of geraniin and tannic acid.

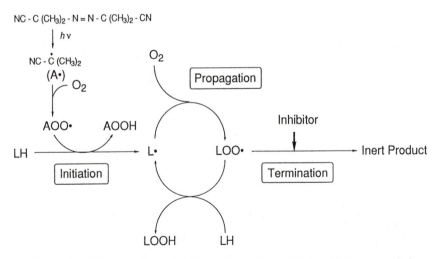

Figure 2. Schemes of autoxidation of methyl linoleate. LH means methyl linoleate.

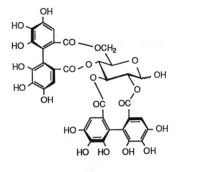

Pedunculagin

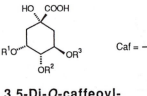

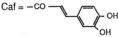

**3,5-Di-*O*-caffeoyl-
quinic acid**: $R^1 = R^3 =$ Caf, $R^2 =$ H
**3,4-Di-*O*-caffeoyl-
quinic acid**: $R^1 =$ H, $R^2 = R^3 =$ Caf
**4,5-Di-*O*-caffeoyl-
quinic acid**: $R^1 = R^2 =$ Caf, $R^3 =$ H

Figure 3.   Chemical structures of pedunculagin and dicaffeoylquinic acids.

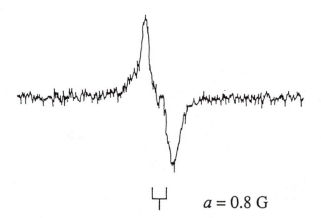

$a = 0.8$ G

Figure 4.   *In situ* ESR signal of geraniin.   This signal was observed upon the photo-irradiation of a mixture of methyl linoleate (0.4 M), AIBN (25 mM) and geraniin (10 mM) in DMSO-water (9:1) in an ESR cell.

molecular weight (~1000) was generally strong, although some small polyphenols of certain structures, such as (-)-epicatechin gallate, also showed strong activity (8).

Upon the reaction of the DPPH radical with various alkyl gallates, dialkyl hexahydroxydiphenates were produced and were isolated in high yields (Figure 5). These products, most probably from mutual coupling of C-centered galloyl radicals, and the ESR spectra of alkyl gallates in alkaline DMSO, verified that the elimination of the DPPH color is due to the scavenging activity of these polyphenols which formed stable free radicals (8).

Inhibition of Lipoxygenase-dependent Lipid Peroxidation.   Tannins and related polyphenols also inhibited the autoxidation of linoleic acid, which was initiated by abstraction of its hydrogen with soybean lipoxygenase (9).

Among these compounds, vescalagin and casuarinin (Figure 6), having two HHDP moieties in each molecule, exhibited stronger inhibition of the autoxidation than the other compounds such as geraniin having an HHDP group (9).   This observation is in accord with the stronger radical-scavenging effect of ellagitannins than that of the other types of polyphenols (5).

Several caffeic acid derivatives remarkably inhibited the peroxidation of linoleic acid.   Among them, 3,5-di-O-caffeoylquinic acid, isolated from *Artemisia* species, and rosmarinic acid, so-called labiataetannin which is widely distributed in plants of Labiatae family, having two caffeic acid (or equivalent) moieties, showed stronger inhibition than chlorogenic acid and caffeic acid, and the effect of ferulic acid was the lowest.   This order of the inhibitory effect among these compounds on the peroxidation was the same as that of the scavenging effect on the DPPH radical. Ferulic acid showed no ESR signal in alkaline DMSO, while all other four compounds (caffeic acid and its esters) showed the signals of their stable radicals under the same condition.   These results show that the inhibition of the autoxidation of linoleic acid by these polyphenols is due to their activity scavenging lipid peroxide radicals (9).   Analogous mechanism can be assumed for the inhibition in the other systems of enzyme-dependent lipid-peroxidation, including that caused by ADP and NADPH in rat liver.

Effects on Arachidonic Acid Metabolism.   Inhibition of the peroxidation catalyzed by a lipoxygenase was also observed in the products from arachidonic acid metabolism.   Several tannins and related polyphenols affected the enzyme-dependent peroxidation in arachidonic acid metabolism in polymorphonuclear leukocytes (10,11).

5-Lipoxygenase in leukocytes catalyzes peroxidation of arachidonic acid, producing 5-hydroperoxy-6,8,11,14-eicosatetraenoic acid (5-HPETE), and this product is further converted into 5-hydroxy-6,8,11,14-eicosatetraenoic acid (5-HETE) and leukotrienes.   On the other hand, cyclooxygenase catalyzes formation of prostagrandin G_2 (PGG$_2$) from arachidonic acid, and subsequent reactions in leukocytes give various prostagrandins and related products, including 6-ketoprostagrandin F_{1a} (6KF), thromboxane B_2 (TXB$_2$) and 12-hydroxy-5,8,10-heptadecatrienoic acid (HHT).   Geraniin and corilagin appreciably inhibited the formation of 5-HETE in a dose-dependent manner, while they inhibited the formation of 6KF, TXB$_2$ and HHT only at much higher concentrations (10).

Caffeic acid and its esters also inhibited the peroxidation catalyzed by 5-lipoxygenase, but they stimulated the formation of prostagrandin E_2 (PGE$_2$).   Since some radical scavengers stimulate the formation of PGE$_2$, these polyphenols may be radical scavengers in the stimulation processes (11).

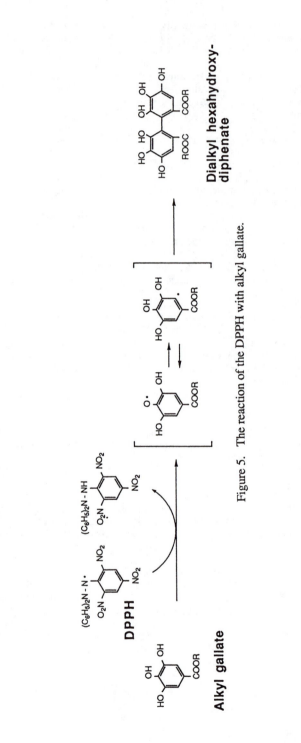

Figure 5.   The reaction of the DPPH with alkyl gallate.

Inhibition of Oxidative Damages Associated with Lipid Peroxidation

Inhibition of Lipid Metabolic Injury by Oral Administration of Polyphenols. Lipid metabolic injury in rat was also inhibited by oral administration of tannins or tannin-rich plant extracts (*12-14*). The elevation of lipid peroxide levels in serum and in liver, and that of the serum transaminases [glutamic-oxaloacetic transaminase (GOT) and glutamic-pyruvic transaminase (GPT)] levels, induced by feeding peroxidized corn oil to rats, were inhibited by oral administration (300 mg/kg/day) of the extracts of *Geranium thunbergii* [containing geraniin (11.7 or 18.4 %) or corilagin (3.0-7.6 %)]. The levels of total cholesterol and free fatty acids in serum, elevated by peroxidized corn oil, were lowered, too (*12*). Oral administration of geraniin showed similar effects even in a dose of 50 mg/kg/day (*12*).

Analogous effects were also observed upon administration of extracts of *Artemisia* species containing 3,5-, 3,4- and 4,5-di-*O*-caffeoylquinic acids (9.4-21.6 % in total) and chlorogenic acid (3.4-3.9 %), and upon administration of caffeic acid, and of chlorogenic acid (*13*). Extracts of fermented and non-fermented tea leaves also showed some improving effects against lipid metabolic injury (*14*).

Inhibition of Cytotoxicity in Primary Cultured Hepatocytes. Primary cultured rat hepatocytes have been used for searching compounds with liver-protecting activity. Addition of carbon tetrachloride to the culture medium of the hepatocytes causes increase of the GPT activity in the medium, due to its cytotoxicity (*15*). This cytotoxic effect was inhibited by several hydrolyzable tannins [such as corilagin, pedunculagin, granatin A and gemin A (Figure 7)], galloylated condensed tannins (procyanidin B2 3'-*O*-gallate, procyanidin B2 3,3'-di-*O*-gallate) and related polyphenols with low molecular weight (pyrocatechol, gallic acid, epicatechin gallate and epigallocatechin gallate) (*16*). Since the hepatotoxicity of carbon tetrachloride is attributable to the lipid peroxidation caused by •CCl$_3$ radical (*15*), tannins and related polyphenols may act as radical scavengers against this lipid peroxidation. Galactosamine-induced cytotoxicity in cultured rat hepatocytes were also inhibited by several tannins and related polyphenols (*16*).

Inhibition of Oxidative Damage of Ocular Lens. Opacification of ocular lens in human senile cataract and diabetic cataract, and several experimental models for cataracts, has been correlated with lipid peroxidation in lens. Geraniin, penta-*O*-galloyl-β-D-glucose, (-)-epigallocatechin gallate and several polyphenols, inhibited lipid peroxidation in lens. The lipid peroxidation in intact lens, which was induced by incubating the lens in a medium containing xanthine, xanthine oxidase, ADP and FeCl$_3$, was inhibited by further incubation of the lens in a medium containing each polyphenol, to various extent depending on the structure of each polyphenol (*17*). The increase of the Na$^+$/K$^+$ ratio, and the decreases of the glutathione level and of the activities of glutathione reductase and Na,K-ATPase, which are accompanied by this lipid peroxidation, were also restored by the incubation of the lens with these polyphenols. The action site of the polyphenols was the plasma membrane of the lens (*17*).

Inhibition of the generation of superoxide anion radical in the hypoxanthine-XOD system was exhibited by tannins and related compounds (*18*). The generated superoxide radical in this system was detected by ESR spectrometry in the presence of 5,5-dimethyl-1-pyrroline-*N*-oxide (DMPO). The ESR signal due to the DMPO adduct of superoxide radical decreased its amplitude upon addition of polyphenolic compound in the reaction mixture, in a dose-dependent manner (Figure 8). The signals of two new radical species assignable to the DMPO adducts of hydrogen (DMPO-H) and of a *C*-centered radical, then appeared. The latter signal shows production of phenoxy radical from polyphenol upon scavenging the superoxide

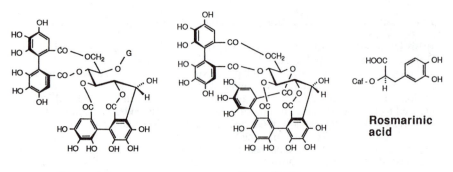

Casuarinin   **Vescalagin**

Figure 6.   Chemical structures of casuarinin, vescalagin and rosmarinic acid.

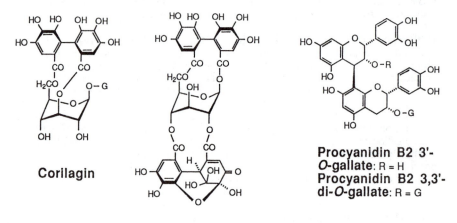

Corilagin

Granatin A

Procyanidin B2 3'-O-gallate: R = H
Procyanidin B2 3,3'-di-O-gallate: R = G

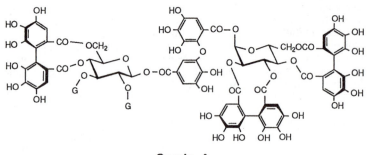

Gemin A

Figure 7.   Chemical structures of corilagin, granatin A, procyanidin O-gallates and gemin A.

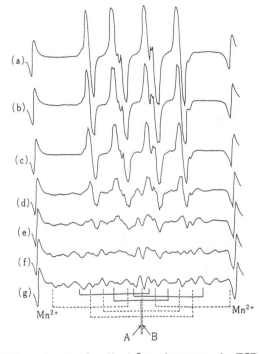

Figure 8. Effect of penta-O-galloyl-β-D-glucose on the ESR signal of the DMPO adduct of superoxide anion radical. The spectra were recorded in the absence [(a)] and in the presence [(b) 1.0×10^{-6} M, (c) 2.5×10^{-6} M, (d) 5.1×10^{-6} M, (e) 1.0×10^{-5} M, (f) 2.5×10^{-5} M, (g) 1.0×10^{-5} M of penta-O-galloyl-β-D-glucose. The solid (A) and broken (B) lines below the spectrum (g) indicate the assignments of the hyperfine splitting patterns for the DMPO adduct of a C-centered radical and the hydrogen adduct of DMPO, respectively.

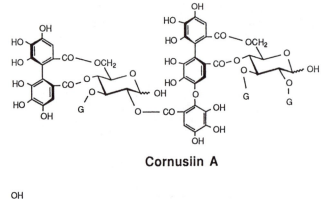

Cornusiin A

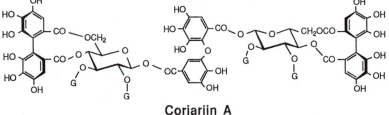

Coriariin A

Figure 9.   Chemical structures of cornusiin A and coriariin A.

radical.   Strong scavenging activity has been found for the polyphenolic compounds with *ortho*-trihydroxy (pyrogallol) structure (*18*).

The compounds which strongly inhibited generation of superoxide radical in the hypoxanthine-XOD system [such as geraniin, cornusiin A, coriariin A (Figure 9), epigallocatechin and epigallocatechin gallate], however, showed only weak inhibition of the XOD activity catalyzing the formation of uric acid.   This result substantiated that their inhibitory effect on the generation of the superoxide radical is ascribable to direct scavenging of the radical (*19*).

Conclusion

The recent findings concerning the antioxidant effects of polyphenolic compounds have revealed the possibility that the effects of these compounds, considerable amount of which is taken in by the people in the world, are underlying various favorite health effects of foods and medicinal plants.   The recent investigations have also shown marked difference of these effects due to the difference of the structure of each polyphenolic compound and of the oxidation system.   Further detailed investigations of their effects, particularly those occurring after they are taken as food and medicine, must be carried out.

Acknowledgments

The authors thank Prof. Y. Fujita and Prof. A. Mori (Okayama University), the late Prof. S. Arichi and Dr. Y. Kimura (Kinki University), Prof. H. Okuda (Ehime University), the late Prof. H. Hikino (Tohoku University), the late Prof. S. Iwata (Meijo University), Prof. T. Noro (University of Shizuoka) and their co-workers for their collaboration.

Literature Cited

1. Okuda, T.; Yoshida, T.; Hatano, T.; Fujita, Y. In *Free Radical and Cino-Japanese Medicine*; Okuda, T.; Yoshikawa, T., Eds.; Kokusai-isho-shuppan: Tokyo, 1990; pp 42-70.
2. Okuda, T.; Fujita, Y.; Yoshida, T.; Hatano, T. In *Free Radicals in Clinical Medicine*; Kondo, M.; Oyanagi, Y.; Yoshikawa, T., Eds.; Nihon-igakukan: Tokyo, 1990, Vol. 4; pp 19-30.
3. Yoshida, T.; Koyama, S.; Okuda, T. *Yakugaku Zasshi* **1981**, *101*, 695.
4. Fujita, Y.; Komagoe, K.; Sasaki, Y.; Uehara, I.; Okuda, T.; Yoshida, T. *Yakugaku Zasshi* **1987**, *107*, 17.
5. Fujita, Y.; Komagoe, K.; Uehara, I.; Okuda, T.; Yoshida, T. *Yakugaku Zasshi* **1988**, *108*, 528.
6. Okuda, T.; Kimura, Y.; Yoshida, T.; Hatano, T.; Okuda, H.; Arichi, S. *Chem. Pharm. Bull.* **1983**, *31*, 1625.
7. Kimura, Y.; Okuda, H.; Okuda, T.; Hatano, T.; Agata, I.; Arichi, S. *Planta Med.* **1984**, *50*, 473.
8. Yoshida, T.; Hatano, T.; Okumura, T.; Uehara, I.; Komagoe, K.; Fujita, Y.; Okuda, T. *Chem. Pharm. Bull.* **1989**, *37*, 1919.
9. Fujita, Y.; Uehara, I.; Morimoto, Y.; Nakashima, M.; Hatano, T.; Okuda, T. *Yakugaku Zasshi* **1988**, *108*, 129.
10. Kimura, Y.; Okuda, H.; Okuda, T.; Arichi, S. *Planta Med.* **1986**, *52*, 337.
11. Kimura, Y.; Okuda, H.; Okuda, T.; Hatano, T.; Arichi, S. *J. Nat. Prod.* **1987**, *50*, 392.
12. Kimura, Y.; Okuda, H.; Mori, K.; Okuda, T.; Arichi, S. *Chem. Pharm. Bull.* **1983**, *31*, 2501.
13. Kimura, Y.; Okuda, H.; Okuda, T.; Hatano, T.; Agata, I.; Arichi, S. *Chem. Pharm. Bull.* **1985**, *33*, 2028.
14. Kimura, Y.; Okuda, H.; Mori, K; Okuda, T.; Arichi, S. *Nippon Eiyo Shokuryo Gakkai Shi* **1984**, *37*, 223.
15. Kiso, Y.; Tohkin, M.; Hikino, H. *Planta Med.* **1983**, *49*, 222.
16. Hikino, H.; Kiso, Y.; Hatano, T.; Yoshida, T.; Okuda, T. *J. Ethnopharmacology*, **1985**, *14*, 19.
17. Iwata, S.; Fukaya, Y.; Nakazawa, K.; Okuda, T. *J. Ocular Pharmacol.* **1987**, *3*, 227.
18. Hatano, T.; Edamatsu, R.; Hiramatsu, M.; Mori, A.; Fujita, Y.; Yasuhara, T.; Yoshida, T.; Okuda, T. *Chem. Pharm. Bull.* **1989**, *37*, 2016.
19. Hatano, T.; Yasuhara, T.; Yoshihara, R.; Agata, I.; Noro, T.; Okuda, T. *Chem. Pharm. Bull.* **1990**, *38*, 1224.

RECEIVED December 17, 1991

Chapter 8

Soybean (Malonyl) Isoflavones

Characterization and Antioxidant Properties

Y. Fleury, D. H. Welti, G. Philippossian, and D. Magnolato

Nestlé Research Centre, Nestec, Ltd., CH–1000 Lausanne 26, Switzerland

Two isoflavone glycosides representing 67% of total isoflavones have been isolated for the first time from soybean. They were identified by spectroscopic techniques as malonylgenistin and malonyldaidzin. Antioxidant activities of soybean isoflavones were checked by accelerated oxidation of chicken fat (Rancimat), chicken olein (storage test) and inhibition of the oxidation of linoleic acid in presence of beta-carotene subjected to irradiation with UV-light, or heated.

Soybeans are known to contain several isoflavones (daidzein, glycitein, genistein) and isoflavone glucosides (daidzin, glycitein-7-0-glucoside, genistin) which have been reported to have estrogenic, antifungal, and antioxidant properties (1).
   Glycitein and glycitin (glycitein-7-0-glucoside), the 6-methoxy derivatives of daidzein and daidzin respectively were found in soybean by Naim et al (2). 6"-0-acetylgenistin and 6"-0-acetylgenistin were first reported by Ohta et al. (3,4) and quantified in 4 soybean varieties (5), 6"-0-acetylglycitin was only found in soybean embryos (6). Phenolic compounds such as anthocyanins (7-9), flavonols (9,10), flavones (11), isoflavones (12), chalcones (13), were acylated with caffeic (14), coumaric (15-17), p-hydroxybenzoic (18,19), ferulic (20,21), malonic, succinic (22), hydroxycinnamic (23), oxalic (24), acetic (25), gallic (26,27) and sinapic (28) acids.
   Acylation with malonic acid was described for anthocyanin (29-34), flavonols (35,36), flavones (37,38), isoflavones (39-43), and chalcones (25). O- and N-malonyl compounds are very important in plant metabolism. Besides phenolic compounds, many different malonyl conjugates have been isolated with amino acids (44,45), intermediary and end products of pesticides degradation in plants (46,47), riboflavin (48) and ethylene precursors (49-51).

0097–6156/92/0507–0098$06.00/0
© 1992 American Chemical Society

Isoflavones acylated with malonic acid were found in Chickpea (Cicer arietinum) (*52-57*) where they constitute 85-90% of the biochanin and formononetin. The malonyl group is always linked on position 6 of the glucose moiety present on C-7. Malonylgenistin was found in submediterranean clover leaves (*58,59*) and the use of malonyldaidzin from Pueraria lobata as aldose reductase inhibitor was patented (*60*).

Malonylation of the sugar moiety seems to be the last step in the biosynthesis of flavonoid glucosides in parsley, where two malonyltransferases were isolated (*61*). It is hypothesized that malonylation may facilitate the transport of flavonoid glycosides through the tonoplast into the vacuole (*62*).

Extraction conditions are essential to obtain the genuine composition of plant constituents. Our purpose was to analyze isoflavones extracted under mild conditions to identify and characterize new molecules in soybean and to check their antioxidative properties. This was accomplished by performing four tests on fats or on emulsions based on heat or UV-light induced oxidation. The environmental versus varietal effects on isoflavone content will also be briefly reported.

Chemical Characterization

Malonylisoflavone Isolation. Milled soybean embryos (100g) obtained from Perikan Co Ltd. were defatted with petroleum ether 60-80 (3 x 750 ml) for 1 hour at room temperature. Defatted flour (89g) was extracted with 80% aqueous methanol (3 x 750 ml) for 1 hour at room temperature. The concentrated extract (380 ml) was diluted to 750 ml with water before being slurried with Amberlite XAD-4 (450 ml). Amberlite was filtrated and washed with water before a crude extract (5.2g) containing 49% isoflavones was desorbed with methanol (750 ml). The crude extract was composed of malonyldaidzin (27.4%), malonylglycitin (7.4%), malonylgenistin (4.9%), glycitin (4.6%), daidzin (3.8%), and genistin (1.0%). Purified isoflavones were obtained by reverse phase chromatography on a Lobar RP-18 column (Merck 10625) successively eluted with aqueous ethanol 10% (400 ml), 20% (200 ml), and 30% (200 ml).

HPLC Separation. Figure 1 shows the HPLC chromatogram of a 80% aqueous methanol extract of soybean embryos obtained with a diode array detector, a Nucleosyl RP-18 column of 5 μm eluted with a gradient of acidified water - acetonitrile (10-50% B) in 40 minutes with a flow rate of 1.0 ml/min. Soybean embryos were selected rather than whole beans due to their high content of glycitin and malonylglycitin. Neither acetylisoflavones nor isoflavone aglycones were found in these extracts.

NMR Analysis. The NMR analysis of the isoflavonoids was performed on a Bruker AM 360 spectrometer equipped with 5mm proton, multinuclear and "inverse" probeheads, at 360.13 MHz for protons and 90.56 MHz for carbon-13. The probe temperature was normally ca. 21°C and TMS was used as an internal standard. The samples were prepared in a nitrogen gas flooded glove box to exclude oxygen and moisture, using 99.95% deuteriated DMSO-d_6 (Dr. Glaser AG, Basel, Switzerland) as a solvent. The following molecules were analysed by various one- and two-dimensional NMR techniques : genistein, daidzein (Plantech LTD., Reading, UK), genistin, daidzin, glycitin, 6"-O-malonylgenistin, and 6"-O-malonyldaidzin (see Figure 2).

There is no space for presenting all these results here, so we limit ourselves to the 13C and 1H spectra of the last two, given in Tables I and II. Representative plots of malonyldaidzin 13C and 1H spectra are shown as Figures 3 and 4. In Figure 4 the very broad signals of the exchangeable protons are indicated. The vertical scale for the 12.0 - 8.6 ppm spectrum inset (but not for the integral) has been increased by a factor 10.

The assignements of all resonances have been carefully elaborated, starting with genistein and daidzein, where two-dimensional INADEQUATE spectra (63) yielded unambiguous 13C assignments. This was necessary to clarify a few remaining ambiguities in the literature (64-66), partly due to solvent effects or accidental line overlap. From the non-decoupled 13C spectra and from two-dimensional long range heteronuclear correlation spectra (HETCOR) (67), a wealth of heteronuclear coupling information could be obtained. This allowed us to assign the aglycone resonances in all three 7-O-beta-D-glucosides. Their sugar protons were assigned based on two-dimensional homonuclear proton correlation spectra (COSY) (68). If necessary, D_2O was added to clarify overlapping sugar proton resonances. The sugar carbon assignments were derived from the protons, via direct heteronuclear correlation. The COSY spectra also confirmed the 7-O-substitution of the isoflavones by the sugar.

The identity of the malonyl group could be confirmed and the site of its linkage unequivocally determined by two-dimensional inverse detected heteronuclear long range correlation spectra (69). Correlation signals were found from both the carbonyls at ca. 169.2 ppm and 168.4 ppm to the AB system of the malonyl CH_2 protons at ca. 3 ppm, but from the carbonyl resonance at ca. 169.2 ppm only to both 6"-CH_2 protons. These findings agree with the chemical shifts arguments used earlier for the site of substitution (70) and also permit an unequivocal determination of the ester and acid carbonyl signal of the malonyl residue.

The spectral characterization by NMR of the malonyl isoflavonoids is somewhat hampered by their chemical instability. From both malonylgenistin and malonyldaidzin, three or four different reaction products were obtained after hours to days at room temperature in DMSO solution. The predominant ones were their 6"-O-acetyl analogs. This was noted earlier, with other flavonoids (70). The other products have not been identified.

The 13C shift value of the malonyl signals seems to be somewhat variable, probably depending upon the conformational, protonation or hydration state of the molecules. This is largely determined by the chromatographic and sample preparation methods used. For example, 48.5 ppm (60) and 41.4 ppm (71) have been found in other laboratories for the malonyl CH_2 shift.

For both malonylisoflavonoids the proton signals of the malonyl CH_2 and the 6"-H's showed an unusual concentration dependence of the

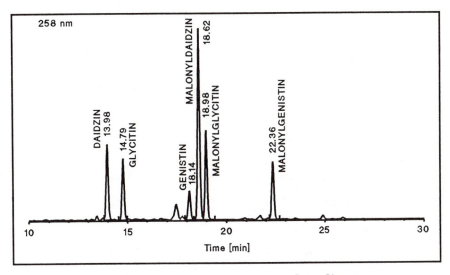

Figure 1 :   RP-HPLC chromatogram of Isoflavones

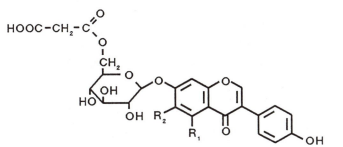

	R₁	R₂
Malonyldaidzin	H	H
Malonylglycitin	H	OCH₃
Malonylgenistin	OH	H

Figure 2 :   Chemical structure of malonylisoflavones

Table I. 90.56 MHz 13C NMR Spectrum of 6"-O-malonylgenistin (*M.G.*) and 6"-O-malonyldaidzin (*M.D.*) in DMSO-d$_6$ at ca. 21°C

Aglycone Carbon	M.G.	M.D.	Glucose and Malonic acid Carbon	M.G.	M.D.
2-CH	154.06	153.20	1"-CH	99.69	99.72
3-C	122.63	123.57	2"-CH	72.99	73.04
4-C=O	179.92	174.70	3"-CH	76.04	76.08
5-C..	161.67	126.91	4"-CH	69.75	69.69
6-CH	99.43	115.32	5"-CH	74.03	74.01
7-C	162.52	161.04	6"-CH$_2$	63.09	63.05
8-CH	94.35	103.42			
9-C	157.20	156.86	1'''-COOR	169.25	169.15
10-C	106.30	123.57	2'''-CH$_2$	45.39	45.16
			3'''-COOH	168.43	168.35
1'-C	120.79	121.96			
2',6'-CH	129.97	129.89			
3',5'-CH	115.04	115.32			
4'-C	157.70	157.43			

Table II. Typical[a] 360.13 MHz 1H NMR Spectrum of 6"-O-malonyl-genistin and 6"-O-malonyldaidzin in DMSO-d$_6$ at ca. 21°C

proton	6"-O-malonylgenistin ppm	mult.[b]	J(Hz)	6"-O-malonyldaidzin ppm	mult.[b]	J(Hz)
2-H	8.36	s		8.35	s	
5-H	-			8.05	d	8.9
6-H	6.48	d	1.8	7.15	dd	8.9, 2.3
8-H	6.69	d	1.9	7.23	d	2.3
2',6'-H	7.37	"d"	8.5	7.39	"d"	8.6
3',5'-H	6.83	"d"	8.6	6.82	"d"	8.6
1"-H	5.06	d	7.4	5.13	d	7.2
2"-H	3.30	"AB,d"		3.33	"AB,d"	
3"-H	3.36	"AB,d"		3.37	"AB,d"	
4"-H	3.20	"t"		3.23	"t"	
5"-H	3.71	"ddd"		3.73	ddd	
6"-H$_2$	4.30	"d"		4.31	d(d)	
	4.06	dd		4.08	dd	
2'''-H$_2$	3.04	AB	14.2	3.06	AB	14.3
	3.00	AB	14.2	3.02	AB	14.3

[a]Most of the shift values are slightly concentration dependent (see text). Exchangeable protons are not listed.
[b]Multiplicities: s singlet, d doublet, t triplet, AB strongly coupled AB system, ".." approximate description.

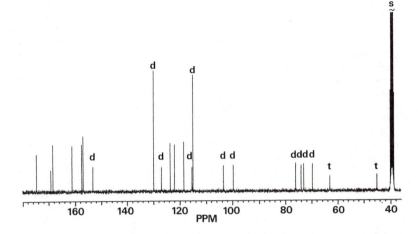

Figure 3 :   90.56 MHz 13C NMR spectrum of 6"-O-malonyldaidzin in DMSO-d$_6$/TMS. The number of directly attached protons is indicated above the lines: d = CH, t = CH$_2$, s = solvent

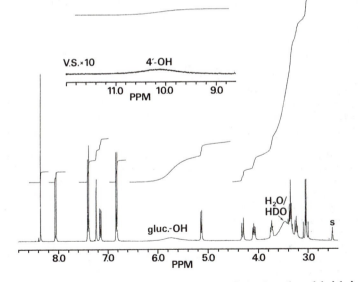

Figure 4 :   360.13 MHz 1H NMR spectrum of 6"-O-malonyldaidzin in DMSO-d$_6$/TMS, with integrals

chemical shifts. Also, the malonyl CH_2, which appears as a singlet at a concentration of 0.5 - 4 mg/ml was seen as a strongly coupled AB system at 30-40 mg/ml. This may indicate an intermolecular association process which restricts the internal rotation of the malonyl residue. At low concentration, e.g. 0.4 mg/ml for malonylgenistin, the individual exchangeable protons which produce broad signals at high concentration could be distinguished and mostly assigned.

The third malonylisoflavonoid, 6"-O-malonylglycitin, has not been examined by NMR in our laboratory, but reasonably compatible data can be found elsewhere (71).

In order to confirm the identity of the glycoside, malonylgenistin was subjected to 18 hours of heating in water under refluxing. Subsequent chromatography of the water solution showed that more than 95% of the malonylgenistin has been converted into genistin (and not 6"-O-acetylgenistin). Correspondingly, in the 13C NMR spectrum of a DMSO-d_6 extract of the lyophilized decomposition product, the predominant signals detected were those of genistin and malonic acid. The latter were found at 171.52 ppm for COOH/COO$^-$ and at 38.73 ppm for CH_2 , as verified by DEPT spectroscopy (72, 73) for the line hidden under the solvent signal. To unambiguously identify these two signals a spiking experiment with malonic acid (168.35 and 41.84 ppm) was done and the new shifts obtained were 169.26 and 40.93 ppm. Thus, heating in DMSO or water yields different decomposition products. It is interesting to note that the significant broadening of the genistin 2-C, 4-C, 5-C, 6-C, 8-C, 9-C, 10-C and 1'-C signals observed in the decomposition product disappeared when the malonic acid was added.

Mass Spectrometry   Both malonylgenistin and malonyldaidzin were analyzed by positive and negative fast atom bombardment (FAB) mass spectrometry on a Finnigan/MAT 8430 sector field instrument using two different matrices (glycerol and "magic bullets"). By these ionization methods rather weak but clean complementary spectra were produced, showing the molecular and some fragment ions.

For malonylgenistin, the protonated molecular ion was obtained in the positive FAB mode at m/z 519. Sodium adduct ions were obtained at m/z 541, 563 and 585. The protonated aglycone was found at m/z 271 and the protonated molecule minus the malonyl moiety at m/z 433. The negative FAB spectra showed the corresponding deprotonated molecular peak at m/z 517 and a sodium adduct ion at 539. The deprotonated aglycone was seen at m/z 269 (plus sodium at m/z 291) and the deprotonated molecule minus the malonyl moiety at m/z 431 ("magic bullet").

For malonyldaidzin, the positive FAB mode yielded the protonated molecular ion at m/z 503, sodium adduct ions at m/z 525, 547, 563 and the protonated aglycone at m/z 255. The negative FAB spectra showed the deprotonated molecular ion at m/z 501, a sodium adduct at m/z 523, and the deprotonated aglycone at m/z 253. A small signal of the deprotonated molecule minus the malonyl substituent was obtained at m/z 415 with glycerol as matrix. All these findings are in agreement with the proposed molecular structures. Malonylglycitin was not analyzed by mass spectrometry.

Isoflavone Accumulation in Soybean

Biosynthesis of Isoflavones during Development of the Beans.   Maple arrow is an early variety of soybean (Glycine max. Mer.) characterized by a 00 maturity type widely grown in Canada. Plants were grown from seeds under controlled conditions in a phytotron. Flowers were tagged every day and pods were harvested at various stages of maturation between 25 and 60 days after flowering. Figure 5 shows isoflavone accumulation in whole beans between 35 and 60 days after flowering.

Malonylgenistin and genistin contents increased during the late development of beans whereas malonyldaidzin and daidzin accumulate during the whole period. Minor isoflavone glycosides, i.e. malonylglycitin and glycitin, doubled during the considered period. No isoflavone aglycones were detected.

Sixty six percent of the isoflavones, which represented 0.8% of the dry weight of mature beans, are present as malonylated derivatives.

Environmental and Varietal Effect on Isoflavone Content.   The 5 soybean varieties C (Corsoy), E (Evans), MA (Maple arrow), V (Vinton) and LF (Lipoxygenase-free) characterized by different precocity levels and protein contents were grown in tropical (Mississippi) and continental (Indiana) climatic areas in order to determine the varietal versus physiological effects on isoflavone content. Values reported in Figure 6 show that the environmental factors were more important than varietal factors since locations induced differences in a ratio 4 to 1 against only 2 to 1 for varieties.

Values reported (74) confirmed the tendency observed here with a variation in the ratio of 3 to 1 within varieties and of 4.2 to 1 for the same variety grown in different locations.

Antioxidant Properties

Soybean, defatted soybean flour, soybean protein concentrates, and soybean protein isolates have appreciable antioxidant activity in lipid-aqueous systems (75). The antioxidant activity is due to compounds such as tocopherols, phospholipids, amino acids and peptides, (76) but also to phenolic compounds such as isoflavonoids (77), chlorogenic acid isomers, caffeic acid, and ferulic acid occurring mainly as glucosides (78-80).

Antioxidant activity of soybean isoflavones was shown in the **in vitro** systems by measuring the extent of inhibition of lipoxygenase (77) and by the Rancimat method (80) on lard at 100°C.

In this work antioxidative activities of soybean isoflavones were compared by four methods differing in the oxidation induction mode and composition of the reaction medium.

Rancimat (81).   Autoxidation of chicken fat (dried Micana, 5g) was induced by a stream of air (17.5 1/h) at 100°C in a model 679 Rancimat (Metrohm AG, Switzerland).The antioxidative efficiency expressed as the ratio of induction time of the fat containing the

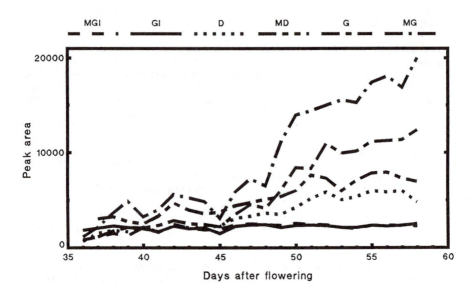

Figure 5 :   Isoflavones accumulation during maturation of Maple
              Arrow soybean seeds

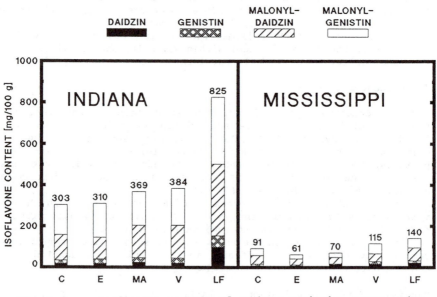

Figure 6 :   Isoflavone content of soybean varieties grown under
              continental (Indiana) and tropical (Mississippi)
              conditions

antioxidant compared to the induction time of the fat alone is reported in Table III.

All isoflavones considered have no antioxidant properties in this system in comparison to BHA and BHT. The rapid decomposition of malonylisoflavones at 100°C could explain their inactivity. Isoflavone glucosides and aglycones, which are more stable at this temperature, are also inactive and do not seem to behave as radical scavengers.

Table III : Rancimat antioxidative efficiency of soybean isoflavones compared to BHA and BHT

Compound	500 PPM	1000 PPM
Malonyldaidzin	1.06	1.16
Malonylglycitin	1.13	1.16
Malonygenistin	1.09	1.05
Daidzin	1.02	1.05
Genistin	1.02	1.01
Daidzein	1.05	1.04
Genistein	1.09	1.16
BHA	4.75	5.84
BHT	1.76	1.96

Storage Test (*81* modified).   Chicken olein (50 mg) composed of oleic acid (49%) and linoleic acid (13%) was stored at 37°C in vials. Triglycerides oxidation was followed at 234 nm by HPLC using a Nucleosil 100 column isocratically eluted with 0.9% isopropanol in hexane. Figure 7 shows the stability of chicken olein alone and in the presence of malonylgenistin (500 and 1000 ppm) stored at 37°C. Malonylgenistin approximately doubled the induction time of chicken olein. Similar results were obtained with malonyldaidzin and malonylglycitin. A weaker antioxidant activity was observed with isoflavone glucosides and isoflavone aglycones.

Heat-induced Oxidation (*82*).   Oxidation of aqueous emulsion system of beta-carotene (1.2%) and linoleic acid (2.0%) was induced by heating at 50°C. The oxidation reaction was followed by discoloration of the beta-carotene measured by spectrophotometry at 450 nm. Figure 8 shows the rate of discoloration of beta-carotene in the presence of different concentrations of malonyldaidzin or 1000 ppm of BHA.

Malonylisoflavones had no antioxidant activity in this system. Their heat lability was therefore not responsible for this inactivity since malonylisoflavones were not decomposed at the end of the test. As already mentioned, isoflavones are hydrophilic molecules which tend to dissolve in the aqueous phase rather than in the lipid phase of the emulsion. Moreover, malonylisoflavones were also inactive in another heat-induced oxidation measured with the Rancimat.

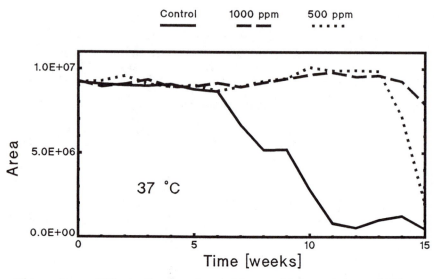

Figure 7 :   Effect of malonylgenistin on chicken olein stability
(storage test)

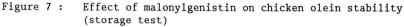

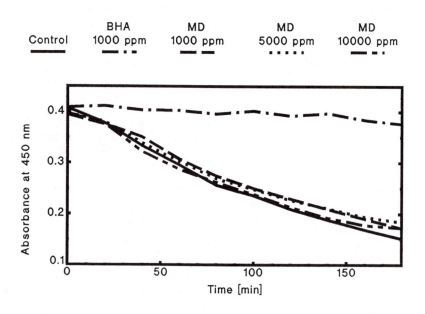

Figure 8 :   Heat induced oxidation of β-carotene/linoleate
system

UV-light induced Oxidation. Oxidation of a methanolic solution beta-carotene (0.0007% w/v) / linoleic acid (0.007% w/v) was induced by UV light (Philips TUV 30 W, G30T8, 360 nm). Irradiation was performed for 1 minute between each spectrophotometric measurement at 450 nm. Figure 9 shows the rate of discoloration of beta-carotene in the presence of malonylisoflavones (12.5 μg/ml). Antioxidant indices corresponding to the ratio of the rate of bleaching of the control to the rate of bleaching when a test compound was added to the system are reported in Table IV. UV light induced oxidation proceeds through a mechanism involving singlet oxygen scavengers. BHA and BHT, not active in this system, seem to proceed through another mechanism than isoflavones.

Table IV : Antioxidant index of isoflavones

Malonyldaidzin	1.88
Malonylgenistin	2.02
Malonylglycitin	1.65
Daidzin	1.73
Genistin	2.05
Daidzein	1.92
Genistein	2.07

Summary of Antioxidant Evaluation. Table V summarizes the antioxidative activities of soybean isoflavones determined with Rancimat, storage test, heat or UV light induced oxidation. In the Rancimat test performed at 100°C, all isoflavones considered were inactive. This agrees with the heat lability of malonylated derivatives but would also suggest that isoflavones may not act as radical scavengers.
Malonylisoflavones, which are good antioxidants in the storage test carried out at 37°C were totally transformed into glucosides at the end of the test. In the heat-induced oxidation of the beta-carotene/linoleate system, only BHA and BHT had a protective effect. Malonylisoflavones were stable at 50°C and their inactivity is probably linked to their very low solubility in the lipidic phase. In the UV light induced oxidation of the same system in methanolic solution, all isoflavones have a protective effect, behaving as photo-antioxidants, in contrast to BHA and BHT acting as radical scavengers.

Legend of Symbols D:Daidzin; G:Genistin; Gl:Glycitin; MD:Malonyldaidzin; MG:Malonylgenistin; MGl:Malonylglycitin; C:Corsoy; E:Evans; MA:Maple Arrow; V:Vinton; LF:Lipoxygenase free; BHA:Butylated hydroxyanisole; BHT:Butylated hydroxytoluene.

Acknowledgments We thank U. Richli and A. Kappeler for the MS spectra, F. Arce Vera for technical assistance with the NMR spectra, J. Perrinjaquet for analytical support, E. Prior for advice on antioxidant tests, C. Desponds for typing the manuscript.

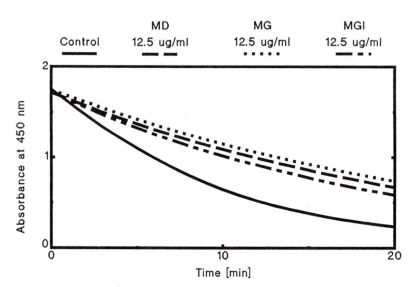

Figure 9 :   UV-light induced oxidation of β-carotene/linoleate system

Table V: Antioxidative properties of Isoflavones
(comparative methods)

Compound	Rancimat 100°C	Storage 37°C	UV/MeOH Room T.	Heating/H_2O 50°C
Malonyldaidzin	-	+++	+	-
Malonylglycitin	-	+++	+	
Malonylgenistin	-	+++	+	-
Daidzin	-	++	+	
Glycitin		+		
Genistin	-	+	+	
Daidzein	-	-	+	
Glycitein		-		
Genistein	-	++	+	
BHA	+		-	+
BHT	+		-	+

Literature cited

1. Eldridge, A.C. *J.Agric.Food Chem.* **1982**, *30*, 353.
2. Naim, M.; Gestetner, B.; Zilkah, S.; Birk, Y.; Bondi, A. *J.Agric.Food Chem.* **1974**, *22*, 806.
3. Ohta, N.; Kuwata, G.; Akahori, H.; Watanabe, T. *Agric. Biol.Chem.* **1979**, *43*, 1415.
4. Ohta, N.; Kuwata, G.; Akahori, H.; Watanabe, T. *Agric. Biol.Chem.* **1980**, *44*, 469.
5. Farmakalidis, E.; Murphy, P.A. *J.Agric.Food* **1985**, *33*, 385.
6. Kudou, S.; Shimoyamada, M.; Imura, T.; Uchida, T.; Okubo, K. *Agric.Biol.Chem.* **1991**, *55*, 859.
7. Maccarrone, E.; Maccarrone, A.; Rapisarda, P. *Ann.Chim. (Rome)*, **1985**, *75*, 79.
8. Yoshida,K.; Kondo, T.; Kameda, K.; Goto, T. *Agric.Biol. Chem.* **1990**, *54*, 1745.
9. Asen,S.; Griesbach, R.J.; Norris, K.H.; Leohhardt, B.A. *Phytochem.* **1986**, *25*, 2509.
10. Wald, B.; Wray, V.; Galensa, R.; Herrmann, K. *Phytochem.* **1989**, *28*, 663.
11. Mizuno, M.; Kato, M.; Iinuma, M.; Tanaka, T.; Kimura, A.; Ohashi, H.; Sakai, H. *Phytochem,* **1987**, *26*, 2418.
12. Hinderer, W.; Koester, J.; Barz, W. *Arch.Biochem. Biophys.* **1986**, *248*, 570.
13. Harborne, J.B.; Greenham, J.; Eagles, J.*Phytochem.* **1990**, *29*, 2899.
14. Lofty, S.; Fleuriet, A.; Ramos, T.; Macheix, J.J. *Plant Cell Rep.* **1989**, *8*, 93.
15. Saleh, N. A. M.; Mansour, R.M.A.; Markham, K.R. *Phytochem.* **1990**, *29*, 1344.
16. Brasseur, T.; Angenot, L. *Phytochem.* **1988**, *27*, 1487.
17. Brasseur, T.; Angenot, L. *Phytochem.* **1987**, *26*, 3331.
18. Saxena, V.K.; Jain, A.K. *Fitoterapia* **1989**, *60*, 85.
19. Franke, A.; Markham, K.R. *Phytochem.* **1989**, *28*, 3566.
20. Morimoto, S.; Nonaka, G.; Nishioka, I. *Chem.Pharm.Bull.* **1986**, *34*, 643.
21. Yoshida, T.; Saito, T.; Kadoya, S. *Chem.Pharm.Bull.* **1987**, *35*, 97.
22. Takeda, K.; Kumegawa, C.; Harborne, J.B.; Self, R. *Phytochem.* **1988**, *27*, 1228.
23. Teusch, M.; Forkmann, G.; Seyffert, W. *Phytochem.* **1987**, *26*, 991
24. Strack, D.; Busch, E.; Wray, V.; Grotjahn, L.; Klein, E. *Z.Naturforsch.C:Biosci.* **1986**, *41*, 707.
25. Hoffmann, B.; Hoelzl, J. *Planta Med.* **1988**, *54*, 450.
26. Iwagawa, T.; Kawasaki, J.;Hase, T.; Sako, S.; Okubo, T.; Ishida, M.; Kim, M. *Phytochem.* **1990**, *29*, 1013.
27. Pichon-Prum, N.; Raynaud, J.; Debourchieu, L.; Joseph, M.J. *Pharmazie* **1989**, *44*, 508.
28. Gluchoff-Fiasson, K.; Jay, M.; Viricel, M.R. *Phytochem.* **1989**, *28*, 2471.
29. Takeda, K.; Harborne, J.B.; Self, R. *Phytochem.* **1986**, *25*, 2191.
30. Takeda, K.; Enoki, S.; Harborne, J.; Eagles, J. *Phytochem.* **1989**, *28*, 499.
31. Saito, N.; Toki, K.; Honda, T.; Kawase, K. *Phytochem.* **1988**, *27*, 2963.

32. Cheminat, A.; Brouillard, R.; Guerne, P.; Bergmann, P.; Rether, B. *Phytochem.* **1989**, *28*, 3246.
33. Kreuzaler, F.; Hahlbrock, K. *Phytochem.* **1973**, *12*, 1149.
34. Harborne J.B.; Self, R. *Phytochem.* **1987**, *26*, 2417.
35. Boerger, G.; Barz, W. *Phytochem.* **1988**, *27*, 3714.
36. Stein, W.; Zinsmeister, H.D. *Z.Naturforsch.C:Biosci.* **1990**, *45*, 25.
37. Veit, M.; Geiger, H.; Czygan, F-C.; Markham, K.R. *Phytochem.* **1990**, *29*, 2555.
38. Stein, W.; Anhut, S.; Tinsmaister, H.D.; Mues, R.; Barz, W.; Koester, J. *Z.Naturforsch.C:Biosci.* **1985**, *40C*, 469.
39. Feeny, P.; Sachdev, K.; Rosenberry, L.; Carter, M. *Phytochem.* **1988**, *27*, 3439.
40. Koester, J.; Bussmann, R.; Barz, W. *Arch.Biochem. Biophys.* **1984**, *234*, 513.
41. Kessmann, H.; Barz, W. *Plant Cell Rep.* **1987**, *6*, 55.
42. Anhut, S.; Zinsmeister, H.D.; Mues, R.; Barz, W.; Mackenbrock, K.; Koster, J.; Markham, K.R. *Phytochem.* **1984**, *23*, 1073.
43. Matern, U.; Reichenbach, C.; Heller, W. *Planta* **1986**, *167*, 183.
44. Anhut, S.; Zinsmeister, H. D.; Mues, R.; Barz, W.; Mackenbrock.; Koester, J.; Markham, K.R. *Phytochem.* **1984**, *23*, 1073.
45. Berlin, J.; Witte, L.; Koester, J.; Markham, K.R. *Planta* **1984**, *155*, 244.
46. Pokorny, M.; Marcenko, E.; Keglevic, M. *Phytochem.* **1970**, *9*, 2175.
47. Frear, D.S.; Swanson, H.R.; Mansager, E.R. *Pestic. Biochem. Physiol.* **1983**, *20*, 299.
48. Larson, J.D.; Lamoureux, G.L. *J.Agric.Food* **1984**, *32*, 177.
49. Ghisla, S.; Mack, R.; Blankenhorn, G.; Hemmerich, P.; Krienitz, R.; Kuster, T. *Eur.J.Biochem.* **1984**, *138*, 339.
50. Amrhein, N.; Schneebeck, D.; Skorupa, H.; Rephof, S.; Stoeckigt, J. *Naturwiss.* **1981**, *68*, 619.
51. Hoffmann, N.E.; Yang, S.F.; McKlon, T. *Biochem. Biophysiol. Res. Commun.* **1982**, *104*, 765.
52. Hoffmann, N.E.; Fu, J.R.; Yang, S.F. *Plant Physiol.* **1983**, *71*, 197.
53. Hoffmann, N. E.; Liu, Y.; Yang, S.F. *Planta* **1983**, *157*, 518.
54. Hinderer, W.; Koester, J.; Barz, W. *Z. Naturforsch.* **1987**, *4*, 251.
55. Hinderer, W.; Koester, J.; Barz, W. *Arch. Biochem. Biophys.* **1986**, *248*, 570.
56. Koester, J.; Strack, D.; Barz, W. *Planta Medica* **1983**, *48*, 131.
57. Jaques, U.; Koester J.; Barz, W. *Phytochem.* **1985**, *24*, 949.
58. Francis, C.M. *J.Sci.Food Agric.* **1973**, *24*, 1235.
59. Beck, A.B.; Knox, J.R. *Aust. J. Chem.* **1971**, *24*, 1509.
60. Hirakura, K. Jpn Kokai Tokkyo JP 01,146,894.
61. Matern, U.; Potts, J.R.M.; Hahlbrock, K. *Arch. Biochem. Biophys.* **1981**, *208*, 233.
62. Teusch, M.; Forkmann, G. *Phytochem.* **1987**, *26*, 2181.
63. Mareci, T.H.; Freeman, R. *J. Magn. Resonance* **1982**, *48*, 158.
64. Markham, K.R.; Chari, V.M.; Mabry, T.J. In *The Flavonoids: Advances in Research*; Harborne J.B. and Mabry T.J., Eds.; Chapman and Hall: London, Great Britain, **1982**, p 30 (table 2.8), and spectrum 84.

65. Agrawal, P.K.; Bansal, M.C. In *Carbon-13 NMR of Flavonoids*; Agrawal, P.K., Ed.; Studies in Organic Chemistry 39; Elsevier: Amsterdam, Netherlands, 1989, pp 194-198.
66. Lane, G.A.; Newman, R.H. *Phytochem.* 1985, *26*, 295.
67. Bax, A.; Morris, G. *J. Magn. Resonance* 1981, *42*, 501.
68. Aue, W.P.; Bartholdi, E.; Ernst, R.R. *J. Chem. Phys.* 1976, *64*, 2229.
69. Bax, A.; Summers, M.F. *J.Am.Chem.Soc.* 1986, *108*, 2093.
70. Horowitz, R.M.; Asen, S. *Phytochem.* 1989, *28*, 2531.
71. Kudou, S.; Fleury Y.; Welti D.; Magnolato, D.; Uchida, T.; Kitamura, K.; Okubo K. *Agric. Biol. Chem.* 1991, *55*, 2227.
72. Doddrell, D.M.; Pegg, D.T.; Bendall, M.R. *J. Magn. Res.* 1982, *48*, 323.
73. Doddrell, D.M.; Pegg, D.T.; Bendall, M.R. *J. Chem. Phys.* 1982, *77*, 2745.
74. Eldridge, A.C.; Kwolek, W.F. *J.Agric.Food Chem.* 1983, *31*, 394.
75. Pratt, D.E.; Birac, P.M. *J. Food Sci.* 1979, *44*, 1720.
76. Hayes, R.E.; Bookwalter, G.N.; Bagley, E.B. *J.Food Sci.* 1977, *42*, 1527.
77. Naim, M.; Gestetner, B.; Bondi, A.; Birk, Y. *J. Agric. Food Chem.* 1976, *24*, 1174.
78. Pratt, D.E. In *Autoxidation in Foods and Biological Systems*; Simic, M.G.; Karel, M.; Eds. Plenum Press, New York ,1980, 283-93.
79. Pratt, D.E.; Di Pietro, C.; Porter, W.L.; Giffee, J.W. *J. Food Sci.* 1981, *47*, 24.
80. Dziedzic, S.Z.; Hudson, B.J.F. *Food Chem.* 1983, *11*, 161.
81. Ragnarsson, J.O.; Labuza, T.P. *Food Chem.* 1977, *2*, 291.
82. Marco, G. *J. J. Amer. Chem. Soc.* 1968 *45*, 594.

RECEIVED January 13, 1992

Chapter 9

Catechins of Green Tea
Antioxidant Activity

T. L. Lunder

Nestlé Research Centre, Nestec, Ltd., CH–1000 Lausanne 26, Switzerland

Green tea contains about 30% polyphenyls and the most
important are the flavanols or tea catechins. Among
these substances the Epigallocatechingallate (EGCG)
which represents 50-60% of the total catechins seem
to play an outstanding role in the antioxidant activity. Our preliminary analytical findings show that
the antioxidant index (AI) could be related to the
content of the EGCG.

A Chinese legend places the introduction of tea drinking in the
reign of the mythological emperor, Shen Nung, at about 2737 B.C.,
but the earliest credible mention is 350 A.D. when a tea handbook
entitled *Ch'a Ching* was discovered.
      A favorite tale of the origin of tea concerns Daruma, a Budd-
hist priest who, having fallen asleep over his devotions, cut off
his eyelids and upon awakening throws them on the ground where they
take root and grow up as a bush whose leaves, when dried and infused
in hot water, produce a beverage that would suppress sleep *(1)*.
      The Dutch brought the first tea to Europe in approximately
1610. It reached Russia in 1618, Paris in 1648 and England and
America in about 1650.

Botany

The tea plant was originally classified by linnaeus under the bi-
nomial system in 1753 as *Thea sinensis*, but the current name, according to the international code of botanical nomenclature, is
Camellia sinensis (family of *Theaceae*). There are many synonyms,
including *Thea Bohea, Thea viridis, Thea cantonensis, Thea assamica,
Camellia Thea* and others.
      Taxonomically speaking, two basic varieties of the tea plant
are recognized: the northern (China) form, *Camellia sinensis*, and
the southern (Assam) form, *Camellia asamica*.

0097–6156/92/0507–0114$06.00/0

The commercial tea belt of the world is largely confined to a ring of mountains around or near the Equator. The quality of tea is at its best at different seasons in the various tea countries. Generally speaking, there are three main classes of tea:
- unfermented or green tea
- semifermented or oolong tea
- fermented or black tea

Chemistry

Like all plant-leaf material, the tea leaf contains the enzymes, biochemical intermediates and structural elements normally associated with plant growth and photosynthesis. In addition, it contains substances that are known to be responsible for the unique properties of tea (2). A representative analysis of green and black tea is given in Table I.

Table I. Approximate Analysis of Green and Black Tea
(% w/w)

Component	Green Tea	Black Tea
Proteins	15	15
Fiber	30	30
Pigments	5	5
Caffeine	4	4
Amino Acids	4	4
Mineral Salts	5	5
Carbohydrates	7	7
Flavanols (catechins)	17-30	5
Flavonol Glycosides	3-4	2-3
Oxidized Polyphenols	0	25

Polyphenolic Substances. The polyphenols or tea tannins, as they are called, constitute the major portion of fresh leaf, green tea and black tea. Their approximate amounts in fresh leaf, green and black teas are in the range of 30 to 35%, 10-25% and 8 to 21%, respectively (3). The polyphenolic composition of tea is shown in Table II.
The most important are the flavanols (catechins), which include (+)-Catechin, (-)-Epicatechin, (+)-Gallocatechin, (-)-Epigallocatechin (-)-Epicatechingallate and (-)-Epigallocatechingallate. Figures for the individual catechins are given in Table III.

Conversion of Green Tea to Black Tea. Catechins play the most significant role in the manufacture of black tea which is the most widely consumed form of tea in Western countries. Black tea is the result of promoting the oxidation of fresh leaf catechins by atmospheric oxygen through catalysis by tea polyphenoloxidase.

Nutritional Value. The nutritional significance of tea (green and black) seems to be confined to some vitamins and minerals. Vitamin C is normally present in green tea leaves in excessively large amounts but, as a consequence of the drying process, this vitamin is almost completely destroyed, or its amount largely reduced (4). More interesting is the Folic Acid whose amounts could represent, in carefully dried green teas, up to 13% of the daily requirements.

Tea (black and green) has a relatively important content of minerals and Table IV gives the quantities provided following an intake of five cups per day.

Table II. Polyphenolic Composition of Tea
(% w/w)

Unprocessed Tea (Green)	
Flavanols (Catechins)	17-30
Flavonols + Flavonol Glycosides	3-4
Phenolic Acids	5
Fermented Tea (Black)	
Thearubigins	15-20
Theaflavins	1-2
Bisflavonols	2-4
Phenolic Acids	4
Unchanged Flavanols	1-3
Flavonols + Flavonol Glycosides	2-3

Table III. Catechin Content of Green Tea Leaves

Catechin	g/100g
Epigallocatechin	2.35
Gallocatechin	0.37
Epicatechin	0.63
Catechin	0.35
Epigallocatechingallate	10.55
Epicatechingallate	2.75

Table IV. Minerals Provided by Black and Green Tea

Minerals	Average Intake from 5-6 Cups (Milligrams)
Sodium	6
Potassium	982
Magnesium	18
Calcium	4
Strontium	0.06
Manganese	1.8
Iron	0.02
Copper	0.2
Zinc	2.4
Phosphorus	10

Physiological Properties

We must point out that tea is the most popular beverage after water and that its excellent reputation is justified. Its healthful ef-

fects cannot be ascribed only to the presence of caffeine, but a great importance is given to catechins (flavanols). Tea is believed to exhibit the following physiological properties (4-5):
- stimulant
- antidepressant
- antiinflammatory
- enhancement of Vitamin C activity
- antihypertensive
- antiatherosclerosis
- hypocholesterolemic
- protection against sunburn
- antioxidant
- retards tooth decay
- protection against nitrosamines
- protection against radioactivity

Tea as an Antioxidant

We focused our attention on the antioxidant properties of green tea extracts because the interest in natural antioxidants is growing (6-9). To determine the Antioxidant Index (AI), tea leaves were not suitable and infusions were difficult to handle. To overcome this drawback, all samples were in the form of spray-dried aqueous extracts.

The AI was carried out in the Rancimat (10) using a ternary mix especially developed for water-soluble substances. The Rancimat used to carry out these measures was the Metrohm Model 617 (Metrohm, SA, Switzerland). The composition of the ternary mix was the following:

Topcithine (trademark for a lecithin)	32%
Vitamin E	3%
Propylenglycol	58%
Green Tea Extract	7%

In the referenced ternary mix, Vitamin C replaced the tea extract at the same level. The fat used in the Rancimat was the chicken fat, Micana III, which is a fat devoid of Vitamin E. The AI is given by Induction Time, i.e., the time necessary to get the fat rancid in the presence of tea divided by the time required for the fat alone to turn rancid. Results are expressed in graphic form and shown in Figure 1. The codes appearing at the bottom of each bar are explained in Table V. Tea having the code MSCOP has the same composition as teas 4204/E and 3859/E2, which are mixtures of green and black tea in ratios from 4:1 to 3:1. The MSCOP was extracted under pressure to get more solids, whereas the other two were extracted at atmospheric pressure (11). The mixture, Green:Black, gives a cold water-soluble powdered extract because catechins have the capability of breaking down the complex formed by caffeine with polyphenols. No chemicals or enzymes are therefore needed.

Relationship between AI and EGCG Content. Catechins, and especially epigallocatechingallate (EGCG) which represent 50-60% of the whole catechin group, seem to be responsible for the antioxidant activity. We tried to find a relationship between the main catechin, i.e., the EGCG and the AI. The EGCG content of each sample was determined by

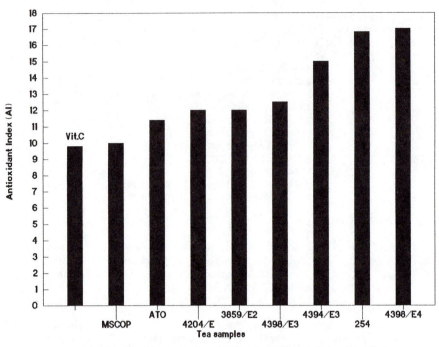

Figure 1. Antioxidant index (AI) of different types of tea.

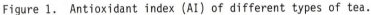

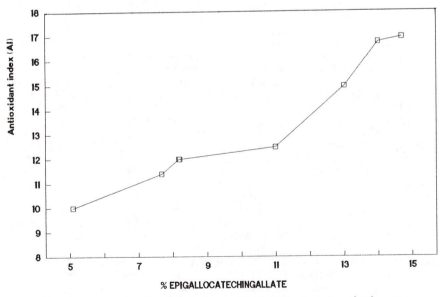

Figure 2. Relationship between Antioxidant index (AI) and Epigallocatechingallate (EGCG) content in tea.

HPLC (*12*) and we found that there was a good correlation between the AI and the EGCG content.
    Table V gives the values for both the AI and EGCG for several experimental tea extracts, and Figure 2 shows that an increase in EGCG content is accompanied by an increase in the AI.

Table V.   Relationship between Antioxidant index (AI) and Epigallocatechingallate (EGCG) content of different types of tea

Sample	Antioxidant Index (AI)	Epigallocatechingallate (% EGCG)
Green Teas		
ATO	11.4	7.68
254	16.8	14.7
4398/E3	12.5	11.0
4394/E3	15.0	13.0
4398/E4	17.0	14.0
Green-Black Tea Mixtures		
4204/E	12.0	8.50
3859/E2	12.0	8.20
MSCOP	10.0	5.10

ATO = Japanese Green Tea
254, 4394, 4398 = Experimental Green Tea
4204, 3859 = Mixtures of Green and Black Teas
MSCOP = Mixture of Green and Black Tea Extracted
           Under Pressure

Conclusion

Green tea extracts, as well as extracts containing definite mix-tures of green and black tea (these extracts have the properties to be cold-water soluble), exhibit interesting antioxidant properties which seem to correlate well with the EGCG content. This substance would be especially useful in evaluating the antioxidant properties of tea infusions because the Rancimat test cannot be carried out on samples in liquid form.

Acknowledgement. We are fully indebted to Ms. E. Prior for having carried out the tests on the Rancimat.

Literature Cited

1.   Graham, H.   In *Encyclopedia of Chemical Technology, 3rd Ed;* John Wiley & Sons: New York, **1978,** Vol. *22;* pp 628-44.
2.   Sanderson, G. W.   *Recent Adv. Phytochem.* **1972,** *5,* 247-315.
3.   Das, D. N.; Ghosh, J. J.; Bhattacharyya, K. C.; Guha, B. C. *Indian J. Appl. Chem.* **1965,** *28(1),* 15-40.
4.   Stagg, G. V.; Millin, D. J.   *J. Sci. Fd. Agric.,* **1975,** *26,* 1439-59.
5.   Graham, H. N.   In *The Methylxanthine Beverages and Foods: Chemistry, Consumption and Health Effects;* A. R. Liss: New York, **1984,** 29-74.
6.   Hara, Y.   *New Food Ind.* **1990,** *32(2),* 33-8.

7.   Herrmann, K. *Fette, Seifen u. Anstrichmit.* **1973,** *75(18),* 499-504.
8.   Bokuchava, M. A.; Skobeleva, N. I. *CRC in Food & Nutrition,* **1979,** *12(4),* 303-70.
9.   Loeliger, J. Swiss Patent 0 326 829, **1989**
10.   Lunder, T. L., Swiss Patent 0 201 000, **1986**
11.   Hoefler, A. C.; Coggon, P. *J. Chromatog.* **1976,** *129,* 460-63.

RECEIVED June 15, 1992

CHEMICAL AND BIOLOGICAL ACTIVITIES OF PHENOLIC ANTIOXIDANTS

Chapter 10

Antioxidative Defense Systems Generated by Phenolic Plant Constituents

T. Osawa, N. Ramarathnam, S. Kawakishi, and M. Namiki

Department of Food Science and Technology, Nagoya University, Chikusa, Nagoya 464—01, Japan

Recently, we have been much interested in the endogenous plant antioxidants that are expected to inhibit lipid peroxidation and protect against oxidative damage to membrane functions. A new type of natural antioxidant, β-diketones have been isolated from *Eucalyptus* leaf wax and identified. The β-diketones may protect endogenous essential oils from oxidative degradation by chemical as well as physical means. Further investigation led to the isolation of novel tocopherol conjugates, named prunusols A and B, from the *Prunus* leaf wax. We have also been investigating whether plant seeds, in particular, rice hulls exhibit any antioxidative defense systems that protect the ability of mature rice seed to germinate during storage As the results of large scale purification and isolation of antioxidative components in the *Indica* rice hull, isovitexin has been identified. These informations lead us to obtain a novel type of antioxidant, tetrahydrocurcumin, which contains both β-diketone and phenolic hydroxy moieties in the structure.

All physiological processes in living systems involve complex combinations of oxidation-reductuion reactions governed by a variety of endogenous compounds including enzymes. Any changes in the normal redox equilibrium, in particular, excess production of oxygen radical species such as hydrogen peroxide, superoxide anion radical, hydroxy radical, and other radicals are speculated to cause the damages in the cells. This damage can result in diseases in accelerated aging syndromes, and eventually, in extreme cases, in death (*1-4*). Most living organisms are, however, speculated to have extremely efficient defense and protective systems in a cell that are essential for defending the organism against the oxidative stress caused by oxygen radical species. Such systems include enzymatic inactivation as well as nonenzymatic protection. Recent publications indicated that endogenous antioxidants play an important role in antioxidative defense mechanism in biological systems since oxidative damage to cell components is apparently one of several factors which contributes to aging (*5-9*).

0097—6156/92/0507—0122$06.00/0

Antioxidant compounds have been already found in numerous plant materials such as oil seeds, crops, vegetables, fruits, leaves and leaf wax, bark and roots. Spices, herbs and crude drugs are also important sources of natural antioxidants, but the chemical properties and physiological role of the active components have not fully been understood. Our work has focused on isolation and identification of plant antioxidants from plant leaf waxes and hull of rice seed hulls.

ANTIOXIDATIVE COMPONENTS IN THE LEAF WAXES

It had been postulated that the stability against oxidative degradation of *Eucalyptus* oil could be due to the presence of a system in a leaf waxes that provides both physical and chemical protection. We have screened seventy-six different kinds of plant leaf waxes, and we found strong antioxidative activity in the leaf waxes extracted from *Eucalyptus* and *Prunus* species of plants (*10*). Based on these data, we have started our new project for isolation and identification of antioxidative components extracted with chloroform.

Model systems composed of chloroform extracts and linoleic acid as a substrate were analysed for antioxidant activity using thiocyanate method. As shown in Fig.1, more than half of the crude chloroform extracts in seventeen diferent type of *Eucalyptus* leaf wax showed quite strong antioxidant activity. Following large scale isolation and identification of antioxidative substances in the leaf waxes of *Eucalyptus viminalis* and *Eucalyptus grobulus*, we were able to identify two active components.

As shown in Fig. 2, *n*-tritriacontane-16,18-dione (*11*) and 4-hydroxy-tritriacontane-16,18-dione (*12*) have been isolated and identified. In order to get more information on structure-activity relationship, many different types of β-diketones were prepared and assayed by the thiocyanate methods, however, no antioxidant activity was noted for the simplest β-diketone, acetyl acetone, the cyclic β-diketone, syncarpic acid, the cunjugated β-diketone and curcumin. The study of structure-activity relationships of β-diketone analogues indicated that long hydrocarbon side chains on both sides of the β-diketones are essential for antioxidative activity. The enol form of the β-diketone moiety seems to be essential for the role as a radical scavenger as well (Fig. 3).   β-Diketones have been reported to be one of the main components in common plant leaf waxes such as acasia, rhododendron, barley, and oat etc. Strong antioxidative activity has also been detected in the chloroform extracts of cherry leaf waxes (*Prunus* species), especially, in the leaf wax of *Prunus grayana* Maxim., usually found as the wild cheery tree in the mountain area in Japan (*13*).

In the course of our investigation of the antioxidative components in the chloroform extracts of the leaf wax of *Prunus grayana* Maxim., it was suggested that the leaf wax of *Prunus grayana* Maxim. contains antioxidative components which were different from tocopherols. Large scale purification of the antioxidative components from the leaf wax of *Prunus grayana* Maxim. led us to isolate two novel phenylpropanoid-substituted tocopherol derivatives, which we named prunusols A and B, and three additional tocopherol dimers. The antioxidative activity of these isolated compounds was examined by several model systems, and prunusols A and B were found to possess distinct antioxidative activity as shown in Fig. 4. Tocopherol dimers have been detected in several vegetable oils, including cottonseed oil, corn oil and soybean oil. These tocopherol derivatives can be formed artificially during the purification of soybean oil, and can also be formed during the course of thermal oxidation of triglycerides. However, these tocopherol dimers have not been previously identified as endogenous components in the leaf wax. By HPLC analysis of these tocopherol derivatives in the leaf wax without

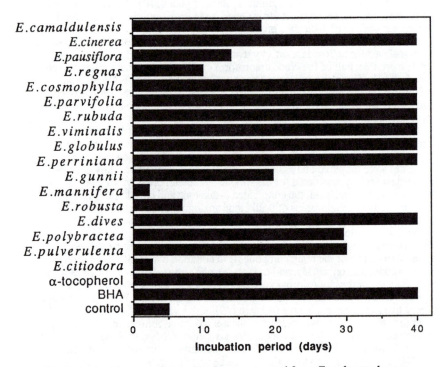

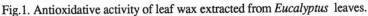

Fig.1. Antioxidative activity of leaf wax extracted from *Eucalyptus* leaves.

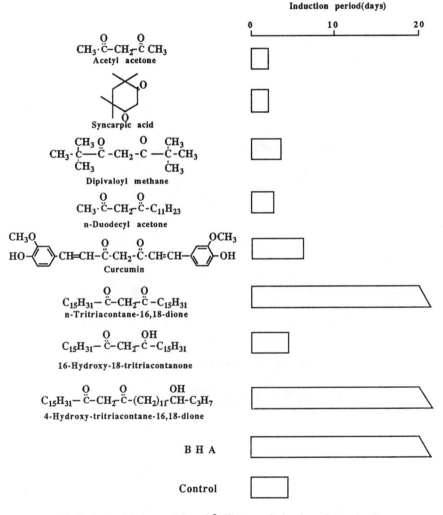

Fig.2. Antioxidative activity of β-diketone derivatives determined by the thiocyanate method.

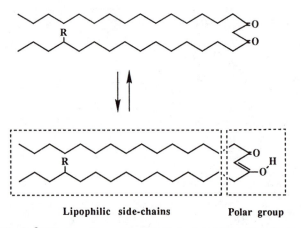

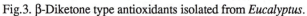

Lipophilic side-chains   **Polar group**

Fig.3. β-Diketone type antioxidants isolated from *Eucalyptus*.

R=H : *n*-tritriacontan-16,18-dione
R=OH : 4-hydroxy-tritriacontan-16,18-dione

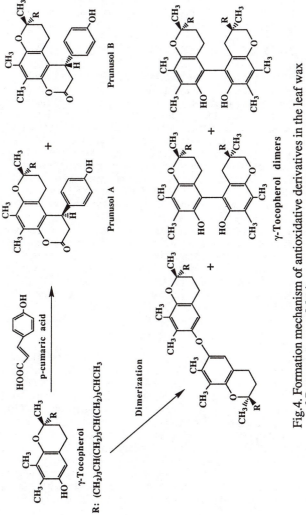

Fig. 4. Formation mechanism of antioxidative derivatives in the leaf wax of *Prunus grayana* Maxim.

purification, it was found that tocopherol dimers are present in a large quantity in the leaf wax of *Prunus grayana* Maxim (total tocopherol dimers; 3.01 mg/g fresh weight). This is the first proof that tocopherol dimers can be produced biogenetically.   Based on NMR analysis, the structures of prunusols A and B have been proposed as shown in Fig. 5. Prunusols A and B are the new class of γ-tocopherol derivatives substituted at the chroman ring with *p* -coumaric acid. The synthesis of the γ-tocopherol dimers has been reported to be achieved by oxidation in the presence of *p* -benzoquinone.   On the other hand, synthesis of prunusols A and B has been carried out by the reaction of γ-tocopherol with *p* -coumaric acid in the presence of *p* -toluenesulfonic acid.

ANTIOXIDATIVE COMPONENTS IN THE RICE HULL

Rice (*Oryza sativa* Linn.) is the principle cereal food in Asia and the staple food of nearly half of the world's population. Recently, the authors investigated the germination potentials of rice seeds during long storage. Differences between the two variety, *Japonica* and *Indica*, after aging could be due to the difference in the effectiveness of defence systems in the rice hull that has been believed until now to offer only physical protection to rice grain.   When *Japonica* and *Indica* rice seeds were stored at room temperature, *Indica* (long-life) rice seeds (KATAKUTARA and CENTURY PATNA) were found to maintain viability even after storage for 1 year but all of the *Japonica* (short-life) rice seeds (KUSABUE, SACHIWATARI, HIMENOMOCHI and HATAKINUMOCHI) except KOSHIHIKARI lost viability after storage for 1 year at the same temperature as shown in Fig. 6 (*14*).

Model experiments were conducted to induce accelerated aging by initiation of lipid peroxidation by hydroxy radicals generated during γ-irradiation of a biological system. It was demonstrated that for rice grain irradiated without intact hulls, the TBA value and radical formation (ESR signal intensity) increased more rapidly than for rice grain irradiated with an intact hull.   There are also some difference in lipid peroxidation between the long- and short-life grains; short-life grain irradiated with an intact hull showed a greater increase both in the TBA value and in radical formation than long-life grain (Fig.7) (*15*).   These data indicate that the hull fraction of long-life rice seeds have antioxidative constituents.   Further experiments showed that the hull of the long-life grain is richer in phenolic constituents than the hull of short-life rice cultivars.

It has long been known that germination of rice seeds is influenced by components present in the hull. Momilactones A and B, inelactone, S-(+)-dehydrovomifoliol, momilactone C, and p-coumaric acid were found as the growth and germination inhibition factors; however, various factors may be responsible for aging, loss of vigor and death of seeds during long-term storage (*16*).   Accelerated aging effects induced by γ-irradiation were investigated on the fatty acid composition of lipids and on the content of the endogenous antioxidants of long-life and short-life rice seeds with and without intact hull. While the linoleic acid content of the phospholipids decreased gradually with the increase in irradiation doses, there was a corresponding increase in the linoleic acid content of the fatty acids. Such changes were drastic, especially, in the case of short-life rice seeds irradiated without intact hull.   However, neutral lipids were found to be resistant to γ-irradiation.

The α-tocopherol content of irradiated rice seeds was found to decrease markedly, particularly, in the short-life rice seeds.   As shown in Table 1, only traces of α-tocopherol could be detected both in short- and long-life rice seeds irradiated with and without intact hull, but oryzanol, a relatively weaker antioxidant, was more resistant to oxidative damage (*16*).   These data indicated that an antioxidant defense system exists in rice hull that offer effective chemical

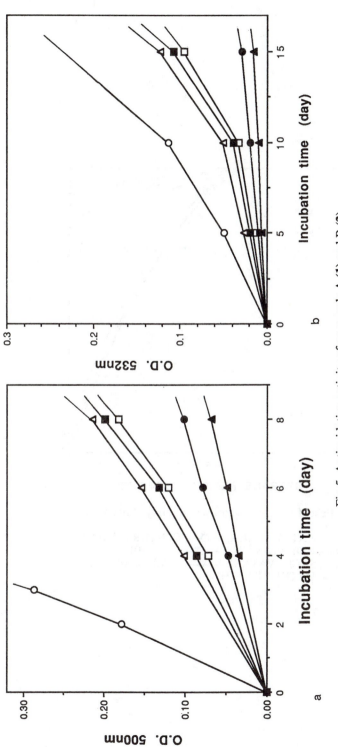

Fig.5. Antioxidative activity of prunusols A (**1**) and B (**2**).

(a) Determined by the thiocyanate method.
(b) Determined by the thiobarbituric acid (TBA) method.

——○——, control; ——●——, BHA; ——△——, α-tocopherol;
——▲——, γ-tocopherol; ——□——, **1**; ——■——, **2**.

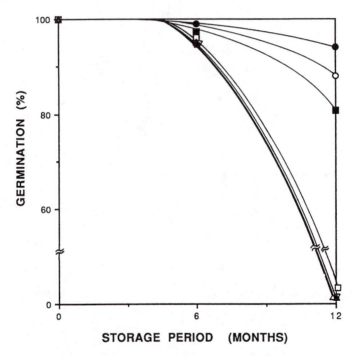

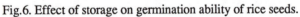

Fig.6. Effect of storage on germination ability of rice seeds.

——●——, KATAKUTARA; ——○——, CENTURY PATNA;
——■——, KOSHIHIKARI; ——□——, KUSABUE;
——▲——, SACHIWATARI; ——▽——, HIMENOMOCHI;
——△——, HATKINUMOCHI

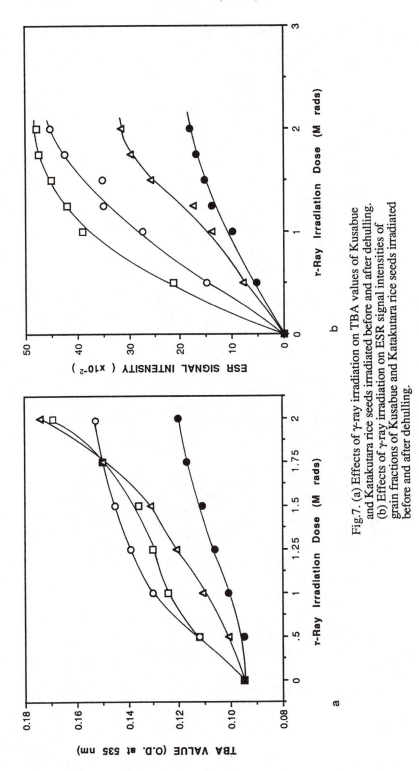

Fig.7. (a) Effects of γ-ray irradiation on TBA values of Kusabue and Katakutara rice seeds irradiated before and after dehulling. (b) Effects of γ-ray irradiation on ESR signal intensities of grain fractions of Kusabue and Katakutara rice seeds irradiated before and after dehulling.

——□——, dehulled Kusabue; ——○——, dehulled Katakutara; ——▲——, Kusabue with hull; ——●——, Katakutara with hull.

TABLE 1. Effects of γ-irradiation on α-tocopherol and oryzanol contents of lipids from rice seeds irradiated with and without hull[a]

	α-tocopherol (μg/g lipid)				Oryzanol (μg/g lipid)			
VAR	OkGy	5kGy	10kGy	15kGy	OkGy	5kGy	10kGy	15kGy
KAT	407	257	114	tr	151	134	108	105
		(110)	(tr)	(tr)		(128)	(104)	(93)
CP	385	227	93	tr	246	227	182	170
		(100)	(tr)	(tr)		(207)	(172)	(153)
K-184	392	216	98	tr	178	161	125	116
		(94)	(tr)	(tr)		(146)	(120)	(105)
IR-8	385	219	85	tr	96	82	75	59
		(89)	(tr)	(tr)		(77)	(62)	(52)
KOS	443	274	115	tr	100	89	74	62
		(128)	(tr)	(tr)		(85)	(69)	(56)
KUS	350	204	88	tr	113	89	78	65
		(95)	(tr)	(tr)		(86)	(69)	(55)
SAC	427	231	94	tr	145	119	103	92
		(107)	(tr)	(tr)		(116)	(97)	(73)
HIM	319	195	76	tr	115	96	81	70
		(84)	(tr)	(tr)		(86)	(71)	(63)

[a]Results are averages of three replicates.
Figures in parentheses refer to values, when rice seeds were irradiated without intact hull.
VAR, cultivar; KAT, Katakutara; CP, Century Patna; KOS, Koshihikari; KUS, Kusabue; SAC, Sachiwatari; HIM, Himenomochi.
SOURCE: Adapted from ref. 16.

protection to both the rice grain and the germ, which is mainly responsible for germination. This protective effect could be supplementary to the endogenous rice antioxidants, α-tocopherol and oryzanol. Though extensive work has been done on the growth and germination inhibiting factors in the rice hull, no attempt has been made yet to investigate the chemical basis of the antioxidative defense system in rice hull. Our objective, therefore, was to isolate and identify the antioxidative components in rice hull.

In order to purify and isolate antioxidative components in the hull of long-life rice seeds, hull of KATAKUTARA (*Indica*) variety of rice seeds was extracted, and fractionated by column chromatography (Amberlite XAD-2) and preparative HPLC techniques. The active components were characterized by FAB-MS, ^{1}H-NMR and ^{13}C-NMR techniques, and were found to be flavonoid substances. One of the flavonoid substances exhibited antioxidant properties as strong as those of α-tocopherol was identified as isovitexin, a C-glycosyl flavonoid (Fig. 8) (*17*). Further investigation on detailed mechanisms of these hull antioxidants for protection against oxidative damages is now in progress.

NOVEL TYPE OF ANTIOXIDANT, TETRAHYDROCURCUMIN

Although many different types of natural antioxidants have been isolated and identified, the major natural antioxidants are plant phenols. Recently, we succeeded in obtaining tetrahydrocurcumin, a novel type of antioxidant which contains both a phenolic moiety and a β-diketone moiety in the same structure. Tetrahydrocurcumin (Fig. 9) is derived from curcumin, the main component of turmeric, by hydrogenation. Curcumin possesses antioxidant activity, and is also responsible for the yellow color of curry cooking. It has been reported to inhibit the microsome-mediated mutagenicity of benzo(a)pyrene and of 7,12-dimethylbenz(a)anthracene, and, more recently, Huang et al. (1988) reported that

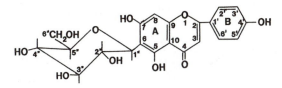

Fig.8. Structure of isovitexin.

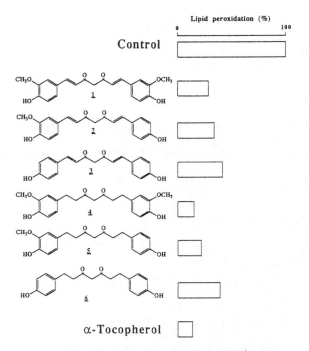

Fig.9. Structure of tetrahydrocurcumin.

curcumin acts as a strong inhibitor of tumor promotion (*18*). These effects roughly parallel its antioxidant activity. At the present stage, we have compared the antioxidative activity of curcumin derivatives in a food model system as well as in an erythrocyte ghost system. Tetrahydrocurcumin (4) was found to be the strongest as shown in Fig. 10. Tetrahydrocurcumin possess both a β-diketone and a phenol moiety in its structure, and has almost the same heat-stability as sesaminol. Correlation between antioxidative activity and antimutagenicity is currently being investigated.

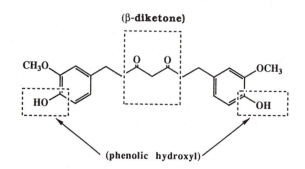

Fig.10. Antioxidative activity of curcumin derivatives determined by erythrocyte ghost system.

LITERATURE CITED

1. Aust, S.D.; Svingen, B.A. In *Free Radicals in Biology*, Vol.V; Pryor, W.A., Ed.; **1982**, Academic Press, New York
2. Balentino, D.B. *Pathology of Oxygen Toxicity*, **1982**, Academic Press, New York.
3. McBrian, D.C.H.; Slater,T.F. Eds.; *Free Radicals, Lipid Peroxidation and Cancer*, **1982**, Academic Press, New York
4. Cutler, R.G. In *Free Radicals in Biology*, Vol.VI; Pryor, W.A., Ed.; **1984**, Academic Press, New York, p. 371-428.
5. McCay, P.B.; King, M.M. In *Vitamin E: A Comprehensive Treatise*, L.J.; Mcchlin, L.J., ed.; **1980**, Marcel Dekker, Inc., New York, p.289-317.
6. Ames, B.N.; Cathcart, R.; Schwiers, E.; Hochstein, P. *Proc. Natl. Acad. Sci., USA* **1981**, *78*, 6858-6862
7. Kohen, R.; Yamamoto, Y.; Cundy, K.C.; Ames, B.N. *Proc. Natl. Acad. Sci., USA* **1988**, *85*, 3175-3179.
8. Simic, M.G. *Mutation Res.* **1988**, *202*, 377-386.
9. Osawa, T.; Namiki; M.;Kawakishi, S. In *Antimutagenesis and Anticarcinogenesis Mechanisms II* ; Kuroda, Y.; Shankel, D.M.; M.D.Water, M.D., Ed.; **1990**, Plenum Press, New York, p.139-153.
10. Osawa, T.; Kumazawa, S.; Kawakishi, S.; Namiki, M. In *Medical, Biochemical and Chemical Aspects of Free Radicals* ; Hayaaishi, O., Niki, E.; Kondo, M.; Yoshikawa, T., Eds.; **1989**, Elsevier, Tokyo, p.583-586.
11. Osawa, T.; Namiki, M. *Agric. Biol. Chem.* **1981**, *45*, 735-939.
12. Osawa, T.; Namiki, M. *J. Agric. Food Chem.* **1985**, *33*, 777-780.
13. Osawa, T.; Kumazawa. S.; Kawakishi, S. *Agric. Biol. Chem.* **1991**, *55*, 1727-1731.
14. Ramarathnam, N.; Osawa, T.; Namiki, M.; Tashiro, T. *J. Sci. Food Agric.* **1986**, *37*, 719-716.
15. Osawa, T.; Ramarathnam, N.; Kawakishi, S.; Namiki, M.; Tashiro, T. *Agric. Biol. Chem.*, **1985**, *49*: 3085-3087.
16. Ramarathnam, N.; Osawa, T.; Namiki, M.; Kawakishi, S. *J. Am. Oil Chem. Soc.*, **1989**, *66*, 105-108.
17. Ramarathnam, N.; Osawa, T.; Namiki, M.; Kawakishi, S. *J. Agric. Food Chem.*, **1989**, *37*, 316-319.
18. Huang, M.-T.; Smart, R.S.; Wong, C.-Q.; Conney, A.H. *Cancer Res.*, **1988**, *48*, 5941-5946.

RECEIVED December 17, 1991

Chapter 11

Phenolic Antioxidants in Dietary Plants as Antimutagens

T. Osawa

Department of Food Science and Technology, Nagoya University, Chikusa, Nagoya 464–01, Japan

In the course of an intensive search for novel antioxidants from plant materials, in particular those used as foods, new types of natural antioxidants have been isolated and identified. Novel lignan type antioxidants have been isolated from sesame seeds and oils, and several phenol glucosides were found to be present in the cultured tissue of the sesame plant. Ellagic acid derivatives and tannins are found to be the main antioxidative components in the herbs, spices and crude drugs. Recently, much attention has been focused on dietary phenolic antioxidants which may be effective in protection from peroxidative damages, especially, from mutagenicity induced during lipid peroxidation. Polyphenol type antioxidants which have o-dihydroxy moiety in the structures, such as ellagic acid derivatives, (-)-epicatechin gallate and (-)-epigallocatechin gallate were found to inhibit not only the free-radical chain reaction of cell membrane lipids but also mutagenicity and DNA damaging activity. Moreover, these polyphenols were found to inhibit DNA single-strand cleavage induced by N-hydroxy-2-amino-naphthalene.

Recently, many reports have suggested the importance of protective defense systems in the living cells against damage caused by active oxygens and free radicals. It was also suggested that several physiological and biological antioxidants like vitamin E, vitamin C, β-carotene, uric acid, bilirubin and carnosine play an important role in nonenzymatic protection (*1-3*). In addition to these natural defense mechanisms, there is increasing interest in the protective biochemical function of dietary antioxidants, which can be a candidate for prevention of mutagenesis and carcinogenesis and, in some cases, extend the life span of animals (*4-5*).

Lipid peroxidation is known to be a free radical chain reaction which takes place

0097–6156/92/0507–0135$06.00/0

in *in vivo* and *in vitro* (*6-7*) and forms many secondary products such as alkanes, alcohols, acids and carbonyls (*8*). Many of these secondary products, especially malondialdehyde (MDA), are highly reactive and have been shown to interact with DNA and exhibit mutagenic activity in a strain of *Salmonella typhimurium* as well as *E.coli* . The mutagenicity of many carbonyl compounds related to lipid peroxidation has also been tested using *S.typhimurium* TA104 (*9*). One of the main aldehydes, 4-hydroxynonenal was also recognized to be the probable candidate for mutagenicity formed during the lipid peroxidation. Akasaka and Yonei (*10*) detected mutation induction in *E.coli* incubated in a reaction mixture containing NADPH-dependent lipid peroxidation of rat liver microsome. The induced mutation frequencies increased in correlation with lipid peroxidation as determined by thiobarbituric acid although identification of the mutagenic products has not been carried out.

From this background, we have been involved in an intensive search for antioxidant compounds from numerous plant materials, in particular those used as foods, and novel type of natural antioxidants have been isolated and identified. Novel lignan type antioxidants have been isolated from sesame seeds and oils, and several phenol glucosides were found to be present in the cultured tissue of sesame plant. Ellagic acid derivatives, flavonoids and other polyphenols were found to be present as the main antioxidative components in herbs, spices, crude drugs and teas.

LIGNAN TYPE ANTIOXIDANTS IN SESAME SEEDS

Sesame seeds and sesame oils have been used traditionally in Japan, China, Korea, and other eastern countries. Sesame oils, in particular roasted sesame oils are widely used in Chinese and Japanese dishes, which has been evaluated as being highly antioxidative. Sesame seeds also have good germination activity even after long-term storage. However, the kind of chemical constituents that are responsible for their wholesomeness and antioxidative activity has not yet been clearly identified.

For a long time, it has been speculated that the component is responsible for the stability of sesame oil is sesamol which is formed from sesamolin (one of the main non-antioxidative components as shown in Fig.1) during the manufacturing process. Sesamol, however, is unstable in the sesame oil, and the quantity of sesamol is not sufficient to exhibit the strong antioxidative activity in sesame oil. This prompted us to investigate other antioxidative components in the sesame seeds, because we speculated that antioxidative components in sesame seeds must be important for antioxidative defense. We have done a large-scale isolation and identification of antioxidative components in sesame seeds and were able to isolate and identify two types of antioxidants, lipid-soluble and water-soluble antioxidants. Two novel lipid-soluble lignan type antioxidants, sesamolinol and sesaminol (Fig. 2) were found to be present only in sesame seeds, together with other lipid-soluble lignan derivatives and water-soluble lignan glucosides (*11*).

FORMATION OF SESAMINOL IN UNROASTED OIL

Usually, sesame oils are manufactured in two different ways, roasted oil and salad oil (unroasted sesame oil). Salad oil is extracted by an expeller and refined by alkaline treatment, water washing, bleaching with acid clay and a deodorizing process. These refining processes are almost the same as for other vegetable oils such as corn, soybean, safflower and sunflower oils. By quantification using HPLC, the sesaminol content of sesame oil was found to be dramatically increased during the manufacturing process, in particular bleaching process (*12*).

These results suggested that sesaminol is chemically formed during the bleaching process. It was found that sesaminol is produced by intermolecular transformation from sesamolin (*13*). Four stereoisomers exist and all of these isomers have quite

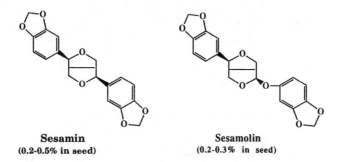

Sesamin
(0.2-0.5% in seed)

Sesamolin
(0.2-0.3% in seed)

Fig. 1. Main components in sesame seeds.

Lipid-soluble Antioxidants

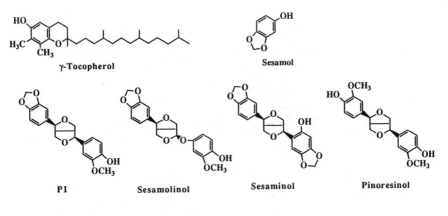

γ-Tocopherol

Sesamol

P1

Sesamolinol

Sesaminol

Pinoresinol

Water-soluble Antioxidants

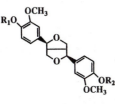

K₁; R₁=H, R₂=Glc-Glc-Glc
K₂; R₁=Glc, R₂=Glc
K₃; R₁=H, R₂=Glc-Glc
K₄; R₁=H, R₂=Glc-Glc

Fig. 2. Structure of lipid- and water-soluble lignan type antioxidants in sesame seeds.

strong antioxidative activity (Fig.3). When the amounts of sesaminol were quantified by HPLC in commercially available sesame oils, the total amount of sesaminol isomers was about four times that of γ-tocopherol in the most commercially available sesame seed oils. Therefore, we concluded that sesaminol is the main antioxidative component present in sesame oil. Sesaminol is an unique antioxidant because it has a superior heat stability, and is able to effectively depress the degradation of tocopherols in corn oil. It is speculated that sesaminol has a synergistic effect on tocopherols during thermal oxidation and inhibits the production of lipid peroxides in the oils, although the detailed mechanisms are still not clear.

ANTIOXIDANT ACTIVITY OF SESAMINOL

In order to determine the antioxidative activity, we have used several *in vitro* lipid peroxidation systems, in particular, the rabbit erythrocyte ghost and rat microsome systems. *t* -Butylhydroperoxide was used to induce autoxidation of membrane lipids of erythrocyte ghosts and enzymatic lipid peroxidation of rat microsome has been induced by ADP-Fe+++/NADPH and ADP-Fe+++/NADPH/EDTA-Fe+++.

All lignans effectively inhibited lipid peroxidation induced in both the rat liver microsomes and the erythrocyte ghost membrane systems (Table 1). Although other lignan-type components have comparable antioxidant activity, the role of sesaminol is very important because it is the main antioxidative component in sesame oil and has superior heat stability.

The protective role of sesaminol against oxidative damage has also been reported using cultured human diploid fibroblasts at various ages *in vitro*. The constitutive level of lipid peroxidation was higher in cells at late passages (75% of lifespan) than those at early passages (26% of life span). Lipid peroxidation in response to the addition of *t* -butylhydroperoxide was more marked in cells at late passages, the presence of sesaminol effectively reduced this peroxidation (*14*).

Recently, it was found that sesaminol inhibited the oxidative damage caused by carbon tetrachloride using SAM (sesnescent accelerated mouse) mice as the *in vivo* system (Fig.4), and long term feeding experiments of sesame seeds have been carried out. These data indicated the potential of sesaminol for inhibition of the oxidative damage caused in *in vivo* systems, and a long term feeding experiment of sesaminol is now ongoing (*15*).

In order to obtain the sesaminol in a large quantity, we have started a new project to increase the seed oil and minor components by breeding of sesame plants to increase the concentration of sesamolin, which is the precursor of antioxidant sesaminol. In the preliminary experiments, we have quantified the amount of sesamolin in 42 different varieties of sesame plants. As shown in Table 2, the sesamolin content of the oil ranged from 0.02-0.48%, avarageing 0.27%. The relationship between the ratio of sesamolin to sesamin and the oil content of *Sesamum* strains with different seed color types are shown in Fig. 5. Brown strains and one of the white-seeded strains (Type A) had relatively high oil content and also high ratio of sesamolin to sesamin. We are now looking for novel types of sesame strains which have a high oil content as well as a high concentrations of sesamolin (*16*).

TANNIN-TYPE ANTIOXIDANTS IN HERBS AND CRUDE DRUGS

Herbs and crude drugs prepared from plant materials are traditionally used medicine and their pharmacological effects have been extensively studied. However, only a few reports are available on their antioxidative components in herbs and crude drugs. From this background, we have started screening two-thirty different types

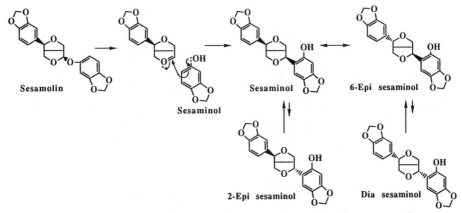

Sesamolin

Sesaminol

Sesaminol   6-Epi sesaminol

2-Epi sesaminol   Dia sesaminol

Fig. 3. Scheme for the mechanism of formation of sesaminol from sesamolin.

Table 1.   Antioxidant activity of sesame lignans using rat liver microsome and erythrocyte ghost systems

	ADP/NADPH/Fe^{3+}	ADP-Fe^{3+}/EDTA-Fe^{3+}/NADPH	Erythrocyte ghost
Control	100	100	100
P1	14.9	13.2	20.2
Sesamolinol	4.6	6.3	24.7
Sesaminol	8.6	10.3	22.8
Pinoresinol	17.2	14.4	18.4
Sesamol	24.1	19.0	19.0
α-Tocopherol	9.2	19.0	24.7

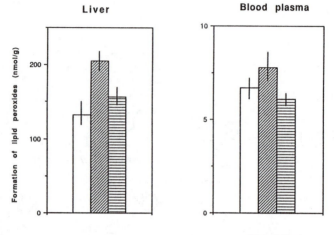

(Rat, 24hrs)

Fig. 4. Effect of sesaminol on formation of lipid peroxides in the rats after feeding CCl4.

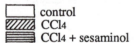

control
CCl4
CCl4 + sesaminol

Table 2.   The seed oil content and the contents of minor components in oil for 42 strains of *Sesamum indicum* L

	Oil (%)	Sesaminol (% in oil)	Sesamolin	100 seedweight (mg)	Hull[a] (%)
Mean	52.4	0.36	0.27	279.8	9.1
Range	43.4-58.8	0.07-0.61	0.02-0.48	218.7-390.9	3.5-23.2
SD[b]	3.9	0.16	0.13	42.8	4.7
CV[c]	7.4	44.2	47.5	15.3	51.1

[a]Ratio of the hull to the whole seed by weight.
[b]Standard deviation.
[c]Coefficient of variation.

SOURCE: Adapted from ref. 16.

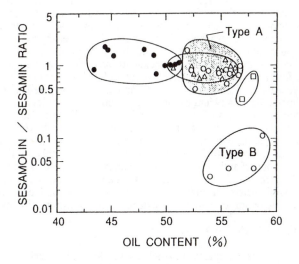

Fig.5. Relationship of the ratios of sesamolin content to sesamin content and to the oil content in different strains.

○ , white-seeded strain; △ , brown-seeded strain;
● , black-seeded strain; □ , yellow-seeded strain.

Adapted from ref. 16.

of herbs and crude drugs obtained from the domestic markets and drug stores in Taipei. Twenty-two species of herbs and crude drugs showed quite strong antioxidative activity. Of course, vitamin E seems to be one of the most important antioxidative components and may be responsible for much of the antioxidative activity in many of the extracts of herbs and crude drugs. The amounts of tocopherol derivatives have been quantified using HPLC and it was concluded that *Osbeckia chinensis* has no tocopherols, while, other herbs and crude drugs were found to have tocopherols as the antioxidative components. These results prompted us to isolate and identify the active principles in the extracts of *Osbeckia chinensis*.

The whole plant of *Osbeckia chinensis* has long been commonly used as an anodyne, anti-inflammation agent and antipyretic in Taiwan, Japan, and China. After large scale purification, many types of antioxidative components have been isolated and identified (*17*). Most of the antioxidative activity in the extracts of *Osbeckia chinensis* is assumed to be due to tannin-type antioxidants (Fig.6). The relative antioxidative activity of the isolated tannins has been examined using the rabbit erythrocyte system, and all of the isolated tannins have quite strong antioxidative activity. These isolated tannins contain the hexahydroxydiphenic acid moiety in their structures, and ellagic acid itself has almost the same antioxidant activity. These results suggest that the hexahydroxydiphenic acid moiety must be responsible for the antioxidative activity of the isolated tannins (*18*). In fact, antioxidative activity of three hydrolyzable tannins were compared, and it was found that the o-diphenolic moiety and lactone ring are both essential for antioxidative activity in the ellagic acid struture as shown in Table 3. Ellagic acid was also found to exhibit the strong (more than 50 times than α-tocopherol) antioxidative protection against oxidative damage in the rat microsome caused by adriamycin, one of the most popular anticancer antibiotics (*19*). Adriamycin is a member of the anthracycline group of antineoplastic agents and is widely used in the medical treatment of carcinoma. It has been reported to cause cardiotoxicity as one side effects due to lipid peroxidation induced by oxygen radicals derived from the quinone structure of adriamycin, but the detailed mechanism is still unknown.

POLYPHENOLS IN TEA LEAVES

Two catechin gallates, (-)-epicatechin gallate and (-)-epigallocatechin gallate, are the main components of tea polyphenols (Fig.7). Antioxidative activity of tea polyphenols was evaluated using *in vitro* lipid peroxidation systems, including the NADPH-dependent microsomal lipid peroxidation system (*5*). (-)-Epicatechin gallate and (-)-epigallocatechin gallate effectively inhibited lipid peroxidation induced by lipid hydroperoxide-dependent initiation although (-)-epigallocatechin has been shown to have strong antioxidative activity only against a perferyl ion-dependent initiation step. (-)-Epicatechin gallate and (-)-epigallocatechin gallate also effectively inhibited rat liver microsomal lipid peroxidation induced by adriamycin. Because (-)-epicatechin gallate and (-)-epigallocatechin gallate exhibited more than 10 times stronger activity than α-tocopherol, we believe there is a potential use of (-)-epicatechin gallate and (-)-epigallocatechin gallate in prevention of the side effects of chemotherapeutic agents such as adriamycin.

Because red blood cell membranes contain high concentration of polyunsaturated lipids, we have carried out an evaluation on *in vitro* antioxidative activity of tea catechins by the simple and convenient antioxidative assay system using erythrocyte membrane ghosts. Eythrocyte ghosts prepared from human and rabbit red blood cell membranes both showed almost the same degree of lipid peroxidation when induced by *t*-butylhydroperoxide, therefore, the evaluation of *in vitro* antioxidative effects of tea polyphenols was carried out using the rabbit erythrocyte membrane ghost system. (-)-Epicatechin gallate and (-)-epigallocatechin gallate showed the

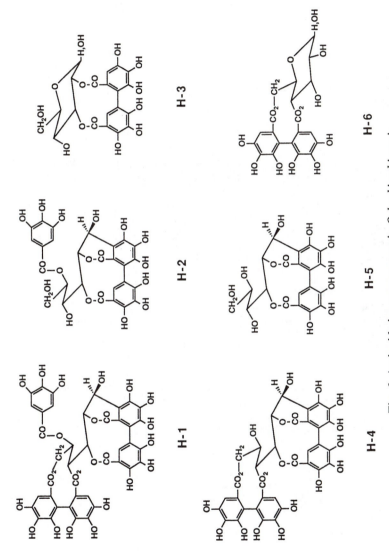

Fig. 6. Antioxidative components in *Osbeckia chinensis*.

Table 3. Inhibition of lipid peroxidation in the rat liver microsome system by ellagic acid and two derivatives

compound	IC_{50}^a		
	ADP-Fe^{2+}/ NADPH	ADP-Fe^{3+}/ EDTA-Fe^{3+}/ NADPH	Adriamycin
α-Tocopherol	>100	0.80+0.02	5.5+0.2
Propyl gallate	>100	20+3	0.30+0.02
Ellagic acid	20+1	23+2	0.10+0.01
Hexahydrodiphenic acid	>100	>100	7.0+0.5
Ellagic acid tetraacetate	>100	>100	31+3

$^a IC_{50}$=concentration (μM) for 50% inhibition of lipid peroxidation in model systems. Reported values are mean + SD (n=3).

SOURCE:Adapted from ref. 19.

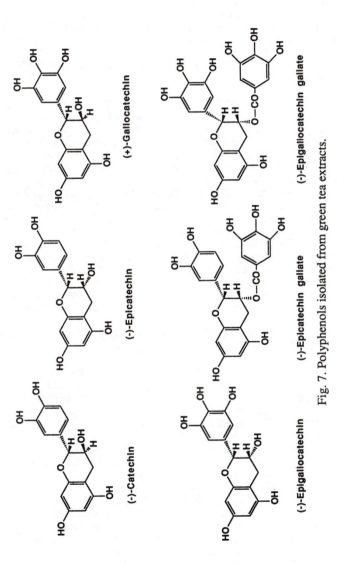

Fig. 7. Polyphenols isolated from green tea extracts.

strongest protection from lipid peroxidation among tea catechins, and were more active than the standard antioxidants α-tocopherol and propyl gallate. These data suggest that dietary antioxidants such as tea polyphenols may play an important role in inhibition of the lipid peroxidation promoted by the heme compounds associated with the phospholipids of erythrocyte membranes. (-)-Epicatechin gallate and (-)-epigallocatechin gallate demonstrate the strong inhibition of radical-induced oxidative destruction of membrane lipids both in erythrocyte and microsomal systems, suggesting that the gallic acid moiety effectively enhances antioxidative activity, although the detailed mode of action is unknown.

INHIBITORY EFFECTS OF ANTIOXIDATIVE COMPONENTS ON MUTAGENESIS

There are many indications that lipid peroxidation plays an important role in carcinogenesis, although there is no definitive evidence. The peroxidative breakdown of the membrane polyunsaturated fatty acids is known to produce a complex mixture of many different types of secondary products such as malondialdehyde, 2-hexenal, 4-hydroxypentenal, 2,4-hexadienal and 4-hydroxynonenal, which are suspected to be the main mutagenic secondary products (Fig.8) (20). Recently, we observed that mutagenic lipid peroxidation products can be formed in the erythrocyte ghost membranes although the mutagenic products were not identified. Because oxygen radical scavengers such as superoxide dismutase, catalase, and mannitol were not effective inhibitors, it is assumed that oxygen radical species such as superoxide anion, hydroxy radical and hydrogen peroxide do not induce both lipid peroxidation and mutagenicity. Both lipid-soluble antioxidants such as α-tocopherol and sesaminol, and polyphenol type antioxidants such as ellagic acid derivatives, (-)-epicatechin gallate and (-)-epigallocatechin gallate inhibited effectively lipid peroxidation and mutagenicity. In addition, these polyphenol type antioxidants were also found to diminish the mutagenicity of secondary metabolites formed during lipid peroxidation(5, 21).

EFFECTS OF POLYPHENOL TYPE ANTIOXIDANTS ON DNA DAMAGES

Recently, much interest has been focused on the oxygen radicals formed during the metabolic activation of mutagenic carcinogens. N-Hydroxy-2-aminonaphthalene is an active form of the mutagenic and carcinogenic 2-aminonaphthalene. It is proposed that cause mutation by binding covalently with DNA and cause DNA damage by forming hydroxy radical by iron-catalyzed Haber-Weiss reactions during the oxidative metabolic process (Fig. 9) (22).

Both ellagic acid derivatives and catechin gallates inhibited the mutagenicity of N-hydroxy-2-aminonaphthalene effectively using *Salmonella typhimurium* TA 104. This result showed that these polyphenol type antioxidants effectively inhibited the formation of covalent binding of active form of 2-aminonaphthalene (N-hydroxy-2-aminonaphthalene) to DNA (step 1 in Fig. 9). These polyphenol type antioxidants which have o-dihydroxy moiety were also found to have a protective effect on breakage of plasmid DNA caused by treatment with N-hydroxy-2-amino-naphthalene. The single-strand cleavage in φX174FRI DNA observed during incubation with N-hydroxy-2-aminonaphthalene (I: supercoiled, II: circular, and III: linear DNA plasmid forms, respectively) clearly inhibited by addition of (-)-epicatechin gallate and (-)-epigallocatechin gallate (23). These data suggest that polyphenol type antioxidants scavenge oxygen free radicals formed during the degradation of N-hydroxy-2-aminonaphthalene (step 2 in Fig. 9) although it is not clear what kind oxygen radicals can be inhibited by addition of polyphenol type antioxidants.

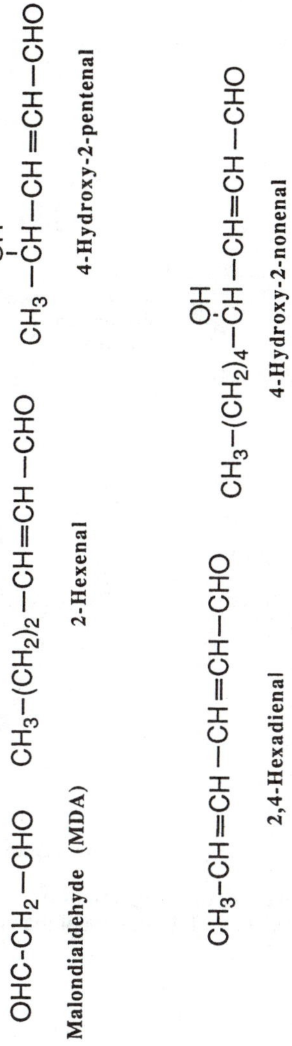

OHC-CH$_2$—CHO

Malondialdehyde (MDA)

CH$_3$—(CH$_2$)$_2$—CH=CH—CHO

2-Hexenal

$$CH_3—\overset{\overset{\displaystyle OH}{|}}{CH}—CH=CH—CHO$$

4-Hydroxy-2-pentenal

CH$_3$–CH=CH—CH=CH–CHO

2,4-Hexadienal

$$CH_3-(CH_2)_4-\overset{\overset{\displaystyle OH}{|}}{CH}—CH=CH—CHO$$

4-Hydroxy-2-nonenal

Fig. 8. Mutagenic aldehydes formed during the lipid peroxidation process.

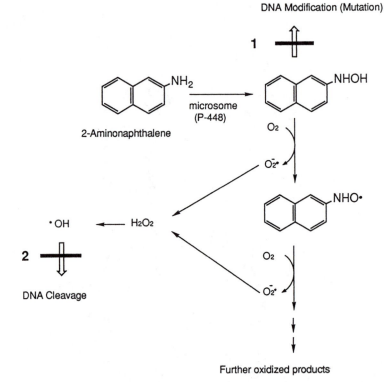

Fig. 9. Proposed mechanism of formation of oxygen radicals during the oxidative degradation of 2-aminonaphthalene.

LITERATURE CITED

1. Ames, B.N.; Cathcart, R.; Schwiers, E.; Hochstein, P. *Proc. Natl. Acad. Sci., USA*, **1981**,*78*, 6858-6862
2. Kohen, R.; Yamamoto, Y.; Cundy, K.C.; Ames, B.N. *Proc. Natl. Acad. Sci., USA*, **1988**, *85*, 3175-3179.
3. Stocker, R.; Yamamoto, Y.; Mcdonagh, .F.; Glazer, A.N.; Ames, B.N. *Science*, **1987**, *235*, 1043-1046.
4. Pziezak, J.D. *Food Tech.*, **1986**, *40*, 94-102.
5. Osawa, T.; Namiki, M.; Kawakishi, S. *In Antimutagenesis and Anticarcinogenesis Mechanisms II* ; Kuroda, Y.; Shankel, D.M.; Water, M.D., Eds.; **1990**, Plenum Press, New York, p.139-153.
6. McBrian, D.C.H.; Slater, T.F., Eds. *Free Radicals, Lipid Peroxidation and Cancer*, **1982**, Academic Press, New York
7. Simic, M.G. *Mutation Res.*, **1988**, *202*, 377-386.
8. Esterbauer, H. *In Free radicals, Lipid Peroxidation, and Cancer*; Mcbrien, D.C.H.; Slater, T.F., Eds.; **1982**, Academic Press, New York, p.101-128.
9. Marnett, L.J.; Hurd, H.K.; Hollstein, M.C.; Levin, D.E.; Esterbauer, H.; Ames, B.N. *Mutation Res.*, **1985**, *148*, 25-34.
10. Akasaka, S.; Yonei, S. *Mutation Res.*, **1985**, *149*, 321-326.
11. Osawa, T.; Nagata, M.; Namiki, M.; Fukuda, Y. *Agric. Biol. Chem.*, **1985**, *49*, 3351-3352.
12. Fukuda, Y.; Nagata, M.; Osawa, T.; Namiki, M. *J. Am. Oil Chem. Soc.*, **1986**, *63*, 1027-1031.
13. Fukuda, Y.; Isobe, M.; Nagata, M.; Osawa, T., Namiki, M. *Heterocycles*, **1986**, *24*, 923-926.
14. Shima, A. (1988) *In Food Functionalities*, Fujimaki, M., Ed.; **1988**, Gakkai Press Center, Tokyo, pp. 227-231.
15. Yamashita, K.; Kawagoe, Y.; Nohara, Y.; Namiki, M.; Osawa, T.; Kawakishi, S. *Nihon Eiyou Shokuryou Gakkaishi*, **1990**, *43*, 445-449.
16. Tashiro, T.; Fukuda, Y.; Osawa, T.; Namiki, M. *J. Am. Oil Chem. Soc.*, **1990**, *67*, 508-511.
17. Su, J.-De; Osawa, T.; Namiki, M. *Agric. Biol. Chem.*, **1986**, *50*, 199-203.
18. Su, J.-De; Osawa, T.; Kawakishi, S.; Namiki, M.; *Phytochem.*, **1988**, *27*, 1315-1319.
19. Osawa, T.; Ide, A.; Su, J.-De; Namiki, M. *J. Agric. Food Chem.*, **1987**, *35*, 808-812.
20. Marnett, L.J.; Hurd, H.K.; Hollstein, M.C.; Levin, D.E.; Esterbauer, H.; Ames, B.N. *Mutat. Res.*, **1985**, *148*, 25-34.
21. Osawa, T.; Kumon, H.; Namiki, M.; Kawakishi, S. *In Mutagens and Carcinogens in the Diet*, Pariza, M.W.; Aeschbacher, H.-U.; Felton, J.S.; Sato, S., Eds.; **1990**, Wiley-liss, Inc., New York, p.223-238.
22. Nakayama, T.; Kaneko, M.; Kodama, M.; Nagata, C. *Carcinogenesis* (Lond.), **1983**, *4*, 765-769.
23. Osawa, T.; Kumon, H.; Nakayama, T.; Kawakishi, S.; Hara, Y. *submitted*.

RECEIVED February 11, 1992

Chapter 12

Phenolic Antioxidants as Inducers of Anticarcinogenic Enzymes

Hans J. Prochaska[1] and Paul Talalay[2]

[1]Molecular Pharmacology and Therapeutics Program, Memorial Sloan–Kettering Cancer Center, 1275 York Avenue, New York, NY 10021
[2]Department of Pharmacology and Molecular Sciences, The Johns Hopkins School of Medicine, 725 North Wolfe Street, Baltimore, MD 21205

The induction of electrophile-processing (Phase II) enzymes is a major mechanism whereby a surprisingly wide variety of compounds can inhibit the development of carcinogen-induced neoplasms. By using the induction of NAD(P)H:quinone reductase (EC 1.6.99.2; QR) in Hepa 1c1c7 murine hepatoma cells as an indicator of Phase II enzyme regulation, we tested a series of analogues of BHA [2(3)-tert-butyl-4-hydroxyanisole] as inducers. Whereas all 1,2- and 1,4-diphenols examined were inducers, 1,3-diphenols were completely inactive. Since 1,2-and 1,4-diphenols are chemically related in that they can undergo facile oxidations to quinones, whereas 1,3-diphenols cannot, we concluded that the signal for induction is chemically mediated. Extension of these structure-activity studies to other chemical classes of anticarcinogenic inducers showed that the presence or acquisition of a Michael acceptor function or equivalent electrophilic center is responsible for the inductive signal. This insight has permitted the prediction of structures of compounds with inductive activity. These generalizations hold promise for the design of more potent and less toxic anticarcinogenic enzyme inducers.

Phenolic antioxidants such as BHT (3,5-di-tert-butyl-4-hydroxytoluene) and BHA [2(3)-tert-butyl-4-hydroxyanisole] were first recognized to protect rodents against carcinogen-induced tumors more than twenty years ago (1-4). Wattenberg is primarily responsible for stimulating scientific interest in these commonly-used food additives by demonstrating that these compounds protected against the toxic and neoplastic effects of a surprisingly wide variety of carcinogens in a number of target tissues (5-8).

Phenolic Antioxidants as Inducers of Phase II Enzymes.

Several mechanisms have been proposed to account for the anticarcinogenic activity of phenolic antioxidants. These include the possibility that they act as

0097–6156/92/0507–0150$06.00/0

antioxidants (which was presumably the original rationale for their use as chemoprotective agents) or as non-specific nucleophiles. The role of Phase I[#] enzyme induction was also examined since: a) phenolic antioxidants dramatically alter the metabolic fate of carcinogens (*10-13*); b) Phase I enzymes play an important role in carcinogen activation and detoxication (*14*); and c) other recognized anticarcinogens such as β-naphthoflavone are known to be very potent inducers of selected Phase I enzymatic activities (*15-17*). However, only minor and variable alterations of Phase I enzymatic activities occurred with BHA treatment (*11,12,18,19*).

The finding that hepatic and extra-hepatic Phase II[#] enzymes and glutathione pools were coordinately (and sometimes dramatically) induced by phenolic antioxidant treatment provided a plausible mechanism for the anticarcinogenic activity (*18,20-22*). Earlier, Huggins and Pataki (*23*) demonstrated that the induction of quinone reductase by azo dyes correlated well with the protection against the toxic and neoplastic effects of 7,12-dimethylbenz(a)anthracene. Thus, Phase II enzyme induction was observed for these seemingly dissimilar anticarcinogens, and this property has been subsequently associated with a large number of other classes of anticarcinogens (*24-29*).

The causal relationship between Phase II enzyme induction and protection appears to be firmly established since [*see references cited in* (*30*)]: a) virtually all Phase II enzyme inducers, including carcinogens such as 3-methylcholanthrene, can behave as anticarcinogens under the appropriate experimental circumstances; b) the protective effects are non-specific with respect to the carcinogen used or target organ; c) the inducer is most effective if given prior to and during carcinogen challenge; d) *in vivo* and *in vitro* evidence [(including transfection experiments (*31,32*)] has established that increasing the activity of these enzymes can prevent the toxic and neoplastic effects of a large number of compounds; e) inhibition of protein synthesis abolishes the protective effects of anticarcinogenic enzyme inducers; f) Phase II enzymes inactivate electrophiles, which are presumed to be the ultimate carcinogenic species. Perhaps the most compelling evidence for the causal relationship between Phase II enzyme induction and anticarcinogenic activity has been the ability to predict correctly that certain

---

[#] Phase I enzymes, which are also referred to as cytochromes P-450 or mixed-function-oxidases, comprise a superfamily of microsomal enzymes that functionalize endogenous compounds and xenobiotics usually by insertion of an oxygen atom between carbon- or nitrogen-hydrogen bonds. Phase II enzymes are conventionally defined as those enzymes responsible for catalyzing the conjugation of endogenous polar ligands to functionalized xenobiotics (*9*) resulting in a metabolically inactive products that can be processed for excretion. Thus, glutathione *S*-transferases (GST) and UDP-glucuronosyltransferases are examples of Phase II enzymes. We have expanded this definition to include additional enzymes such as NAD(P)H:(quinone-acceptor) oxidoreductase (QR) since they do not introduce functional groups, play an important detoxication role, and are induced coordinately with the above-cited conjugating activities.

compounds (e.g., 1,2-dithiole-3-thiones) would possess anticarcinogenic activity based solely on their ability to induce Phase II enzymes (33,34). Moreover, anticarcinogens have been isolated from natural sources using Phase II enzyme induction as a short term bio-assay (35,36). Thus, Huggins' (23) suggestion that it will be profitable to screen for candidate anticarcinogens by using Phase II enzyme expression as a short-term marker for chemoprotective activity appears correct.

Identification of the Chemical Signal for the Induction of Phase II Enzymes

By the early-1980's, a bewildering array of compounds had been identified that possessed inductive as well as chemoprotective activity. Although some compounds shared a common structural motif and could be characterized as large planar aromatics (i.e., flavonoids, azo dyes, polycyclic hydrocarbons), other inducers defied such simple categorization. The latter included the aforementioned phenolic antioxidants, coumarins and other lactones, cinnamates, dithiocarbamates, 1,2-dithiole-3-thiones, and other sulfur-containing compounds. These compounds possessed no obvious structural pattern and could be distinguished by their inability to induce selected Phase I enzymatic activities. The large planar aromatics, on the other hand, potently elevated the specific activities of selected Phase I enzymes that are directly transcriptionally regulated by the Ah receptor (37). For descriptive purposes we have defined two families of Phase II enzyme inducers (38): *bifunctional inducers* which also induce Ah receptor-dependent Phase I enzymatic activities; and *monofunctional* inducers which have minor and variable effects on Phase I enzymatic profiles. Since the induction of Phase II enzymes by bifunctional inducers has been shown to segregate only with functional Ah receptors in mouse strains and cell lines, it had been originally assumed that Phase II enzymes are regulated in a manner identical to the regulation of Phase I enzymes (39).

Structure-activity Studies of BHA Congeners in Mice. Our initial attempts to understand the mechanism for Phase II enzyme induction involved a systematic examination of the relation between the structure of BHA analogues and their ability to induce GST and QR in mouse tissues. One possible mechanism was that these compounds bound to a conventional cellular receptor, and consequently potency would be expected to be very sensitive to structural changes. A series of 19 congeners of BHA were synthesized with substituents varying as shown below:

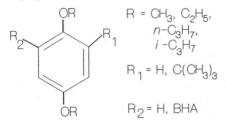

$R = CH_3, C_2H_5,$
$n-C_3H_7,$
$i-C_3H_7$

$R_1 = H, C(CH_3)_3$

$R_2 = H, BHA$

The compounds were administered to female CD-1 mice for 5 days by oral intubation and two to seven tissues were examined for GST and QR activity (40,41). Surprisingly, all compounds tested did not differ greatly in inducer activity in the liver. Thus, there was no apparent relation between structure and activity.

Three clues were critical in the elucidation of the chemical nature of the inductive signal: a) the free phenol, *tert*-butylhydroquinone, was the least tissue-selective inducer; b) all compounds tested contained oxygen atoms in the 1 and 4 position; c) BHA was an exceedingly weak inducer of QR activity in cultured murine hepatoma (Hepa 1c1c7) cells. These results together suggested that BHA and its congeners required *O*-dealkylation to their cognate hydroquinones to be active.

Mechanism for the Induction of Phase II Enzymes by Phenols: A Unifying Proposal.

Why should hydroquinones be inducers of Phase II enzymes?   One obvious possibility is that these "redox labile" species can undergo reversible one- or two-electron oxidations to semiquinone or quinone forms.   If correct, this would imply a novel mechanism of gene regulation: the chemical reactivity of these compounds rather than their steric interaction with a conventional receptor was responsible for inductive activity.   Diphenols provide an ideal group of compounds to test this hypothesis since catechols (1,2-diphenols) and hydroquinones (1,4-diphenols) can undergo such reversible oxidations whereas resorcinols (1,3-diphenols) cannot (Figure 1).

Consequently, we (*42*) tested a variety of substituted diphenols for their ability to induce quinone reductase in the Hepa 1c1c7 murine hepatoma cell line (Figure 2).   The results are unambiguous: all catechols and hydroquinones tested were inducers of this enzymatic activity whereas resorcinols were completely devoid of inductive activity. Moreover, these patterns were preserved when amino groups replaced phenolic groups. Later data *in vivo* have shown identical inductive patterns (*38,44*).

These results suggested that a chemical signal was responsible for induction.   Since 1,2- and 1,4-diphenols are known to: a) "redox cycle" in cells resulting in the generation of reactive oxygen species at the expense of NADPH; and b) form electrophilic quinones (*45*), studies with diphenols could not unequivocally identify the nature of the chemical signal. This problem was solved by the systematic examination of partial structures of coumarin (*46,47*). We found that the critical feature for Phase II enzyme induction by coumarin was the presence (or

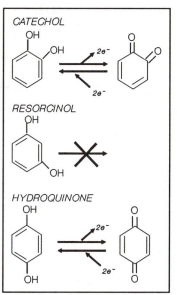

Figure 1. *Oxidation reactions of diphenols.*   Resorcinols cannot give rise to quinones.

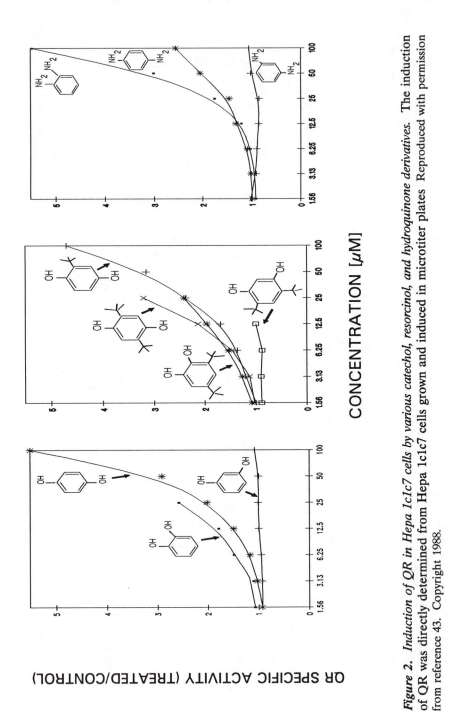

Figure 2. Induction of QR in Hepa 1c1c7 cells by various catechol, resorcinol, and hydroquinone derivatives. The induction of QR was directly determined from Hepa 1c1c7 cells grown and induced in microtiter plates Reproduced with permission from reference 43. Copyright 1988.

acquisition by metabolism) of an olefin conjugated to an electron-withdrawing group, i.e., a Michael reaction acceptor. Indeed, quinones are exceedingly efficient Michael reaction acceptors. Other classes of inducers are not formally Michael reaction acceptors *per se*, but possess similar electrophilic centers (e.g., isothiocyanates, halogenated nitroaromatics, epoxides). Moreover, inductive activity appeared to parallel their efficiency as Michael reaction acceptors and as electrophiles (*46-49*).

Bifunctional Inducers and the Metabolic Cascade. The conclusion that monofunctional inducers act via a chemical signal led us to reexamine the mechanism for Phase II enzyme regulation by bifunctional inducers. Although genetic studies had demonstrated that the induction of Phase II enzymes by these compounds segregated with Phase I inducibility and a functional Ah receptor (*39,50,51*), these findings did not establish direct involvement of the Ah receptor in Phase II enzyme regulation. In 1985, we proposed a model [*see* Figure 3; (*42*)] wherein Phase II enzyme regulation was at least in part the result of a *metabolic cascade*; i.e., the Ah-dependent Phase I activities that were induced by bifunctional inducers catalyze the conversion of bifunctional inducers to metabolites resembling monofunctional inducers (e.g., conversion of polycyclic aromatic hydrocarbons to quinone metabolites and azo dyes to phenylenediamines and aminophenols). The proposal that bifunctional inducers regulate Phase II enzymes via a metabolic cascade (as well as by direct participation of the Ah receptor) received further indirect support by the systematic examination of the dose-response of Phase I and Phase II enzymes to a series of inducers alone or in combination, in wild-type or mutant cell lines defective in aryl hydrocarbon hydroxylase expression (*38*). Recent molecular analyses of the 5'- upstream region of the DNA sequences for GST and QR are consistent with this model (*53-56*).

Prospects for the Future.

The possible contribution to protection of the small quantities of antioxidant food additives consumed in the Western diet is unknown. Experimental protection by these agents against carcinogens is dose-dependent and requires relatively large amounts of these substances (*57-60*). Moreover, BHA causes cancer of the forestomach in rodents (*61*) and BHT is a tumor promoter (*7,8*). However, structure-activity studies with phenolic antioxidants have provided an understanding of the structural requirements that endow compounds with the ability to induce Phase II enzymes. It is now possible to predict structures that have inductive activity, and many novel inducers have been identified (*46-49*). Indeed, some recently identified compounds are potent and non-toxic inducers of Phase II enzymatic activities and are derivatives of normal tissue constituents [e.g., esters of fumaric and maleic acids (*48*)]. It remains to be seen if these novel compounds are also anticarcinogenic, but it seems likely that they will be.

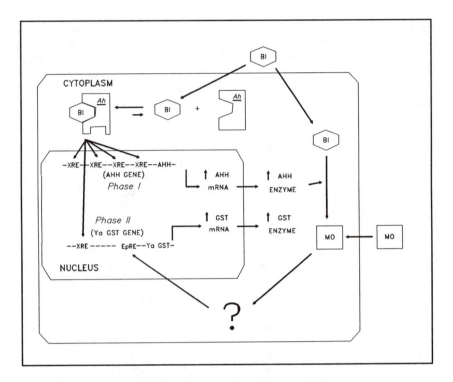

Figure 3. *Mechanisms for the regulation of Phase I and Phase II enzymes by monofunctional and bifunctional inducers [modified from (38,42); enhancer elements added per (52-56)].* Monofunctional inducers (MO) generate an electrophilic signal that stimulates the induction of Phase II enzymes. Although the targets for these electrophiles are unknown, they interact with a novel enhancer element [termed EpRE (*Electrophile Responsive Element*) or ARE (*Antioxidant Responsive Element*)]. Bifunctional inducers (BI) act via two mechanisms: *1)* they bind to the Ah receptor and activate Phase I and II enzymes via XRE's (*Xenobiotic Responsive Elements*); *2)* the dramatically elevated Phase I enzymes convert bifunctional inducers to metabolites resembling monofunctional inducers.

＊　＊　＊　＊　＊　＊　＊

Acknowledgments. This work was supported by a Special Institutional Grant (SIG-3) from the American Cancer Society and grants from the American Institute for Cancer Research and National Cancer Institute (NIH 1 PO1 CA 44530). H.J.P. was supported by the National Institutes of Health Medical Scientist Training Program (GM07903) and Training Grant CA 09243 during the course of these studies. We thank Annette B. Santamaria for performing the experiments shown in Figure 2.

Literature Cited

1. Frankfurt, O. S.; Lipchina, I. P.; Bunto, T. V.; Emanuel, N. M. *Bull. Exp. Biol. Med. (USSR)* **1968**, *8*, 86-88.
2. Wattenberg, L. W. *J. Natl. Cancer. Inst.* **1972**, *48*, 1425-1430.
3. Wattenberg, L. W. *Fed. Proc.* **1972**, *31*, 633.
4. Wattenberg, L. W. *J. Natl. Cancer. Inst.* **1973**, *50*, 1541-1544.
5. Wattenberg, L. W. *Cancer Res.* **1983**, *43*, 2448s-2453s.
6. Wattenberg, L. W. *Cancer Res.* **1985**, *45*, 1-8.
7. Kahl, R. *Toxicology* **1984**, *33*, 185-228.
8. Kahl, R. In *Oxidative Stress: Oxidants and Antioxidants*; Sies, H., Ed.; Academic Press: London, 1991, pp 247-273.
9. Williams, R. T. *Fed. Proc. Fed. Am. Soc. Exp. Biol.* **1967**, *26*, 1029-1039.
10. Speier, J. L.; Wattenberg, L. W. *J. Natl. Cancer. Inst.* **1975**, *55*, 469-472.
11. Wattenberg, L. W.; Speier, J.; Kotake, A. *Adv. Enzyme Regul.* **1976**, *14*, 313-322.
12. Speier, J. L.; Lam, L. K. T.; Wattenberg, L. W. *J. Natl. Cancer. Inst.* **1978**, *60*, 605-609.
13. Anderson, M. W.; Boroujerdi, M.; Wilson, A. G. E. *Cancer Res.* **1981**, *41*, 4309-3415.
14. Miller, E. C.; Miller, J. A. In *Accomplishments in Cancer Research*; Fortner, J. G.; Rhoads, J. E., Eds.; Lippincott: Philadelphia, 1980, pp 63-89.
15. Wattenberg, L. W.; Leong, J. L. *Proc. Soc. Exp. Biol. Med.* **1968**, *128*, 940-943.
16. Wattenberg, L. W.; Leong, J. L. *Cancer Res.* **1970**, *30*, 1922-1925.
17. Diamond, L.; McFall, R.; Miller, J.; Gelboin, H. V. *Cancer Res.* **1972**, *32*, 731-736.
18. Cha, Y.-N.; Bueding, E. *Biochem. Pharmacol.* **1979**, *28*, 1917-1921.
19. Rahimtula, A. D.; Jernström, B.; Dock, L.; Moldeus, P. *Br. J. Cancer* **1982**, *30*, 121-124.
20. Benson, A. M.; Batzinger, R. P.; Ou, S.-Y. L.; Bueding, E.; Cha, Y.-N.; Talalay, P. *Cancer Res.* **1978**, *38*, 4486-4495.
21. Benson, A. M.; Cha, Y.-N.; Bueding, E.; Heine, H. S.; Talalay, P. *Cancer Res.* **1979**, *39*, 2971-2977.
22. Benson, A. M.; Hunkeler, M. J.; Talalay, P. *Proc. Natl. Acad. Sci.* **1980**, *77*, 5216-5220.
23. Huggins, C.; Pataki, J. *Proc. Natl. Acad. Sci.* **1965**, *53*, 791-796.
24. Sparnins, V. L.; Wattenberg, L. W. *J. Natl. Cancer. Inst.* **1981**, *66*, 769-771.
25. Sparnins, V. L.; Venegas, P. L.; Wattenberg, L. W. *J. Natl. Cancer. Inst.* **1982**, *68*, 493-496.
26. Sparnins, V. L.; Chuan, J.; Wattenberg, L. W. *Cancer Res.* **1982**, *42*, 1205-1207.
27. Benson, A. M.; Barretto, P. B. *Cancer Res.* **1985**, *45*, 4219-4223.
28. Benson, A. M.; Barretto, P. B.; Stanley, J. S. *J. Natl. Cancer. Inst.* **1986**, *76*, 467-473.

29.   Fujita, S.; Matsunaga, T.; Matsubuchi, Y.; Suzuki, T. *Cancer Res.* **1988**, *48*, 254-259.
30.   Talalay, P.; De Long, M. J.; Prochaska, H. J. In *Cancer Biology and Therapeutics*; Cory, J. G.; Szentivanyi, A., Eds.; Plenum Press: New York, 1987, pp 197-216.
31.   Manoharan, T. H.; Puchalski, R. B.; Burgess, J. A.; Pickett, C. B.; Fahl, W. E. *J. Biol. Chem.* **1987**, *262*, 3739-3745.
32.   Manoharan, T. H.; Welch, P. J.; Gulick, A. M.; Puchalski, R. B.; Lathrop, A. L.; Fahl, W. E. *Mol. Pharmacol.* **1991**, *39*, 461-467.
33.   Ansher, S. S.; Dolan, P.; Bueding, E. *Hepatology* **1983**, *3*, 932-935.
34.   Wattenberg, L. W.; Bueding, E. *Carcinogenesis* **1986**, *7*, 1379-1381.
35.   Lam, L. K. T.; Sparnins, V. L.; Wattenberg, L. W. *Cancer Res.* **1982**, *42*, 1193-1198.
36.   Wattenberg, L. W.; Lam, L. K. T. In *Coffee and Health. Banbury Report No. 17*; MacMahon, B.; Sugimura, T., Eds.; Cold Spring Harbor Laboratory: CSI, NY, 1984, pp 137-145.
37.   Whitlock, J. P., Jr. *Ann. Rev. Pharm. Toxicol.* **1990**, *30*, 251-277.
38.   Prochaska, H. J.; Talalay, P. *Cancer Res.* **1988**, *48*, 4776-4782.
39.   Nebert, D.W.; Gonzalez, F.J. *Ann. Rev. Biochem.* **1987**, *56*, 945-993.
40.   De Long, M. J.; Prochaska, H. J.; Talalay, P. *Cancer Res.* **1985**, *45*, 546-551.
41.   Prochaska, H. J.; Bregman, H. S.; De Long, M. J.; Talalay, P. *Biochem. Pharmacol.* **1985**, *34*, 3909-3914.
42.   Prochaska, H. J.; De Long, M. J.; Talalay, P. *Proc. Natl. Acad. Sci.* **1985**, *82*, 8232-8236.
43.   Prochaska, H. J.; Santamaria, A. B. *Anal. Biochem.* **1988**, *169*, 328-336.
44.   Talalay, P.; Prochaska, H. J. *Chemica Scripta* **1987**, *27A*, 61-66.
45.   Prochaska, H. J.; Talalay, P. In *Oxidative Stress: Oxidants and Antioxidants*; Sies, H., Ed.; Academic Press: London, 1991, pp 195-211.
46.   Talalay, P.; De Long, M. J.; Prochaska, H. J. *Proc. Natl. Acad. Sci.* **1988**, *85*, 8261-8265.
47.   Talalay, P. *Adv. Enzyme Regul.* **1989**, *28*, 237-250.
48.   Spencer, S. R.; Wilczak, C. A.; Talalay, P. *Cancer Res.* **1990**, *50*, 7871-7875.
49.   Spencer, S. R.; Xue, L.; Klenz, E. M.; Talalay, P. *Biochem. J.* **1991**, *273*, 711-717.
50.   Kumaki, K.; Jensen, N. M.; Shire, J. G. M.; Nebert, D. W. *J. Biol. Chem.* **1977**, *252*, 157-165.
51.   Owens, I. S. *J. Biol. Chem.* **1977**, *252*, 2827-2833.
52.   Fisher, J. M.; Wu, L.; Denison, M. S.; Whitlock, J. P., Jr. *J. Biol. Chem.* **1990**, *265*, 9676-9681.
53.   Rushmore, T. H.; Pickett, C. B. *J. Biol. Chem.* **1990**, *265*, 14648-14653.
54.   Rushmore, T. H.; King, R. G.; Paulson, K. E.; Pickett, C. B. *Proc. Natl. Acad. Sci.* **1990**, *87*, 3826-3830.
55.   Friling, R. S.; Bensimon, A.; Tichauer, Y.; Daniel, V. *Proc. Natl. Acad. Sci.* **1990**, *87*, 6258-6262.
56.   Favreau, L. V.; Pickett, C. B. *J. Biol. Chem.* **1991**, *266*, 4556-4561.

57. Reddy, B. S.; Maeura, Y.; Weisburger, J. H. *J. Natl. Cancer. Inst.* **1983**, *71*, 1299-1305.
58. Cohen, L. A.; Polansky, M.; Furuya, K.; Reddy, M.; Berke, B.; Weisburger, J. H. *J. Natl. Cancer. Inst.* **1984**, *72*, 165-174.
59. Reddy, B. S.; Maeura, Y. *J. Natl. Cancer. Inst.* **1984**, *72*, 1181-1187.
60. Maeura, Y.; Weisburger, J. H.; Williams, G. M. *Cancer Res.* **1984**, *44*, 1604-1610.
61. Ito, N.; Fukushima, S.; Hagiwara, A.; Shibata, M.; Ogiso, T. *J. Natl. Cancer Inst.* **1983**, *70*, 343-352.

RECEIVED December 2, 1991

Chapter 13

Polyphenols from Asian Plants

Structural Diversity and Antitumor and Antiviral Activities

Takuo Okuda, Takashi Yoshida, and Tsutomu Hatano

Faculty of Pharmaceutical Sciences, Okayama University, Tsushima, Okayama 700, Japan

Numerous polyphenolic compounds (tannins), isolated from Asian plants used as food and medicine, can be classified in several classes, most of which are new types; dehydroellagitannins and their metabolites, oligomeric hydrolyzable tannins, complex tannins, and others. Antitumor, antiviral, antioxidant, and other activities, which are sometimes specific to chemical structure of each polyphenol, have also been found.

A large number of plants, which have been used as food and medicine in Asia, often based on the traditional prescriptions in literature, have been supplying spices, beverages and medicines to the other parts of the world. Alkaloids, terpenoids, essential oils, and phenolics, etc., produced by these plants, have been important sources for development of many modern medicines.

Polyphenolic compounds are widely distributed in these plants, occasionally in large quantity. However, because of poor chemical information as to polyphenolic compounds, their presence in these plants was presumed, until recent years, only on the basis of astringent taste and/or the conventional chemical test for "tannin-containing plant". Therefore, attention was usually not given to the differences of tannins contained in different plants, thus inducing misunderstanding of tannins such as that found in some medical literature, which did not differentiate between commercially sold "tannic acid" and the "tannins" contained in various plants. It was also not realized well by many scientists, except for the phenolics specialists, that "tannic acid" (1) is a mixture of many galloyl esters of poor reproducibility, although the crude drugs which provide "tannic acid," such as Chinese gall and Turkish gall, were rather exceptionally well-investigated tannin sources.

Hundreds of new polyphenolic compounds have been isolated recently from numerous "tannin-containing plants," and their chemical structures have been determined, assisted by the developments in new methods of tannin analysis (2). Any discussion about tannins, at least about those contained in the plants associated with health effects (3), must be based on the properties of each isolated polyphenolic compound. Such a situation of each polyphenolic compound (tannin) in the tannin

0097–6156/92/0507–0160$07.00/0

family is similar to that of each alkaloidal compound, *e.g.*, morphine, berberine, caffeine, *etc.*, in the alkaloid family.

Among hundreds of polyphenolic compounds characterized during our investigation, there have been several new types of tannins, particularly of those belonging to hydrolyzable tannins.

Ellagitannins, Dehydroellagitannins, and Their Metabolites

Several dehydroellagitannins (*4*) having a (or more) dehydrohexahydroxydiphenoyl (DHHDP) group which is regarded as an oxidative metabolite of hexahydroxydiphenoyl (HHDP) group, and their analogs were isolated from medicinal plants at the initial stage of our investigation (*5-7*). Some other tannins, having analogous polyphenolic groups producible by further oxidation of DHHDP group, have also been obtained (*8*).

Geraniin (I) (*6*), a key metabolite in biosynthesis of hydrolyzable tannins having the $^1C_4$ glucopyranose core (*1*), was isolated from *Geranium thunbergii* (*9*) which is one of the most popular medicinal plants and an official medicine in Japan, frequently used for the treatment of intestinal disorder, in a way similar to that of tea. It was obtained as yellow crystals, which show almost no astringent taste, in spite of its fairly strong binding to protein and basic compound (*10*). Dehydrogeraniin (III) and furosinin (IV), which have two DHHDP groups in the molecule, were also isolated from this plant (*8*) (Figure 1).

Mallotusinic acid (II) and geraniin were found as the main components in the leaf of *Mallotus japonicus* (*7*) which has been used for the treatment of stomach ulcer in Japan.

An example of the oxidative metabolites of geraniin is chebulagic acid (V) (*11*, *12*) which was isolated from Myrobalans (dried fruits of *Terminalia chebula*), a crude drug used as a laxative and a tonic in India and other parts of Asia, and is one of the important tannin sources in the leather industry. Its analog, geraniinic acid B (VI), was obtained from *G. thunbergii* (*13*) and *Phyllanthus flexuosus* (Yoshida, T. *et al.*, *Chem. Pharm. Bull.*, in press). The latter plant also produced phyllanthusiins A (VII), B (VIII) and C (IX), which are biogenetically regarded as the products of metabolic modifications at the highly reactive DHHDP group in the molecule of geraniin.

Elaeocarpusin (X), isolated from water-soluble portion of the extract of *G. thunbergii* and *Acer nikoense*, was a condensate of geraniin and ascorbic acid (*14*).

***C*-Glycosidic Tannins.**   *C*-Glycosidic ellagitannins, casuarinin (XI) and stachyurin (XII) (Figure 2), which were initially isolated from *Casuarina stricta* and *Stachyurus praecox* (*15*), have been found to occur in a variety of plants used as medicine or food, including *Psidium guajava*, *Syzygium aromaticum* and *Eucalyptus viminalis* (*16*). These tannins are almost inevitably accompanied by ellagitannins such as casuarictin (XIII) and tellimagrandin II (XIV), etc., and are thus regarded as the metabolites formed by an intramolecular phenol-aldehyde condensation in the molecule of XIII or its analog (*16*). This biogenesis, assumed based on the seasonal variation in the *C*-glycosidic tannins and co-existing ellagitannins (XIII and XIV, etc.) in *Liquidambar formosana* leaf, exhibiting increase of XI in place of decrease of XIII, XIV and pentagalloylglucose (*17*), was further supported by the isolation of liquidambin (XV), a plausible biosynthetic intermediate, from the same plant leaf (*18*).

Oligomeric Hydrolyzable Tannins.   Over one hundred oligomeric hydrolyzable tannins (*19*) of the molecular size up to tetramers, having HHDP and/or DHHDP group, have been isolated from various medicinal plants. The first oligomer isolated from plant was agrimoniin (XVI), a dimer from *Agrimonia pilosa* var. *japonica* (*20*,

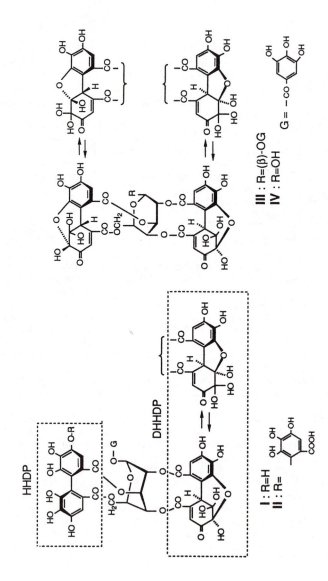

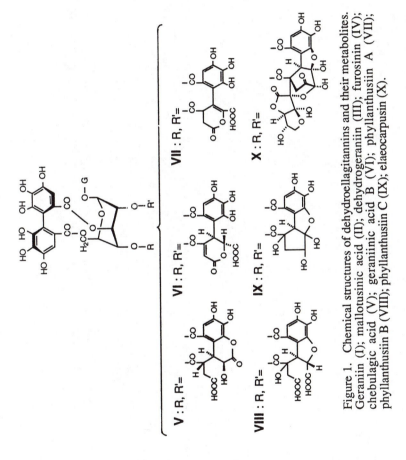

Figure 1. Chemical structures of dehydroellagitannins and their metabolites. Geraniin (I); mallotusinic acid (II); dehydrogeraniin (III); furosinin (IV); chebulagic acid (V); geraniinic acid B (VI); phyllanthusiin A (VII); phyllanthusiin B (VIII); phyllanthusiin C (IX); elaeocarpusin (X).

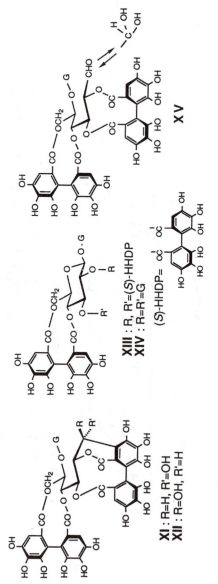

Figure 2.   Chemical structures of some ellagitannins. Casuariin (XI); stachyurin (XII); casuarictin (XIII); tellimagrandin II (XIV); liquidambin (XV).

21), which is used for treatment of diarrhea and hemorrhage in Japan, and also of inflammation and cancer in China. Considerably wide distribution of this dimer in Rosaceous plants (*Agrimonia, Duchesnea, Potentilla* and *Rosa* species) has been revealed. For example, *Rosa laevigata* and *R. davurica*, which are Chinese traditional medicines, produce agrimoniin as their main tannin (*22, 23*). Its analogs, laevigatins B (XVII) and C (XVIII) (from *R. laevigata*) (*22*), and davuriciins D2 (XIX) (dimer) and T1 (XX) (trimer) (from *R. davurica*) (*23*), were also isolated (Figure 3). Rugosins D (XXIV) and E (XXV), which were first isolated from the flower petals of *Rosa rugosa* (*24, 25*) together with rugosins A (XXI), B (XXII), C (XXIII) (monomers), F (XXVI) (dimer) and G (XXVII) (trimer), have also been found in the plant species of various families such as Coriariaceae, Euphorbiaceae, Hamamelidaceae and Stachyuraceae (*19*).

Cornusiins A (XXVIII), C (XXXI) and their analogs (*26*), which were originally isolated from *Cornus officinalis* (Cornaceae) (used as tonic and astringent in China and Japan), and later found in several plants of Trapaceae and Nyssaceae (*19*), are also examples of the oligomeric hydrolyzable tannins occurring in various medicinal plants. Their analogs, trapanins A (XXXII) and B (XXXIII) (*27*) and camptothins A (XXIX) and B (XXX) (*28*), were isolated from *Trapa japonica* and *Camptotheca acuminata*, respectively (Figure 4).

There are many Euphorbiaceous plants which are known to be rich in tannins, and have been used traditionally for various medicinal purposes in Asia. Upon the early survey with HPLC, geraniin was shown to distribute widely in these plants (*29*). Recent investigations on the polyphenolics in *Euphorbia* species revealed the occurrence of dimeric hydrolyzable tannins possessing geraniin as a constituent monomeric unit. The first example of the dimers of this type was euphorbin A (XXXIV) (Figure 5) and its analogs, isolated from *Euphorbia hirta*, a Chinese folk medicine used as antidiarrhea and antihemorrhage (*30*).

Hydrolyzable tannin oligomers of unique macrocyclic structures [oenothein B (XXXV), woodfordin C (XXXVI) (dimers) and oenothein A (XXXVII) and woodfordin D (XXXVIII) (trimers)] have been isolated from the flower of *Woodfordia fruticosa* (Lythraceae) which has been used as a popular crude drug "sidowayah" in Indonesia and Malaysia, for the treatment of dysentry, sprue, rheumatism, dysuria and hematuria (*31, 32*). Oenotheins A (XXXVII) and B (XXXV) were first isolated from *Oenothera erythrosepala* (*33*) and *O. biennis* (*32*), respectively, and are the oligomers characteristic of Onagraceous plants. Camelliin B (XXXIX), an analogous macrocyclic dimer, was obtained from *Camellia japonica* and *C. sasanqua* flowers (*34*). It is noticeable that the leaves of *C. japonica* yielded a dimer of different type, camelliatannin D (XL) (Hatano, T. *et al.*, Okayama University, unpublished data), consisting of an ellagitannin monomer and a complex tannin (described below).

Several oligomers of different type, nobotanins B (XLI), G (XLII), H (XLIII) (dimers) and J (XLIV) (trimer), were isolated from an Indonesian crude drug "daun halendong" [the dried leaves of *Melastoma malabathricum* (Melastomataceae)] which has been used for the treatment of diarrhea and leucorrhea (*35*). These oligomers, which were isolated first from tropical plants, *Tibouchina semidecandra* (36) and *Heterocentron roseum* (Yoshida, T. et al., *Chem. Pharm. Bull.*, in press), have been widely found in the melastomataceous plants such as *Melastoma normale, M. candidum, Schizocentron elegans* and others (Yoshida, T. *et al.*, Okayama University, unpublished data).

Complex Tannins and Analogs.   Complex tannins are condensates of a hydrolyzable tannin with catechin or its analog. The hydrolyzable tannin unit of these tannins is usually *C*-glycosidic tannin such as stachyurin, vescalagin and its analogs, and its C-1 in the open-chain glucose residue binds to C-8 or C-6 of the catechin

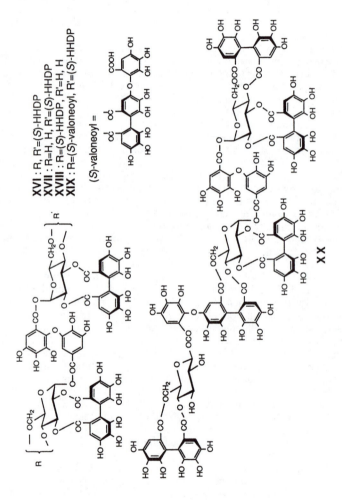

XVI : R, R'=(S)-HHDP
XVII : R=H, H, R'=(S)-HHDP
XVIII : R=(S)-HHDP, R'=H, H
XIX : R=(S)-valoneoyl, R'=(S)-HHDP

(S)-valoneoyl =

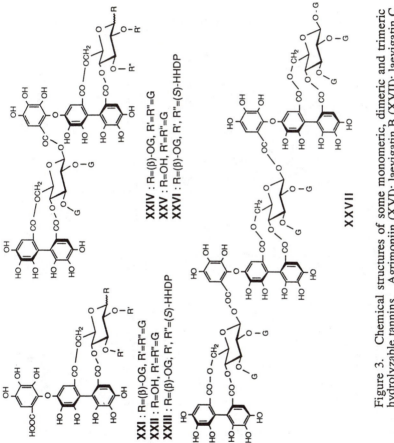

XXIV : R=(β)-OG, R'=R''=G
XXV : R=OH, R'=R''=G
XXVI : R=(β)-OG, R', R''=(S)-HHDP

XXI : R=(β)-OG, R'=R''=G
XXII : R=OH, R'=R''=G
XXIII : R=(β)-OG, R', R''=(S)-HHDP

XXVII

Figure 3. Chemical structures of some monomeric, dimeric and trimeric hydrolyzable tannins. Agrimoniin (XVI); laevigatin B (XVII); laevigatin C (XVIII); davuriciin D2 (XIX); davuriciin T1 (XX); rugosin A (XXI); rugosin B (XXII); rugosin C (XXIII); rugosin D (XXIV); rugosin E (XXV); rugosin F (XXVI); rugosin G (XXVII).

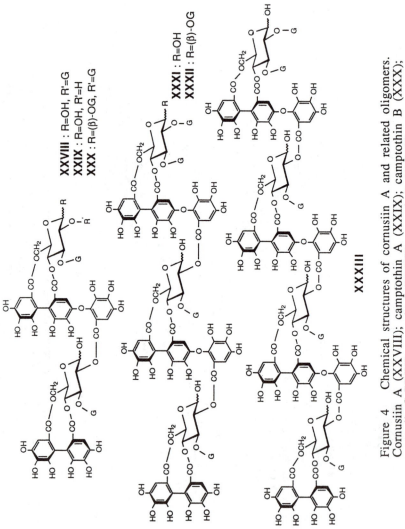

Figure 4   Chemical structures of cornusiin A and related oligomers. Cornusiin A (XXVIII); camptothin A (XXIX); camptothin B (XXX); cornusiin C (XXXI); trapanin A (XXXII); trapanin B (XXXIII).

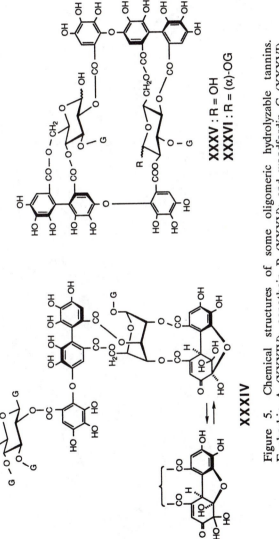

Figure 5. Chemical structures of some oligomeric hydrolyzable tannins. Euphorbin A (XXXIV); oenothein B (XXXV); and woodfordin C (XXXVI). *Continued on next page.*

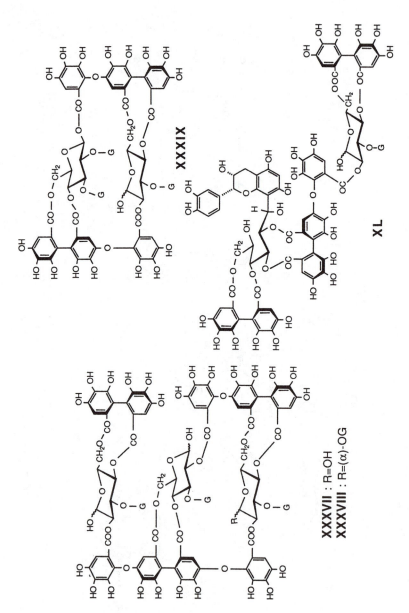

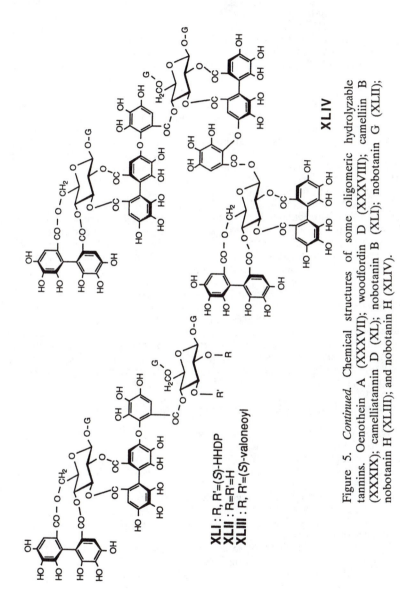

Figure 5. *Continued.* Chemical structures of some oligomeric hydrolyzable tannins. Oenothein A (XXXVII); woodfordin D (XXXVIII); camelliin B (XXXIX); camelliatannin D (XL); nobotanin B (XLI); nobotanin G (XLII); nobotanin H (XLIII); and nobotanin H (XLIV).

XLI : R, R'=(S)-HHDP
XLII : R=R'=H
XLIII : R, R'=(S)-valoneoyl

moiety. Typical tannins of this class are camelliatannin A (XLV) and its C-6 isomer, camelliatannin B (XLVII), isolated from *Camellia japonica* leaves (*37*) (Figure 6). The structures of camelliatannins C (XLVIII) and E (XLIX) are exceptions lacking the C-C bond between C-1 in the open-chain glucose and HHDP group at O-2/O-3 (*38*). Camelliatannin E (XLIX) is one of the monomers composing camelliatannin D (XL), a dimer described earlier. Camelliatannin F (L), which is regarded as a metabolite of camelliatannin A, was also isolated from *C. japonica* leaves (*38*). Galloylated congeners [malabathrins A (XLVI) and E (LI)] of camelliatannins A (XLV) and F (L) were obtained from *Melastoma malabathricum* (*39*).

Caffeic Acid Derivatives. Caffeic acid occurs widely in nature, often as esters of several types. Among these esters, 3,5-dicaffeoylquinic acid (LII) and its positional isomers concerning caffeoyl group, called "caffeetannins," are abundant in *Artemisia* species, several of which are popular as medicine and food plants in Japan and China (*40*). Rosmarinic acid (LIII), a caffeic acid dimer, was first isolated from *Rosmarinus officinalis*, and has been found to occur widely in various Labiatae plants used as herb medicines and spices in Europe (*Melissa officinalis* and *Salvia officinalis*, and also in those used in Japan and China (*Perilla* species and *Prunella vulgaris* var. *lilacina*, etc.) (*41*). Rosmarinic acid, called "labiataetannin," exhibits the activity of tannin evaluated by binding with proteins and basic compounds (astringency relative to that of geraniin) (*10*). Lithospermic acid and lithospermic acid B, possessing a benzofuran moiety, are caffeic acid trimer and tetramer (*42, 43*). A novel tetramer with a lignan skeleton [rabdosiin (LIV)] was isolated from *Rabdosia japonica* (*44*), a folk medicine for gastrointestinal disorders in Japan (Figure 7).

Highly Galloylated Condensed Tannins. Although condensed tannins (proanthocyanidins) distribute widely in herbal and woody plants, the distribution of their congeners, highly galloylated at O-3 of epicatechin units, is limited to the plants of several orders. (-)-Epicatechin gallate dimer (LV), trimer (LVI) and tetramer (LVII) were found in *Saxifraga stolonifera*, a folk medicine for earache and painful hemorrhoids in Japan (*45*) (Figure 8). Rhubarb, a popular Chinese crude drug (dried rhizomes of *Rheum* species), also produces galloylated procyanidin oligomers, the extent of galloylation at O-3 in some of which is *ca.* 70-80% (*46*).

Distribution of Galloyl Esters and Their Metabolites in Plants

Hydrolyzable tannins of various types hitherto isolated, which are most probably metabolites of galloylglucoses, have been found in Choripetalae of Dicotyledoneae in Angiospermae, and in Ericales of the Sympetalae. The orders, to which the plants producing these hydrolyzable tannins belong, were correlated with the plant evolution system (Figure 9) (*47*).

Highly galloylated condensed tannins have been found in the plants of limited families as described above, in contrast with the wide distribution of non-galloylated condensed tannins in the plants, from Pteridophyta to Monocotyledoneae and Dicotyledoneae of Angiospermae.

Although caffeic acid esters distribute in plant species of both Dicotyledoneae and Monocotyledoneae of Angiospermae (*48*), distribution of caffeic acid dimer (LIII), trimer (lithospermic acid) and tetramer (LIV) is limited to Labiatae and a few other families (*44*).

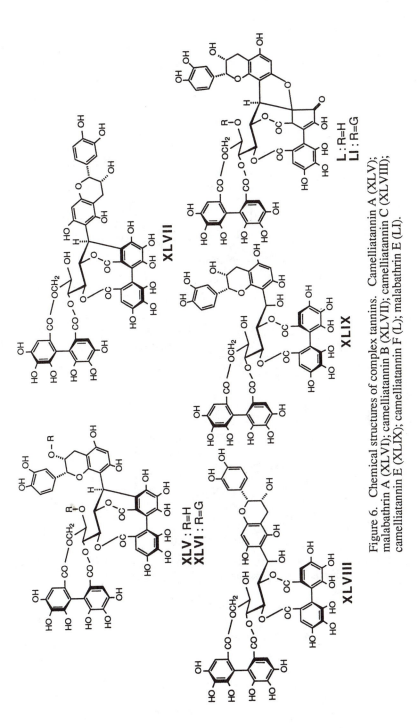

Figure 6. Chemical structures of complex tannins. Camelliatannin A (XLV); malabathrin A (XLVI); camelliatannin B (XLVII); camelliatannin C (XLVIII); camelliatannin E (XLIX); camelliatannin F (L); malabathrin E (LI).

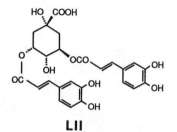

LII

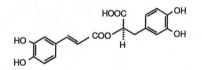

LIII

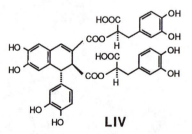

LIV

Figure 7.   Chemical structures of caffeic acid derivatives.   3,5-Di-*O*-caffeoylquinic acid (LII); rosmarinic acid (LIII); rabdosiin (LIV).

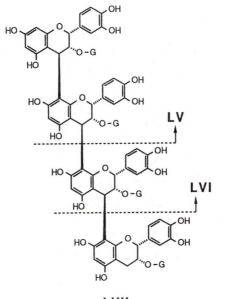

LVII

Figure 8. Chemical structures of (-)-epicatechin gallate dimer (LV), trimer (LVI) and tetramer (LVII).

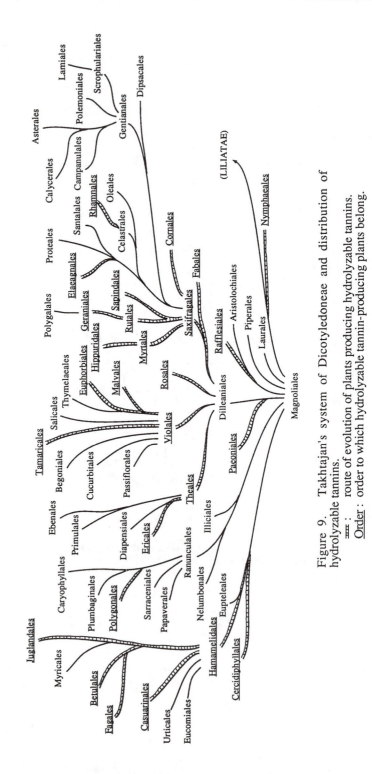

Figure 9. Takhtajan's system of Dicotyledoneae and distribution of hydrolyzable tannins.
────── : route of evolution of plants producing hydrolyzable tannins.
Order : order to which hydrolyzable tannin-producing plants belong.

Biological and Pharmacological Activities Associated with Several Chemical Structures

Polyphenolic compounds inhibit active oxygen species in various experimental systems, and this inhibitory effect is attributable to stable free radical formation by polyphenolic compounds (*49*). The free radicals, formed upon donation of hydrogen radical from the polyphenols to active oxygen species, has been exhibited by ESR measurement (*49*). It is probable that such inhibitory effect underlies several biological and pharmacological activities of polyphenolic compounds, directly or indirectly correlated with the inhibition of active oxygen species. However, since the inhibitory effect of active oxygen species by polyphenolic compounds are reviewed in the chapter "Antioxidant Effect of Tannins and Related Polyphenols," some other main biological and pharmacological activities of polyphenolic compounds are summarized in the present chapter, although not all of them are independent of the suppression of active oxygen species.

Antitumor Activities. (-)-Epigallocatechin gallate (EGCG) (LVIII) is the main polyphenolic compound responsible for the astringent taste of green tea. It potently inhibited tumor promotion in the two-stage carcinogenesis on mouse skin, induced with 7,12-dimethylbenz[*a*]anthracene (DMBA) as initiator and teleocidin as promotor (*50*). In the experiments using 12-*O*-tetradecanoylphorbol-13-acetate (TPA) as promotor, EGCG (LVIII) and some hydrolyzable tannins inhibited the binding of TPA to the receptor prepared from the particulate of mouse skin (*50*).

Oral administration of EGCG (LVIII) also remarkably inhibited the duodenal cancer of rats induced by *N*-ethyl-*N*'-nitrosoguanidine (*51*). Marked antimutagenic effect was exhibited by LVIII and several other polyphenolic compounds (*52*). The detail of these experiments is described in the chapter, "(-)-Epigallocatechin Gallate (EGCG), a Cancer Preventing Agent," in this book.

Upon intraperitoneal administration of each tannin in mice on the 4th day before intraperitoneal inoculation of Sarcoma 180 tumor cell (1 x 10^5), several oligomeric hydrolyzable tannins, including agrimoniin (XVI), oenothein B (XXXV) and coriariin A (LX) (Figure 10), exhibited strong host-mediated antitumor activity, while monomeric hydrolyzable tannins and galloylated condensed tannins, and polyphenols of small molecules were almost inactive (*53*). Most of the macrocyclic dimers, represented by oenothein B (XXXV) and camelliin B (XXXIX), showed strong antitumor activity of this kind (*31, 54*). The increase of life-span and the numbers of survived mice on the 60th day after administration of the active oligomers are summarized in Table I (*54*).

Agrimoniin (XVI) and oenothein B (XXXV) also exhibited similar host-mediated activity against ascites type tumor (MM2), and solid type rodent tumor (Meth A) (*54*). The structures of most of active oligomers were those composed of tellimagrandin II (XIV) and/or tellimagrandin I (LIX). The antitumor activity of these tannins was shown to be elicited by potentiation of the host-immunity through activation of macrophages, followed by that of natural killer cells and lymphocytes (*54*).

Antiviral Activities against HSV and HIV. Hydrolyzable tannin monomers, oligomers and galloylated condensed tannins exhibited antiviral activity (ED$_{50}$ 0.03-0.1 mg/ml) against herpes simplex virus (HSV) *in vitro*, although small molecule polyphenols such as gallic acid were inactive (Table II) (*55, 56*). The antiviral activity of tannins was shown to be due to the inhibition of HSV replication by blocking the virus adsorption to the cultured cells (*55*).

Human immunodeficiency virus (HIV) causes aquired immunodeficiency syndrome (AIDS). Several oligomeric hydrolyzable tannins inhibited the replication of HIV *in vitro*. The anti-HIV activity was evaluated by a selective index (SI = CC$_{50}$

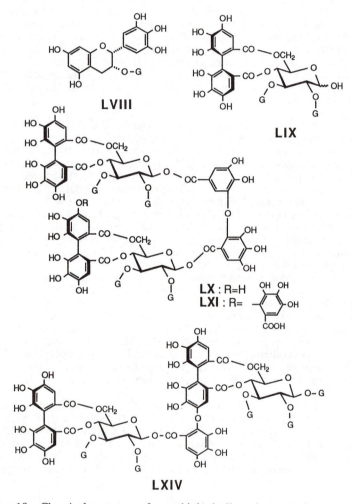

Figure 10.   Chemical structures of some biologically active polyphenols. (-)-Epigallocatechin gallate (LVIII); tellimagrandin I (LIX); coriariin A (LX); coriariin C (LXI); hirtellin B (LXII); tamarixinin A (LXIII); isorugosin D (LXIV); woodfordin E (LXV).

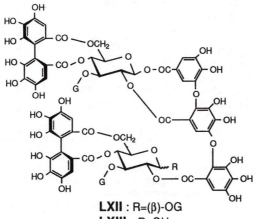

LXII : R=(β)-OG
LXIII : R=OH

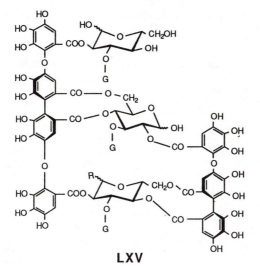

LXV

Figure 10. Continued

Table I. Antitumor Activity of Tannins against Sarcoma-180 in Mice[a]

Tannin	Dose (mg/kg)	%ILS[b]	60-day survivors[c]
Cornusiin A (XXVIII)	10	89.1	2
Hirtellin B (LXII)	10	114.2	3
Rugosin E (XXV)	10	169.5	2
Isorugosin D (LXIV)	5	37.2	2
Agrimoniin (XVI)	10	136.2	3
Coriariin A (LX)	5	110.9	3
Coriariin C (LXI)	10	51.7	2
Nobotanin H (XLIII)	10	-30.2	3
Tamarixinin A (LXIII)	10	75.0	4
Oenothein B (XXXV)	10	196.0	4
Camelliin B (XXXIX)	10	36.5	2
Woodfordin E (LXV)	10	77.4	2
OK-432[d]	100 (KE/kg)	208.4	2

[a] Six female ddY mice (6 weeks old) per group were used.
[b] Percent increase in the life span = {[(mean survival days of the treated group) - (mean survival days of the vehicle control group)] / (mean survival days of the vehicle control group)} x 100. Excluding 60-day survivors.
[c] No tumor take.
[d] Anticancer agent prepared from *Streptococcus pyogenes*. CAS Registry No. 39325-01-4.

Table II. Inhibition of HSV-1 plaque formation by tannins

Tannin or related polyphenol	EC_{50} (μg/ml)	CC_{50} (μg/ml)
Oenothein B (XXXV)	0.036	>30
Coriariin A (LIX)	0.038	>30
Rugosin D (XXIV)	0.034	>30
Cornusiin A (XXVIII)	0.039	>30
Tellimagrandin I (LIX)	0.036	>30
Casuarictin (XIII)	0.044	>30
Penta-O-galloyl-β-D-glucose	0.047	7
Geraniin (I)	0.093	>30
Epicatechin gallate tetramer (LVII)	0.14	>30
Licopyranocoumarin[a]	>10	23
Glycyrrhisoflavone[a]	>10	15
Gallic acid	>10	
Idoxuridine	16	
Sodium dextran sulfate	3.2	

[a] Flavonoids isolated from Licorice. See references (57, 58).

/ EC_{50}), calculated on the basis of 50% effective concentration (EC_{50}) and 50% cytotoxic concentration (CC_{50}) of each tannin, in the assay using MT-4 cells infected with HTLV-IIIB (*56*, Nakashima, H. et al., *Antiviral Res.*, in press).   Among 87 tannins tested, including monomeric and oligomeric hydrolyzable tannins and condensed tannins, camelliin B (XXXIX) (EC_{50} 4.8 µg/ml, CC_{50} 74.3 µg/ml, SI=15), nobotanin B (XLI) (EC_{50} 2.4 µg/ml, CC_{50} 33.7 µg/ml, SI=14) and trapanin B (XXXIII) (EC_{50} 2.7 µg/ml, CC_{50} 39.1 µg/ml, SI=14) showed potent anti-HIV activity.   Slightly lower activity was also exhibited by several other oligomeric hydrolyzable tannins such as nobotanins A, C and F, and camelliatannin A (XLV). However, no significant anti-HIV activity was found in most of the other tannins and small molecule polyphenols related to tannin [gallic acid, ellagic acid, (+)-catechin and (-)-epicatechin, etc.].   The activity of this kind is thus specific to the structures of tannins.   The anti-HIV activity of these oligomers *in vitro* was shown to be evoked by inhibition of adsorption of the virus to the cells (*56*), and also by inhibition of the reverse transcriptase of HIV (*59*).

Conclusion

The remarkable change in the definition of tannins, due to recent advances in the chemistry of natural polyphenolic compounds, has been accompanied by the changes in our recognition of the significance of their existence and utilization, from that for leathering material to that for health effects.   It is noticeable that several biological activities are specific to some structures of polyphenolic compounds, as exemplified by the host-mediated antitumor activity almost exclusively exhibited by some oligomeric hydrolyzable tannins.   Any research hereafter on the health effects of polyphenolic compound should be based on each polyphenolic compound of defined structure.

Acknowledgments

The authors thank Prof. H. Fujiki (National Cancer Center), Prof. R. Koshiura, Prof. K. Miyamoto (Hokuriku University), and Dr. H. Sakagami (Showa University), and their co-workers for their collaboration in the study of biological activities.

Literature Cited

1.   Haslam, E. *Plant Polyphenols, Vegetable Tannins Revisited* Cambridge University Press; U. K. 1989.
2.   Okuda, T.; Yoshida, T.; Hatano, T. *J. Nat. Prod.* **1989**, *52*, 1.
3.   Okuda, T.; Yoshida, T.; Hatano, T. *Planta Med.* **1989**, *55*, 117.
4.   Okuda, T.; Yoshida, T.; Hatano, T.; Koga, T.; Toh, N.; Kuriyama, K. *Tetrahedron Lett.* **1982**, 3941.
5.   Okuda, T.; Yoshida, T.; Nayeshiro, H. *Tetrahedron Lett.* **1976**, 3721.
6.   Okuda, T.; Yoshida, T.; Hatano, T. *J. Chem. Soc., Perkin Trans. 1* **1982**, 9.
7.   Okuda, T.; Seno, K. *Tetrahedron Lett.* **1978**, 139.
8.   Yazaki, K.; Hatano, T.; Okuda, T. *J. Chem. Soc., Perkin Trans. 1* **1989**, 2289.
9.   Okuda, T.; Mori, K.; Hayashi, N. *Yakugaku Zasshi* **1976**, *96*, 1143.
10. Okuda, T.; Mori, K.; Hatano, T. *Chem. Pharm. Bull.* **1985**, *33*, 1424.
11. Haslam, E.; Uddin, M. *J. Chem. Soc. (C)* **1967**, 2381.
12. Yoshida, T.; Fujii, R.; Okuda, T. *Chem. Pharm. Bull.* **1980**, *28*, 3713.
13. Namba, O.; Hatano, T.; Yoshida, T.; Okuda, T. *108th Annual Meeting of the Pharmaceutical Society of Japan* **1988**, p. 339 (Abstr.).

14. Okuda, T.; Yoshida, T.; Hatano, T.; Ikeda, Y.; Shingu, T.; Inoue, T. *Chem. Pharm. Bull.* **1986**, *34*, 4075.
15. Okuda, T.; Yoshida, T.; Ashida, M.; Yazaki, K. *J. Chem. Soc., Perkin Trans. 1* **1983**, 1765.
16. Okuda, T.; Yoshida, T.; Hatano, T.; Yazaki, K.; Ashida, M. *Phytochemistry* **1982**, *21*, 2871.
17. Hatano, T.; Kira, R.; Yoshizaki, M.; Okuda, T. *Phytochemistry* **1986**, *25*, 2787.
18. Okuda, T.; Hatano, T.; Kaneda, T.; Yoshizaki, M.; Shingu, T. *Phytochemistry* **1987**, *26*, 2053.
19. Okuda, T.; Yoshida, T.; Hatano, T. *Heterocycles* **1990**, *30*, 1195.
20. Okuda, T.; Yoshida, T.; Kuwahara, M.; Memon, M. U.; Shingu, T. *J. Chem. Soc., Chem. Commun.* **1982**, 163.
21. Okuda, T.; Yoshida, T.; Kuwahara, M.; Memon, M. U.; Shingu, T. *Chem. Pharm. Bull.* **1984**, *32*, 2165.
22. Yoshida, T.; Tanaka, K.; Chen, X.-M.; Okuda, T. *Chem. Pharm. Bull.* **1989**, *37*, 920.
23. Yoshida, T.; Jin, Z.-X.; Okuda, T. *Phytochemistry* **1991**, *30*, 2747.
24. Hatano, T.; Ogawa, N.; Yasuhara, T.; Okuda, T. *Chem. Pharm. Bull.* **1990**, *38*, 3308.
25. Hatano, T.; Ogawa, N.; Shingu, T.; Okuda, T.; *Chem. Pharm. Bull.* **1990**, *38*, 3341.
26. Hatano, T.; Ogawa, N.; Kira, R.; Yasuhara, T.; Okuda, T. *Chem. Pharm. Bull.* **1989**, *37*, 2083.
27. Hatano, T.; Okonogi, A.; Yazaki, K.; Okuda, T. *Chem. Pharm. Bull.* **1990**, *38*, 2707.
28. Hatano, T.; Ikegami, Y.; Shingu, T.; Okuda, T. *Chem. Pharm. Bull.* **1988**, *36*, 2017.
29. Okuda, T.; Mori, K.; Hatano, T. *Phytochemistry* **1980**, *19*, 547.
30. Yoshida, T.; Chen, L.; Okuda, T. *Chem. Pharm. Bull.* **1988**, *36*, 2940.
31. Yoshida, T.; Chou, T.; Nitta, A.; Miyamoto, K.; Koshiura, R.; Okuda, T. *Chem. Pharm. Bull.* **1990**, *38*, 1211.
32. Yoshida, T.; Chou, T.; Matsuda, M.; Yasuhara, T.; Yazaki, K.; Hatano, T.; Okuda, T. *Chem. Pharm. Bull.* **1991**, *39*, 1157.
33. Hatano, T.; Yasuhara, T.; Matsuda, M.; Yazaki, K.; Yoshida, T.; Okuda, T. *J. Chem. Soc., Perkin Trans. 1*, **1990**, 2735.
34. Yoshida, T.; Chou, T.; Maruyama, Y.; Okuda, T. *Chem. Pharm. Bull.* **1990**, *38*, 2681.
35. Yoshida, T.; Nakata, F.; Hosotani, K.; Nitta, A.; Okuda, T. *110th Annual Meeting of the Pharmaceutical Society of Japan* **1990**, 224 (Abstr. Part 2).
36. Yoshida, T.; Ohwashi, W.; Ishihara, K.; Okano, Y.; Shingu, T.; Okuda, T. *Chem. Pharm. Bull.* **1991**, *39*, 2264.
37. Hatano, T.; Shida, S.; Han, L.; Okuda, T. *Chem. Pharm. Bull.* **1991**, *39*, 876.
38. Han, L.; Hatano, T.; Okuda, T. *111th Annual Meeting of the Pharmaceutical Society of Japan* **1991**, p. 175 (Abstr. Part 2).
39. Nakata, F.; Yoshida, T.; Hosotani, K.; Nitta, A.; Okuda, T. *38th Annual Meeting of the Japanese Society of Pharmacognosy* **1991**, p. 172 (Abstr.).
40. Okuda, T.; Hatano, T.; Agata, I.; Nishibe, S.; Kimura, K. *Yakugaku Zasshi* **1986**, *106*, 894.
41. Okuda, T.; Hatano, T.; Agata, I.; Nishibe, S. *Yakugaku Zasshi* **1986**, *106*, 1108.
42. Kelley, C. J.; Mahajan, J. R.; Brooks, L. C.; Neubert, L. A.; Breneman, W. R.; Carmack, M. J. *J. Org. Chem.* **1975**, *40*, 1804.

43. Wagner, H. in *Biochemistry of Plant Phenolics* (Swain, T.; Harborne, J. B.; Van Sumere, C. F. eds) **1977**, p. 598, Plenum Press, New York.
44. Agata, I.; Hatano, T.; Nishibe, S.; Okuda, T. *Chem. Pharm. Bull.* **1988**, *36*, 3223.
45. Hatano, T.; Urita, K.; Okuda, T. *J. Med. Pharm. Soc. WAKAN-YAKU* **1986**, *3*, 434.
46. Okuda, T.; Yoshida, T.; Hatano, T.; Kuwahara, M.; Iida, S. *Proc. Symp. WAKAN-YAKU* **1982**, *15*, 111.
47. Okuda, T. *Proc. 23rd Symp. Phytochemistry* **1987**, p. 56.
48. Mølgaard, P.; Ravn, H. *Phytochemistry* **1988**, *27*, 2411.
49. Hatano, T.; Edamatsu, R.; Hiramatsu, M.; Mori, A.; Fujita, Y.; Yasuhara, T.; Yoshida, T.; Okuda, T. *Chem. Pharm. Bull.* **1989**, *37*, 2016.
50. Yoshizawa, S.; Horiuchi, T.; Fujiki, H.; Yoshida, T.; Okuda, T.; Sugimura, T. *Phytotherapy Res.* **1987**, *1*, 44.
51. Fujita, Y.; Yamane, T.; Yanaka, K.; Kuwata, K.; Okuzumi, J.; Takahashi, T.; Fujiki, H.; Okuda, T. *Jpn. J. Cancer Res.* **1989**, *80*, 503.
52. Okuda, T.; Mori, K.; Hayatsu, H. *Chem. Pharm. Bull.* **1984**, *32*, 3755.
53. Miyamoto, K.; Kishi, N.; Koshiura, R.; Yoshida, T.; Hatano, T.; Okuda, T. *Chem. Pharm. Bull.* **1987**, *35*, 814.
54. Miyamoto, K.; Koshiura, R.; Hatano, T.; Yoshida, T.; Okuda, T. *8th Symposium on the Development and Application of Naturally Occurring Drug Materials* **1991**, p. 61 (Abstr.).
55. Fukuchi, K.; Sakagami, H.; Okuda, T.; Hatano, T.; Tanuma, S.; Kitajima, K.; Inoue, Y.; Inoue, S.; Ichikawa, S.; Nonoyama, M.; Konno, K. *Antiviral Res.* **1989**, *11*, 285.
56. Sakagami, H.; Nakashima, H.; Murakami, T.; Yamamoto, N.; Hatano, T.; Yoshida, T.; Okuda, T. *8th Symposium on the Development and Application of Naturally Occurring Drug Materials* **1991**, p. 57 (Abstr.).
57. Hatano, T.; Kagawa, H.; Yasuhara, T.; Okuda, T. *Chem. Pharm. Bull.* **1988**, *36*, 2090.
58. Hatano, T.; Yasuhara, T.; Fukuda, T.; Okuda, T. *Chem. Pharm. Bull.* **1989**, *37*, 3005.
59. Asanaka, M.; Kurimura, T.; Koshiura, R.; Okuda, T.; Mori, M.; Yokoi, H. *4th International Conference on Immunopharmacology* **1988**, p. 35 (Abstr.).

RECEIVED April 21, 1992

Chapter 14

Custom Design of Better In Vivo Antioxidants Structurally Related to Vitamin E

L. Hughes[1], G. W. Burton[1], K. U. Ingold[1], M. Slaby[1], and D. O. Foster[2]

[1]Steacie Institute for Molecular Science and [2]Institute for Biological Sciences, National Research Council of Canada, 100 Sussex Drive, Ottawa, Ontario K1A 0R6, Canada

Vitamin E (α-T) owes its bioactivity mainly or entirely to its ability to inhibit lipid peroxidation *in vivo* by trapping peroxyl radicals. The reactivity towards peroxyl radicals is controlled almost exclusively by the chroman head group. Biokinetic studies using deuterated α-T stereoisomers suggest that the differences in biopotencies of the α-T diastereoisomers originate largely from the differences in the chirality at carbon 2. Furthermore, results using the rat curative myopathy pyruvate kinase (PK) bioassay show that the methyl branching in the hydrophobic tail is unimportant. However, the length of the tail is important. The optimum chain length appears to be 11-13 carbon atoms. Increased bioactivity has been observed in an α-T analog in which the heterocyclic ring size was reduced from 6 to 5 atoms. The resultant *all-racemic* dihydrobenzofuran acetate is 1.55 times more bioactive than *all racemic* α-T acetate in the PK bioassay. We believe increased bioactivity is due to stereoelectronic enhancement of radical trapping *in vivo*.

The chemical reaction of an organic substrate with oxygen can lead to a process known as autoxidation (*1*), often referred to as lipid peroxidation in living systems. Lipid peroxidation begins with the formation of a carbon centered free radical (R•), (reaction 1). Rapid reaction of the carbon centered radical with molecular oxygen yields a peroxyl radical (reaction 2) which can attack a lipid molecule (RH) to form a hydroperoxide and a new R•.(reaction 3). The sequence of reactions is repeated until two peroxyl radicals react to form inactive molecular products (reaction 4). This overall process is described as a free radical chain reaction which can be represented as follows;

Initiation:	production of R•	(1)
Propagation:	R• + O_2 → ROO•	(2)
	ROO• + RH → ROOH + R•	(3)
Termination:	ROO• + ROO• → molecular products	(4)

For a free radical chain reaction to occur radicals must be first formed from a molecular precursor. Chain initiation may be caused by heat or light, or by single electron transfer from a reducing agent such as Fe2+ to an acceptor (e,g., a hydroperoxide), or it may be enzymatically catalyzed. The reaction of R• with molecular oxygen to form a peroxyl radical is diffusion controlled (2). In comparison, the attack of ROO• on RH is a much slower reaction (3).

In biological systems, polyunsaturated fatty acids, essential compounds of mammalian biomembranes, are very susceptible to peroxidation. Autoxidation of a biological membrane can ultimately breach its integrity. The consequences for the organism may be disastrous since one of the principal functions of a membrane is to act as a dividing wall and compartmentalize biochemical processes into specific cells and into specific regions within the cell.

Protection of living organisms against oxidative degradation is provided by small quantities of specific compounds called *antioxidants* (4) There are two broad classes of antioxidants. *Preventive* antioxidants, which reduce the rate of chain initiation, and *chain breaking* antioxidants,,which interfere with one or more of the propagation steps. Most phenols are chain breaking antioxidants (5). The major (>90%) lipid soluble antioxidant found in plasma is vitamin E (6) α-Tocopherol (α-T) or 1 is the most biologically active (7) member of the tocopherol family.As a result of the existence of three chiral centers there are eight possible stereoisomers of α-T. The most biologically active is the *RRR* isomer (7)

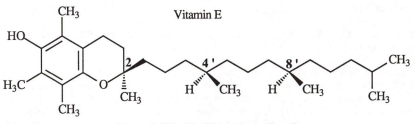

Vitamin E

1   (Natural (2R,4'R,8'R)-a-Tocopherol)

Chain carrying peroxyl radicals are "trapped" by phenols (reaction 5).

$$ROO• + ArOH → ROOH + ArO• \qquad (5)$$

The phenoxyl radical is too unreactive to continue the chain and is eventually destroyed by reaction with a second peroxyl radical (reaction 6).

$$ROO• + ArO• → \text{inactive molecular products} \qquad (6)$$

The effectiveness of a chain-breaking antioxidant, especially in a living organism, depends on its reactivity towards peroxyl radicals. A chain breaking antioxidant must, to be effective, be able to protect a relatively large amount of lipid using a small amount of the antioxidant. The rate constant for reaction 5 must consequently, be far greater than for reaction 3, i.e., $k_5 \gg k_3$.

Determination of Rate Constant for Reaction 5 (k_5)

Rate constants for reaction 5 have been carried out on a custom built autoxidation apparatus (8) designed to measure extremely low rates (~1x10-9 M s-1) of oxygen uptake. The organic substrate is styrene which is thermally autoxidized at a constant and reproducible rate using azobisisobutyronitrile (AIBN) as an initiator. The time during which the rate of oxidation is suppressed after the addition of a phenol is known as the induction period and its duration, τ is related to the concentration of phenol and the rate of chain initiation R_i (equation 7). Most hindered phenolic antioxidants trap two peroxyl radicals, thereby breaking two autoxidation chains

$$\tau = 2[Ar\ OH]/R_i \qquad (7)$$

After the phenol is consumed, the rate of oxidation will, of course return to the value it has when no antioxidant is present (see figure 1).The initial rate of phenol inhibited oxidation is given by

$$\frac{-d[O_2]}{dt} = \frac{k_3[RH]R_i}{2k_5[ArOH]} \qquad (8)$$

where RH represents the concentration of the substrate, in this case styrene, and k_3 the rate constant for chain propagation ,i.e., the rate constant for equation (3). For styrene, k_3 is relatively large (41 M-1 s-1 at 30°) (9). This means that, even with a very good antioxidant, inhibited oxidation can occur at a measurable rate and with a chain length greater than one. That is, the inhibited autoxidation is still a *chain reaction* (which is a necessary condition if k_5 is to be derived from equation 8). All our measurements of k_5 have been carried out at chain lengths > 4. The results in Table 1 give the k_5 values of selected phenolic antioxidants obtained using this inhibited autoxidation of styrene method.

What are the Structural Prerequisites of a Good Chain-Breaking Antioxidant ?

Over thirty years ago Howard and Ingold (*10*) explored the effect of various ring substituents on the k_5 values for simple phenols. Their results showed that the best 4-substituent was a methoxy group and that the best substituents at the four remaining aromatic positions were methyl groups. However, what would be predicted to be the *best* phenol, 4-methoxy-2,3,5,6,-tetramethylphenol, **2**, was not examined until twenty four years later (*11*). We fully expected this compound to have a k_5 similar to vitamin E. As can be seen from Table 1, compound 2 is only ~1/10 as effective (*11*) as vitamin E!

So far, in our model system, the most effective chain-breaking antioxidant we have examined (*12*) has proven to be 5-hydroxy-4,6,7-trimethyl-2,3-benzofuran, **4**. Since steric hindrance to abstraction of the phenolic hydrogen by ROO• must be very similar for all three of the phenols 1, 3, and 4 it was clear that the differences in k_5 values must be due mainly to differences in the exothermicities of reaction 5. That is, the O-H bond strength must decrease in the order $4 < 1 < 2$ and hence the phenoxyl radical derived from 4 must be more stabilized that that derived from 1 or 2.

What then is stabilizing the phenoxyl radical derived from 4 ? The likely answer is a *stereoelectronic effect* involving the oxygen para to the phenolic OH. If the para oxygen p-type lone pair orbital overlaps with the semi-occupied molecular orbital (SOMO) in the radical, then the phenoxyl radical will be further stabilized by *conjugative electron delocalization* (A ↔ B) .

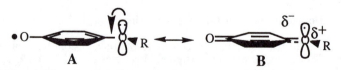

.The extent of this p-type lone pair overlap will depend on the dihedral angle, θ between the p-type orbital on O_1 and a perpendicular to the aromatic plane, an angle

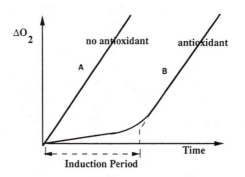

Figure 1. Plot of the consumption of oxygen by an oxidizable substrate (e. g., styrene) in a model system with a constant production of radicals. A: no chain-breaking antioxidant; B: antioxidant added.

Table 1. Values of k_5 and θ' for selected phenols obtained by the inhibited autoxidation of styrene at 30^0

	Phenol	k_5	θ'
3		380	17^0
2		39	89^0
4		570	6^0
1		320	–
5		470	–

which is equal to the dihedral angle θ' between the O_1-C_2 bond and the aromatic plane. Stabilization will be maximized when θ = 0° and will be at a minimum when θ = 90°.

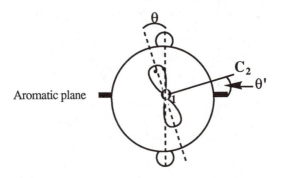

Our proposed theory that stereoelectronic factors are responsible for the increased antioxidant activity of compound 4 is supported by X-ray analyses of the various phenols were carried out (Table 1). The least reactive compound 2 had θ' = 89° and the most reactive compound 4 had θ' = 6°. In 2 the methoxy group is perpendicular to the plane of the aromatic ring. In this position, the p-type lone pair on the ether oxygen atom lies in the plane of the aromatic ring and therefore cannot stabilize the corresponding phenoxyl radical. However, in α-T and its corresponding model compound 3, the second ring adopts a half-chair conformation and holds the ether oxygen in a position corresponding to θ = 17°. The lone pair can thus overlap with the orbital containing the unpaired electron and stabilize the radical.

A *Better* Vitamin E

Since stereoelectronic effects are responsible for the increased activity *in vitro* of compound 4, it was of considerable interest to us to synthesize (*13*) a five-membered ring analog (dihydrobenzofuran-DHBF), 5 of vitamin E.

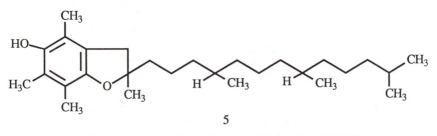

5

All-Racemic Dihydrofuran Analog of Vitamin E   (DHBF)

We wanted to determine its bioactivity relative to α-T in order to see if there was any correlation with our *in vitro* results, i. e.,

In vitro trapping of peroxyl radicals

$$ROO\cdot + ArOH \xrightarrow{k_5} ROOH + ArO\cdot$$

$$\frac{k_5(\text{all-rac-DHBF})}{k_5(\text{a-T})} = \frac{4.7 \times 10^6}{3.2 \times 10^6} = \boxed{1.47}$$

In vivo (rat curative myopathy assay) potency ratio

$$\text{potency ratio} \left\{ \frac{\text{all-rac-DHBT}}{\text{all-rac - α-T}} \right\} = \boxed{?}$$

The synthesis (*13*) (scheme 1) was effected in ten steps, the key step being a Wittig reaction between 2-formyl-2,4,6,7-tetramethyl-5-benzyloxy dihydro-benzofuran, 6, and the phosphonium salt of 3,7,11-trimethyldodecyl bromide, 7.

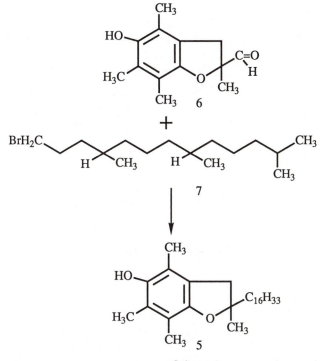

Scheme 1

Also we eventually synthesized both the $2R,4'R,8'R$ *(5-RRR)* and the $2S,4'R,8'R$ *(5-SRR)* stereoisomers (*13*)

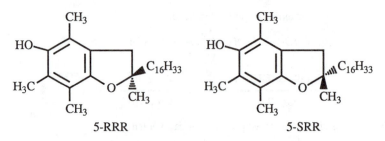

5-RRR   5-SRR

We chose to test the DHBF analogs *in vivo* using the rat curative myopathy bioassay (*14*) to determine relative vitamin E activity. The method is based on the reduction of highly elevated plasma pyruvate kinase (PK) activities in vitamin E deficient rats. In each bioassay the regression of PK activity vs. *ln* (dose) of the test compounds was computed by the method of least squares (see figure 2).

The results of seven bioassay experiments which employed, in their acetate forms, *all-rac-5*, *RRR-5*, *SRR-5*, *all-rac-1*, *RRR-1*, and *SRR-1* are summarized in Table 2. Most startling is the result for *all-rac-5* (DHBF) vs *all-rac-1* (α-T) for which we obtained a potency ratio A:B (Table 2) of 1.55. If we compare this to our *in vitro* result for the same pair of compounds we see remarkable agreement (k_5 (5)/k_5 (1)=1.47) !

Table 2.   Potency ratio of various pure stereoisomers and mixtures of stereoisomers of the acetates of 5 and 1 in reducing plasma pyruvate kinase in the rat curative myopathy bioassay

Test	Pair of compounds		Potency ratio, A:B [a]			
	A	B	Assay 1	Assay 2	Assay 3	Average
1	RRR-5-Ac	RRR-1-Ac	1.09[ns]	0.96[ns]	1.24[ns]	1.10
2	SRR-5-Ac	RRR-1-Ac	0.76[ns]			0.76
3	SRR-5-Ac	SRR-1-Ac	1.16[ns,d]			1.16
4	all-rac-5-Ac	all-rac-1-Ac	1.55[c,d]			1.55
5	SRR-5-Ac	RRR-5-Ac	0.60[c]	0.43[c]		0.51
6	SRR-1-Ac	RRR-5-Ac	0.42[c]	0.47[c]		0.45
7	SRR-1-Ac	RRR-1-Ac	0.44[b]	0.56[c]	0.46[c]	0.49

[a] The superscripts ns, b and c indicate the degree of significance, $p > 0.05$, $p < 0.05$, and $p < 0.01$, respectively, that the test pair of compounds do not differ in potency.
[d] Double assay with 36 rats per test compound rather than the usual 18 rats.

However the results obtained with the pure isomers are quite different. For example the directly measured potency ratio for the *RRR-5-Ac/RRR-1-Ac* pair is 1.10 (test 1), while the potency ratios which may be calculated for this pair by combining the data from tests 2 and 5 and from tests 6 and 7 are 1.49 and 1.09, respectively.

Similarly the directly measured potency ratio for the *SRR*-5-Ac/*SRR*-1-Ac pair is 1.16 (test 3) while the potency ratios which may be calculated for this pair by combining the data from tests 2 and 7 and from tests 5 and 6 are 1.55 and 1.13, respectively. Although both the directly measured and indirectly calculated potency ratios for the *RRR* 1 and 5 pair and the *SRR* 1 and 5 pair are not, statistically speaking, significantly different from 1.0 (see Table 2) the total weight of evidence is that *RRR*-5-Ac and *SRR*-5-Ac are ca. 1.1 to 1.15 times as active as the corresponding stereoisomers of α-T-Ac. We therefore conclude that the statistically greater activity of *all-rac*-5-Ac relative to *all-rac*-1 (Table 2) is most likely due to a significantly higher activity of some (but not all) of the stereoisomers of 5-Ac relative to the corresponding stereoisomers of 1-Ac.

In conclusion, to our knowledge, this represents the first time that the methods of physical organic chemistry have been used in vitamin research to design a compound which improves on nature.

The Stereochemistry at C-2 Controls the Biopotency

Although phenol 3 is an excellent antioxidant (*11*) it has no vitamin E activity. The phytyl tail of tocopherols serves to hold the chemically reactive phenolic head group within biomembranes. The stereochemistry of the phytyl tail is known to affect the bioactivity (*7*) of α-T. Of the eight possible stereoisomers of α-T, the natural isomer, *RRR*-α-T has been shown to be the most bioactive (*7*). The isomer with inverted stereochemistry at C-2, i.e., *SRR*-α-T, has been reported to have only about 30% the activity (*7*) of *RRR*-α-T. There appear to have been no real attempts to explore or explain the origin of the different bioactivities of the α-T stereoisomers. To us, the magnitude of the *RRR/SRR* difference appeared far too small to be due to some enzyme or protein mediated process. We therefore decided to explore *biokinetically* the transport of the α-T isomers over both short and extended periods of time. However, in order to do long term experiments we could not use radioisotopically labelled α-T isomers. We therefore chose to synthesize (*15*) stable, isotopically labelled, isomers of *RRR* and *SRR*- α-T using different degrees of deuteration to differentiate between the two stereoisomers. Thus a competitive uptake study with this pair of compounds, differing in structure only at the 2-position where the chroman ring and the phytyl tail are joined (see structure 1) would provide direct evidence of the importance of the stereochemistry at the ring-tail junction.

We chose as starting materials γ-Tocopherol, 8 and δ-Tocopherol 9, which can be obtained from soybean distillate in large quantities, to synthesize (*15*), as depicted in Scheme 2, multigram amounts of 5-CD_3-*SRR*-α-T,10 (after inversion of the configuration at C-2) and 5,7-CD_3-*RRR*-α-T, 11, respectively.

The competitive biokinetic experiment was carried out as follows: three-week old male rats were raised on a standard vitamin E free diet containing 36 mg of unlabelled *RRR*-α-T-Ac/kg of diet. After four weeks the diet was substituted with a diet in which the d_0-*RRR*-α-T-Ac was replaced with an equimolar quantity of a 1:1 mixture of 18mg d_3-*SRR*-α-T-Ac, 10, and 18 mg d_6-*RRR*-α-T-Ac, 11 by diet. Blood and tissue samples were taken at various times and their lipids were extracted by our sodium dodecyl sulphate (SDS) method (*16*). The experiment was carried out over a period of ~3 months. The α-T extracts were purified by HPLC, silylated, and analyzed by GC/MS.

The appropriate ions are extracted from the single gas chromatographic peak obtained (See figure 3) using the selected ion monitoring mode (SIM). An internal standard (d_9-*all-rac*-α-T) (*17*) was used to quantitate the absolute concentrations of d_0, d_3 and d_6-α-T. Also, corrections were applied for the presence of natural abundance isotopes (^{13}C, ^{29}Si, ^{30}Si). The use of deuterated tocopherols (*15,17*), our SDS method (*16*) and our GC/MS technology permits the detection of as little as 40 pg of biopsy samples as small as 1mg·

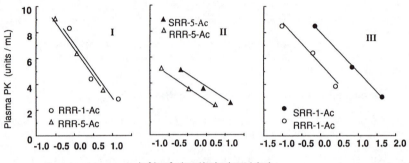

Figure 2. Dose -dependence of the reduction in plasma pyruvate kinase (PK) levels measured on day 5 in vitamin E deficient rats after administration once daily for four consecutive days of I: *RRR*-1-Ac (O) or *RRR*-5-Ac (Δ) (Test 1,Assay 1); II: *RRR*-5-Ac (Δ) or *SRR*-5Ac (▲) (Test 5,Assay 1); and III:*RRR*-1-Ac (O) or *SRR*-1-Ac (●) (Test 7, assay 1). The data points displayed in each graph are the means of data from six animals. The potency ratios found in I (*RRR*-5-Ac:*RRR*-1-Ac), II (*SRR*-5-Ac:*RRR*-1-Ac) and III (*SRR*-1-Ac:*RRR*-1-Ac) were 1.09, 0.60, and 0.44 respectively.

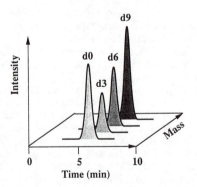

Figure 3. Depiction of resolution by mass spectrometer, of α-tocopherol gas chromatographic peak into component peaks corresponding to α-tocopherols containing 0, 3, 6, and 9 deuterium atoms/ molecule.

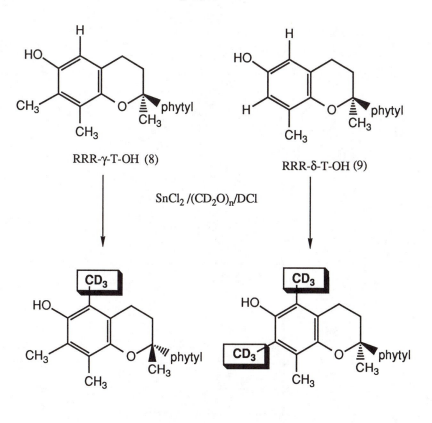

RRR-γ-T-OH (8)

RRR-δ-T-OH (9)

SnCl₂ /(CD₂O)ₙ/DCl

d3-SRR-α-T-OH (10)

d6-RRR-α-T-OH (11)

Scheme 2

   The results of this experiment are summarized in Figure 4. There are three interesting observations. Firstly,we observed from the very first day a discrimination in plasma favoring the *RRR* isomer. The *RRR* vs *SRR* ratio begins on day one at 1.4 and reaches 2.4 after more than a 100 days. Secondly, the liver at the onset shows an initial excess, by a factor of two, of the unnatural *SRR*-isomer. Eventually, at about three weeks a slight preference for the *RRR* isomer is observed.(*RRR/SRR* = ~1.1). This pattern in the liver was maintained until the experiment was terminated after more than 100 days. Thirdly, the brain shows an extraordinary progressive discrimination in favor of the natural isomer with the *RRR/SRR* ratio exceeding a value of five after five months.

   In conclusion, the degree of discrimination depends not only on the tissue but also on the duration of the experiment. To us these results suggest that there can be no simple comparative tests of the biopotency of synthetic antioxidants vs. natural α-T. The results obtained in any test must depend on the nature of the test (e.g., the tissue examined) and the duration of the test. The general assumption that *RRR*-α-T has 1.36 times the activity (7) of synthetic *all-rac*-α-T(which is an equimolar mixture of all eight stereoisomers) must be questioned.

   There are important practical consequences to these biokinetic studies. Thus if the brain needs extra protection against radical damage (because of radiation therapy

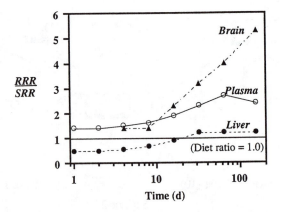

Figure 4. Ratio of uptake of RRR- vs SRR-α-tocopherol in male rats vs time.

after surgery, for example) even massive oral doses of vitamin E will not reach the brain very quickly.

Is Methyl-Branching in α-Tocopherol's Tail Important for its *In Vivo* Activity ?

Despite the overwhelming evidence that α-T owes its bioactivity solely to its ability to trap radicals and hence inhibit lipid peroxidation (*18-20*) other roles for this vitamin have occasionally been suggested. The most intriguing of all these suggestions was made by Lucy, Diplock and their co-workers (*21-28*) who proposed that vitamin E did not stabilize biomembranes by inhibiting lipid peroxidation with its "head" group by trapping peroxyl radicals, but instead stabilized the membranes with its "tail" group via a purely physical interaction with the polyunsaturated fatty acid moities of the phospholipid constituents of the membrane. Based on molecular model building, it was suggested that the 4'-CH$_3$ group of α-T could "fit" into a "pocket" provided by the double bonds at Δ5 and Δ8 in arachidonic acid, and the 8'-CH$_3$ group could fit into a similar "pocket" provided by the Δ11 and Δ14 double bonds. In the "complex" so formed, the 6-OH group and the polar groups of the phospholipids are both in the surface region of the membrane. The 4' and 8' CH$_3$ groups are nested into the pockets created by the cis-double bonds, allowing the methylene groups in the phytyl "tail" of α-T and in the fatty acyl chain of the phospholipids to associate closely, promoting the stability of the complex through London-Van der Waals dispersion-attraction forces.

The hypothesis of Lucy and Diplock must overcome the handicap that the ratio of arachidonic acid residues to α-T in biological membranes may be greater than 500:1 (*22*). To avoid this difficulty, it was suggested that α-T interacts within membranes "in a dynamic manner with a number of phospholipid molecules" (*23*). However this theory has been questioned on theoretical grounds (*29*). Furthermore it has been pointed out (*29*) that Diplock and co-workers' (*23-26*) experimental results with a variety of model systems, which they had taken to support the "physical" "membrane-stabilizing effect of vitamin E, could be equally well explained in "chemical" terms as consequence of reduced lipid peroxidation.

Model systems employed by other workers have also yielded results which have been claimed to support the Lucy-Diplock "physical" membrane stabilizing effect of α-T (*30*) or some related physical effect(s) which influence membrane fludity and/or permeability (*21-35*). However, other model systems have provided fairly convincing evidence that vitamin E does not stabilize membranes via the Lucy and Diplock "complex" nor by any kind of "physical" effect, particularly at physiological concentrations of vitamin E (*42*).

As we have pointed out in another context (the question as to whether or not vitamin C interacts synergistically with vitamin E *in vivo*) (*43*), even the most carefully designed model systems may fail to reflect the *in vivo* reality. A proper check of the relevance of the Lucy -Diplock "complex" to the stabilization of the biological membranes therefore demanded a proper *in vivo* test. The simplest such test appeared to be to employ an approved vitamin E bioassay using compounds in which the methyl branched hydrocarbon "tail" of α-T was replaced by an unbranched, saturated hydrocarbon "tail" of suitable length, i.e. 2RS-2,5,7,8-tetramethyl-6-hydroxychromans,12.

Compounds 12 were synthesized (see scheme 3) in 6 steps, the key linkage being given by Fouquet-Schlosser (*44*) condensation of tosylate (*45*) 13 and the Grignard reagent from the appropriate alkyl bromide 14

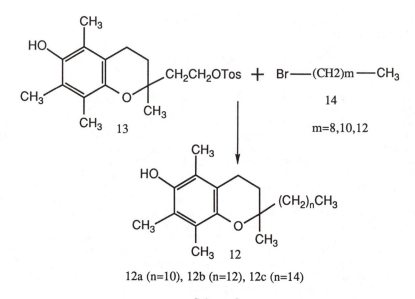

12a (n=10), 12b (n=12), 12c (n=14)

Scheme 3

A search of the literature revealed only two earlier whole animal bioassays for vitamin E activity of compounds with structure 12 (46,47). In 1939 John (46) reported that 12 (n=11) had vitamin E activity in the gestation-resorption bioassay but did not quantify the activity of this compound relative to that of α-T. In 1981 Kingsley and Combs (47) reported that 12 (n=12) in selenium deficient chicks had 50%, 46% and 44% of the activity of all-rac-α-T in supporting growth, in preventing exudative diathesis and for sustaining chick survival, respectively. These results certainly provide support for the relative unimportance of the 4'-CH$_3$ and 8'-CH$_3$ groups to the biological activity of α-T. Nevertheless, we thought that a more thorough study of the bioactivity of 12 with various chain lengths for the "tail" would be worthwhile as an in vivo test of the Lucy- Diplock hypothesis. We chose the rat curative myopathy bioassay (14) to measure the vitamin E bioactivities of compounds 12a (n=10), 12b (n=12), and 12c (n=14). The potency ratios for each of the pair of compounds tested are given in Table 3.

Since all three acetates of 12 which were studied were racemic compounds (i.e., 2RS-12-Ac) it seems more appropriate to compare their bioactivities to that of all-rac-α-T-Ac than to that of RRR-α-T-Ac. Fortunately, this can readily be done since Weiser et all (7) have employed the rat curative myopathy bioassay (14) to measure the potency ratio for RRR-α-T-Ac:all-rac-α-T-Ac pair, for which they obtained a ratio of 1.47:1.0. In those cases in which a 12-Ac compound was compared experimentally with RRR-α-T-Ac we have therefore calculated a 12-Ac:all-rac-α-T-Ac potency ratio by multiplying the measured potency ratio by 1.47 (see final column in Table 3).

The results given in Table 3 show that for the hydrocarbon "tails" with 11- and 13-carbon atoms the bioactivities of the 12-Ac compounds are equal within experimental error to the bioactivity of all-rac-α-T-Ac. There is also an indication that 12a may be more active than all-rac-α-T-Ac which, in turn may be more active thar. 12b, but these differences are not statistically significant. However there can be no doubt that 12c, with a 15 carbon tail, has a much lower bioactivity than the other compounds.

Table 3.   Potency ratios of the acetates of 2-RS-n-alkyl-2,5,7,8-tetramethyl-6-hydroxy-chroman, 12, and α-tocopherol, α-T

Test	Pair of Compounds		Potency Ratios	
	A	B	A:B[a]	A:all-rac-α-T-Ac[b]
1	12a-Ac (n=11)	all-rac-α-T-Ac	1.35[ns]	1.35
2	12b-Ac (n=13)	all-rac-α-T-Ac	0.89[ns]	0.89
3	12b-Ac (n=13)	RRR-α-T-Ac	0.69[ns]	1.01
4	12b-Ac (n=13)[c]	RRR-α-T-Ac	0.61[*]	0.90
5	12c-Ac (n=15)	RRR-α-T-Ac	0.15[*]	0.22
6	12a-Ac (n=11)	12b-Ac (n=13)	1.11[ns]	

[a] ns and * signify the probabilities $p > 0.05$ and $p < 0.01$, respectively, that the test pair do not differ in potency.

[b] For tests of A vs. RRR-α-T-Ac this is the measured A:B potency multiplied by 1.47 (which is the all-rac-α-T-Ac:RRR-α-T-Ac potency ratio determined in the rat curative myopathy bioassay.

[c] Repeat of test 3.

We conclude that methyl branching in the hydrocarbon "tail" of α–tocopherol plays no role (or, at best an insignificant role) in determining the bioactivity of this compound. The present study therefore demonstrates that Lucy and Diplock's hypothesis is invalid *in vivo*. It also invalidates other hypotheses which have been advanced regarding the function of methyl branching in α–tocopherol's "tail"(*48*).

Conclusion

We are now in a position to list the important features of lipid-soluble phenolic antioxidants analogous to the tocopherols which give these compounds vitamin E activity.
(i) The aromatic ring should be fully methylated.
(ii) The size of the heterocyclic ring is important. For example, reduction in ring size from the six atoms of α-tocopherol to the five atoms of the dihydrofurans analogues increases the bioactivity (*49,50*).
(iii) The stereochemistry at the 2-position of both α-tocopherol (*14,51-58*) and a dihydrofuran analog (*50*) should have the 2-R configuration.
(iv) Methyl branching in the hydrophobic "tail" is unimportant but the length of the tail is important, the optimum chain length being ca, 11-13 carbon atoms (*59*).

Acknowledgement

We thank Mrs, S, Lacelle for help with the bioassays, Mrs. H. Burton for help with the animals, Dr. M. Zuker for designing the statistical analysis program, and Mr. D. A.Lindsay for the development of the GC/MS method. We also thank the Association for International Cancer research, the National Foundation for Cancer Research,

Eastman Chemical Products, Inc., Eisai Co., Ltd., Henkel Corporation Hoffman-Laroche (Basle) and the Natural Source Vitamin E Association.

Literature Cited

1. Howard J. A. *Free Radicals*; Kochi J. K.,Ed.; Wiley: New York, N.Y.,1973;Vol 2, 3-62.
2. Maillard, B.; Ingold, K.U.; Scaiano, J, C. *J Am. Chem. Soc.* **1983**, *105*, 5095.
3. Ingold, K. U. *Acc. Chem Res.* **1969**, *2*, 1-9.
4. Ingold, K. U. *Chem. Rev.* **1961**, *61*, 563-589.
5. Ingold, K. U. *Spec. Publ.- Chem. Soc.* **1971**, *24*,285-293.
6. Burton, G. W.; Ingold, K. U.; Joyce A.. *Lancet.* **1982**,327.
7. Weiser, H.; Vecchi, M.; Schlachter, M. *Int. J. Vit. Nutr. Res.* **1985**, *55,*149-158.
8. Howard, J. A.; Ingold, K. U. *Can. J. Chem.* **1969**, *47*, 3809-3815.
9. Howard, J. A. ; Ingold, K. U. *Can. J. Chem.* **1965**, *43*, 2729-2736.
10. See also: (a) Howard, J. A. : Ingold, K. U. *Can. J. Chem.* **1962**, *40*, 1851-1864; (b) **1963**, *41*, 1744-1751; (c) **1963**, *41*, 2800-2806; (d) **1964**, *42*, 1044-1056.
11. Burton, G. W. ; Ingold, K. U. *J. Am. Chem. Soc.* **1981**, *103*, 6472-6477.
12. Burton G. W.; Doba T.; Gabe E.W; Hughes L.; Lee F.L.; Prasad L.; Ingold, K.U. *J. A. Chem. Soc.* **1985**, *105*, 7053-7065.
13. Brownstein, S.; Burton, G. W.; Hughes, L.; Ingold, K.U. *J. Org. Chem.* **1989**, *54*, 560-569.
14. Machlin, L. J.; Gabriel, E.; Brin M. *J. Nutr.* **1982**, *112*, 1437-1440.
15. Ingold, K.U.; Hughes, L.; Slaby, M.; Burton, G. W. *J. Lab. Cmpds. and Radio.* **1987**, *24*, 817-831.
16. Burton, G. W.; Webb, A.; Ingold. *Lipids.* **1985**,*20* ,29-39.
17. Hughes, L.; Burton, G. W.; Slaby, M.; Ingold, K.U. *J. Lab. Cmpds. and Radio.* **1990**, *28*,1049-1057.
18. *Vitamin E A Comprehensive Treatise,*Machlin, L. J.; ed., Dekker, New York, N.Y. **1980**.
19. Burton, G. W.; Joyce, A.; Ingold, K. U. *Arch Biochem. Biophys.* **1983**, *221*, 281-290.
20. Bieri, J. G. *Prog. Chem. Fats, Other Lipids.* **1964**, *7*, 247-266.
21. Lucy, J. A.; *Ann. NY. Acad. Sci.* !972, *203*, 4-11.
22. Diplock, A. T.; Lucy, J. A. *FEBS Lett.* **1973**, *29*, 205-210.
23. Maggio, B.; Diplock, A.T.; Lucy, J. A. *Biochem.* **1977**, *161*, 111-121.
24. Diplock, A. T.; Lucy, J. A.; Verrinder, M.; Zieleniewski, A.*FEBBS Lett.* **1977**, *82*, 341-344.
25. Giasuddin, A. S. M.; Diplock, A. T. *Arch. Biochem. Biophys.* **1979**, *196*, 270-280.
26. Giassudin, A. S. M.; Diplock, A. T. *Arch. Biochem.***1981**, *210*, 348-362.
27. Diplock, A. T. *Biology of Vitamin E.*, Ciba Foundation Symposium . **1983**, Pitman: London; 45-53.
28. *Fat Soluble Vitamins*s;Diplock, A. T., Ed.; Technomic: Lancaster, PA, **1985**; 154-224.
29. Ingold, K. U. Comment on ref. 27, p. 54.
30. Sunamoto, J.; Baba, Y.; Iwamoto, K.; Kondo, H. *Biochem Biophys. Acta.* **1985**, *833*, !44-150.
31. Erin, A. N.; Spinn, M. M.; Tabidge, L. V.; Kagan, V. E. *Biochim Biophys. Acta.* **1984**, *774*, 96-102.

32. Erin, A. N.; Skrypin, V. V.; Kagan, V. E. *Biochim. Biophys. Acta.* **1985**, *815*, 209-214.
33. Stillwell, W.; Bryant, L. *Biochim. Biophys. Acta.* **1983**, *731*, 483-486.
34. Mutchie, E. J.; McIntosh, G. H. **1986**, *J. Nutri. Sci. Vitaminol.*, **1986**, *32*, 551-558.
35. Wassall, S. R.; Thewalt, J. L.; Wong, L.; Gorrissen, H.; Cushley, R. *J. Biochemistry.* **1986**, *25*, 319-326.
36. Ohki, K.; Tadanobu, T.; Yoshinori, N. *J. Nutr. Sci. Vitaminol.* **1984**, *30*, 221-234.
37. Urano, S.; Matsuo. M. *Synthesis and applications of Isotopically Labelled Compounds.* **1985**. Proceedings of the sec ond International symposium, Kansas City, MO, U. S. A.Amsterdam: Elsevier, 517-518.
38. Urano, S.; Iida, M.; Otari, I,; Matsuo, M. *Biochem Biophys, Res. Comm.* **1987**, *146*, 1413-1418.
39. Urano, S.; Yano, K.; Matsuo, M. *Biochem. Biophys. Res. Commun.***1980**, *150*, 469.-475 40. Wassall, S. R.; Phelps, T. M.; Wang, L.; Langsford, C. A.; Stillwell, W. *Prog. Clin. Biol. Res.* **1989**, *292*, 435-444.
41. Sugata, S.; Urano, S.; Matsushima, Y.; Matsuo, M. *Biochim. Biophys. Acta.* **1988**, *962*, 385-386.
42. Takahashi,M.; Tsuchiya, J.; Niki, E.; Urano, S. *J. Nutr. Sci. Vitaminol.* **1988**, *34*, 25-34.
43. Burton, G. W.; Wronska, U.; Stone, L.; Foster, D.O; Ingold, K. U. *Lipids.***1990**, 25, 199.
44. Fouquet, G.; Schlosser, M. *Agew. Chem. Int. Ed. Engl.* **1974**, *13*, 82-83.
45. Cohen, N.; Eichel, W. F.; Lopresti, R. J.; Neukom, C.; Saucy, G. *J. Org. Chem.* **1976**, *41*, 3505-3511.
46. John, W. *Angew. Chem.* X, *52*, 413-419.
47. Kinsley, P. B.; Combs, G. F. *Proc. Soc. Exp. Biol. Med.* **1981**, *166*, 1-5.
48. Ivanov, I. I. *J. Free Radical Biol. Med.* **1985**, *1*, 247-253.
49. Ingold, K. U.; Burton, G. W.; Foster, D. O.; Zuker, M.; Hughes, L.; Lacelle, S.; Lusztyk, E.; Slaby, M. *FEBBS. Lett.* **1986**, *205*, 117-120.
50. Ingold, K. U.; Burton, G. W.; Foster, D. O.; Hughes, L. *FEBS Lett..* **1990**, *267 (1)*, 63.
51. Ames, S. R.; Ludwig, M. I.; Nelan, D. R.; Robeson, C. D. *Biochemistry.* **1963**, *2*, 188-190.
52. Scott, M. L.; Desai, I. D. *J. Nutrr.* **1964**. *83*, 39-43.
53. Witting, L. A.;Horwitt, M. K. *Proc. Exp. Biol. Med.* **1964**,*116*, 655-658.
54. Ames, S. R. *Lipids.* **1971**, *6*, 281-290.
55. Ames, S. R. *J. Nutr.* **1979**,*109*, 2198-2204.
56. Weiser, H.; Vecchi, M. *Internat. J. Vit.Nutr. Res.* **1981**,*51*, 100-113.
57. Weiser, H.; Vecchi. *Internat, J. Vit. Nutr, Res.* **1982**,*52*, 351-370.
58. Ingold, K. U.; Burton, G. W.; Foster, D. O.; Hughes, L.; Lindsay, D. A.; Webb, A. *Lipids.* **1987**, *22*. 163-172.
59. Ingold, K. U.; Burton, G. W.; Foster, D. O.; Hughes, L. *Free. Rad. Biol. and Med.* **1990**, *9*, 205.

RECEIVED November 7, 1991

Chapter 15

Thermal Degradation of Phenolic Antioxidants

M-C. Kim and D. E. Pratt

Department of Foods and Nutrition, Purdue University,
West Lafayette, IN 47907

Chemical decomposition of 2-tert-butyl-hydroxy-anisol,
2-tert-butyl-hydroxytoluene and 2-tert-butyl-
hydroquinone at 175±5°C in the absence of hydroperoxy or
alkylperoxy radicals was studied. 2-tert-butyl-
benzoquinone was identified as the primary and major
oxidation product of TBHQ. Interconversion between TBHQ
and TBBQ was observed to play the most important role in
antioxidant effectiveness and carry through effect of
TBHQ. 3,3'5,5'-tetra-bis-Itert-butyl)-4,4'-dihydroxyl-
1.2-diphenylethane from BHT and 3,3'-bis-(1.1-dimethyl
ethyl)-5,5'-dimethoxy-1,1'-biphenyl-2,2'-diol and its
ethylene glycol type of isomer from BHA were determined
as major decomposed products. 2.6-d9-tert-butyl-4-
hydroperoxy-4-methylcyclohexa-2.3-dienone was
thermolized at frying temperature to simulate and
simplify the real system containing BHT and lipid. The
products included BHT, 2.6-di-tert-butyl-4-hydroxy-4-
methylcyclohexa-2.5-dione, 3.5-di-tert-butyl-4-
hydroxybenzaldehyde, 2.6-di-tert-butyl-4-
hydroxymethylphenol and unidentified compounds having
M.W. 414,442,274 and 278. Trace amounts of 2.6-di-tert-
butylbenzoquinone and quinone methide were detected. In
the model system containing palmitic acid (16;0) and
linoleic acid (18;2), the order of antioxidant
effectiveness was TBHQ>BHA>BHT.

Most commonly used antioxidants today are synthetic chemicals and
possible toxicity of these synthetic antioxidants has not been
completely elucidated. Especially, studies regarding breakdown
pathways of phenolic antioxidant and interactions between secondary
degradation products of oil and antioxidants themselves at frying
temperature are scarce.

Many studies have utilized conditions that are not typical in
food processing systems and use high concentrations of antioxidants

0097–6156/92/0507–0200$06.00/0

to simplify the analysis of the reaction mixtures. The relevance in food processing is nonexistent or very obscure. In the case of phenolic antioxidant, this may be illustrated:

$$ROO\bullet + AH \longrightarrow ROOH + A\bullet$$
$$RO\bullet + AH \longrightarrow ROH + A\bullet$$
$$R\bullet + AH \longrightarrow RH + A\bullet$$
$$OH\bullet + AH \longrightarrow H_2O + A\bullet$$

Antioxidant free radicals A• are not sufficiently active to propagate the reaction chain, so they are usually deactivated by combination with another radical A• or with radicals derived from oxidized lipids, with formation of inactive products. In the case of strong antioxidants, nearly all A• radicals disappear by reaction with A• or ROO• radicals, while in the case of weak antioxidants other reactions may take place. Various side reactions (very slow in comparison with the main reaction) may proceed simultaneously, resulting in chain initiation and thus increasing the reaction rate. Phenolic antioxidants are converted into radicals during protection reaction. If a phenolic antioxidant is to be considered technologically useful, the phenoxy radical A• must not abstract H• from RH by itself and thus contribute to the propagation process. Rather it is converted into stable quinone, forms dimers or generates adducts with the peroxy radical species.

Scott (1) has thoroughly reviewed the rules of substitution at the ortho, meta, and para position in terms of phenolic antioxidant stability. Basically, to be a good antioxidant, the compound must be able to have an effective delocalization of the unpaired electron produced in the reaction with the free radical. The more effective the resonance forms, the better the antioxidant. Therefore, substitution at ortho and para position is much more effective than at the meta position because of the greater number of resonance forms possible. Another important factor is the size of the substitution group. A bulky attachment helps to protect the antioxidant radical and gives more stability towards further reaction, but this also makes it more difficult to react with the peroxy radical.

Pokorny (2) reported that the unsubstituted phenol is practically inactive. Substitution of the phenolic ring with alkyl groups increases electron density on active centers and thus improves the reactivity with ROO• radicals, which, being electron acceptors, have a tendency to react with phenols at the point of maximum electron density. Alkylation usually increases solubility of phenols in the substrate and decreases the tendency toward oxidation with atmospheric oxygen.

Antioxidants escapes from the frying medium by volatilization and steam distillation due to the high temperature characteristic and the large amount of water boiled out. Loss of antioxidant during frying has often been reported as a factor contributing to the loss of antioxidant potency. Freeman et al (4) reported that BHT was not effective in frying operations. Augustin et al (5) have observed ineffectiveness of BHA and BHT under frying conditions, but TBHQ was relatively more effective under the same conditions. Stuckey (4) postulated that some phenolic antioxidants are ineffective during frying due to thermal destruction and loss through steam distillation.

Peled et al (6) also found that the low effectiveness of BHT
in retarding thermal deterioration of the oil in frying could be
attributed to BHT losses due to volatilization and/or destruction by
oxidation. They also found that when the oil was heated in the
presence of air, the antioxidant destruction was lessened by the
presence of water in the system, but a large portion of BHT was lost
through steam distillation.

Warner et al (7,8) determined the extent of volatilization and
decomposition of antioxidants and the extent to which unknown
antioxidant decomposition products are retained by the food by
introducing ring-labelled [^{14}C]BHA, ring-labelled [^{14}C]TBHQ, and [7-
^{14}C]BHT in deep fat frying under actual cooking conditions. BHA or
its decomposition products were largely (over 80%) retained by the
lard. For [7-^{14}C]BHT, the radioactivity decreased markedly with
frying. After the equivalent of four batches of French-fried
potatoes, 50% of the label was retained, while after 12 batches less
than 20% remained in the lard. BHT, which was found to be the most
volatile of the phenolic antioxidants, volatilized to the extent of
28%, while the corresponding losses for BHA and TBHQ were 11% and
6.1%, respectively.

Unlike BHA and BHT, which were stable in unheated lard
throughout the experiment, TBHQ appeared to be unstable in lard even
at room temperature. Furthermore, TBHQ was undetectable in the
heated lard, and was present only in a minute amount (2%) in the
volatiles. Overall, the heat treatment resulted in virtually total
decomposition (99%) of the added TBHQ in lard, and the resultant
products appeared to be less polar than TBHQ itself.

Perhaps the most comprehensive study on the thermal
decomposition of phenolic antioxidant was done recently by Hamama
and Nawar (9). These authors not only identified products of
thermal decomposition but also establish the order of stability for
BHT, BHA, TBHQ, and propyl-gallate.

Labuza (10) reported that other factors, e.g., changes in
reaction pathways and interactions of food components, may play a
major role in influencing antioxidant potency under frying
conditions. Sherwin (11,12) reported that antioxidants absorbed in
food are lost during storage. These decreasing levels of
antioxidants may be attributed to the reaction of antioxidants with
reactive species in the food or the oil.

Thermal Decomposition of Antioxidants

BHT (2.6-di-tert-butyl-hydroxyl-toluene) and BHA (butylated
hydroxyanisol) form adducts with 2 peroxy free radicals at para
position as shown in Figure 1. Ingold (13) has proposed three
possibilities for the mechanism in which 4-alkyl-peroxycyclohexa-
dienone (APCD) was formed through reaction between BHT and two alkly
peroxy free radicals (Figure 1). Peroxy free radicals, like other
free radicals, are electron deficient and will tend to react with
the phenols at regions of high electron density. Therefore, peroxy
radicals will react with the phenolic end of the inhibitor, and for
this reason mechanism can be discarded in spite of the fact that
bulky tert-butyl groups hinder the inhibition process.

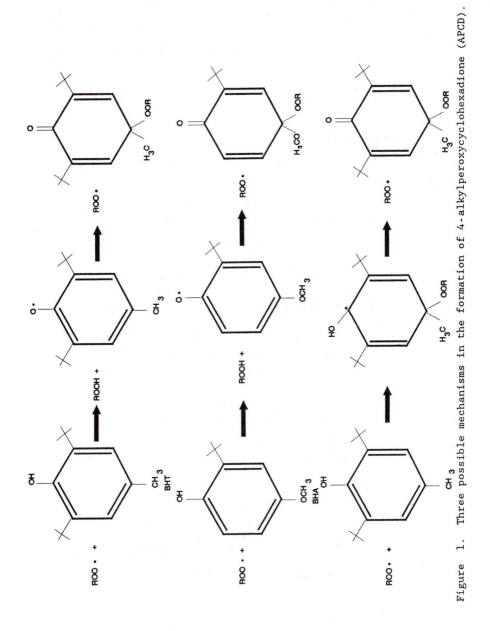

Figure 1. Three possible mechanisms in the formation of 4-alkylperoxycyclohexadione (APCD).

Howard and Yamada (*14*) isolated APCD to determine the
structure of the chain-ropagating species HROO•. APCD is thermally
and photolytically unstable. These reactions would be expected to
play an important role in the nature of the products formed during
thermolysis of BHT. Hodgeman (*15*) found the volatile products from
the thermal decomposition of the APCD at 80-200°C listed in Figure
2. APCD were observed to undergo thermal decomposition by both O-O
bond and C-O bond scission mechanisms, the relative proportion of
the two fragmentation reactions depending on the nature of the 4'-
substituent and on the temperature of thermal decomposition. Most
of the products formed on thermal decomposition of APCD are
accounted for by the reactions in Figure 5. Acetone, tert-butanol
and isobutene are obtained from secondary reactions of the tert-
butoxy and tert-butyl peroxy radicals. 2,6-di-tert-butyl-
benzoquinone (DBBQ) and 2,6-di-tert-butyl-4-methylphenol arise from
O-O and C-O bond scission reactions, respectively. However, they
could not identify the major products of molecular weight 234
obtained on pyrolysis of APCD at 200°C Figure 3.

   Koch (*16*) reported the formation of 2,6-di-tert-
butylbenzoquinone (DBBQ, 3,5-di-tert-butyl-4-hydroxybenzaldehyde
(DBHB), acetone and tert-butanol on thermal decomposition of APCD at
170°C and Buben and Pospisil (17) reported no change in APCD at 65°C
but at 200°C obtained several unidentified products, among which
2,6-di-tert-butylbenzoquinone (DBBQ) was not detected. Warner et al
(*7,8*) determined two thermolysis pathways of 2,6-di-tert-butyl-4-
hydroperoxy-4-methylcyclohexa-2,5-dienone (HBHT), primary oxidation
product of BHT. The first route leads to the predominant product
which is BHT itself. This observation suggests the intriguing
possibility that BHT can capture two chain-propagating alkylperoxy
radicals and in subsequent thermolysis reappear as the parent
antioxidant. The second route of decomposition, involving the
intermediate alkoxy radical, would yield the products given in
Figure 4. Only a trace of DBBQ (mol. wt. 220) was detected. Of the
molecular weight 236 compounds, 2,6-di-tert-butyl-r-hydroxy-4-
methylcyclohexa-2,5-dione (MHCD) was the predominant product. For
BHA, under the well-characterized reaction conditions typical of
organic reaction, BHA is readily oxidized to a coupling product,
3,3'-bis-(1,1-dimethylethyl)-5,5'-dimethoxy-1,1'-biphenyl-2,2'-diol.

   Rosenwald and Chenicek (18) obtained CHDP by the oxidation of
BHA with potassium ferricyanide, in which DHDP was a colorless
compound possessing a high melting temperature (225°C) and low
solubility. It was found that DHDP is not devoid of antioxidant
properties, since it possesses 33% at 0.005% concentration and 50%
at 0.02% concentration of antioxidant effectiveness contrasted to
the antioxidant effectiveness of BHA itself.

   Furthermore, Maga and Monte (*19*) succeeded in isolating and
identifying this DHDP (2',3-di-tert-butyl-2-hydroxy-4',5-dimethoxy-
biphenyl ether) and its structural isomers in fresh corn oil systems
containing 0.02% BHA at the temperature of 218°C as shown in Figure
5. These results are consistent with findings in the model systems
of Kurechi et al (20) previously described (Fig. 5).

   This substance was not detected in heated fats by TLC and
autoradiography (*7,8*). DHDP was formed in 2% BHA-containing rat

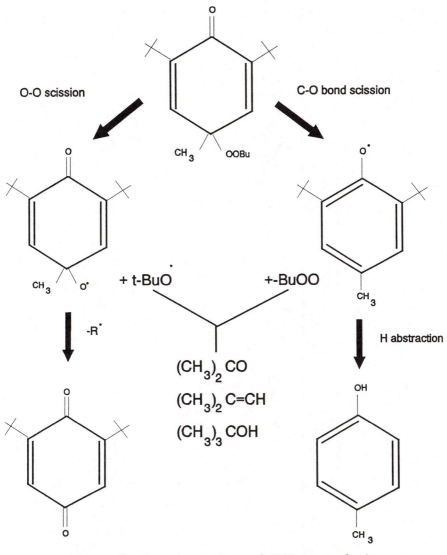

Figure 2. Reaction pathway of APCD on pyrolysis.

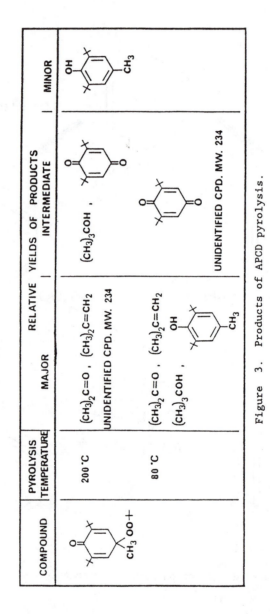

Figure 3. Products of APCD pyrolysis.

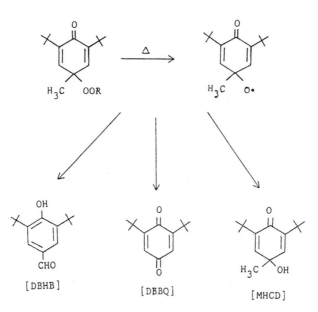

Figure   4.   Reaction of the Alkoxy radical formed by cleavage of
the oxygen bond in 2.6-di-tertbutyl-4-hydroperoxy-4-
methylcyclohexa-2,5-dienone to form products of
molecular weight 234, 220 or 236.

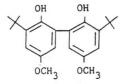
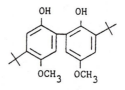

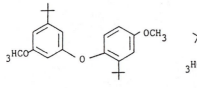
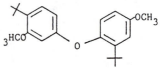

FROM   BHA

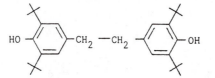

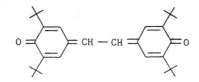

FROM   BHT

Figure   5.   Dimer formation from BHA and BHT.

chow which had been subjected to accelerated stability studies at
50°C for 4 days.
Costcove and Waters (21) reported that BHT could be oxidized
to dimers 1.2-bis-(3.5-di-tert-butyl-4-hydroxyphenyl)ethane and
3,3'-5,5'-tetra-tert-butylstilbene-4,4'luinone.   Daun et al (22)
isolated and identified these substances from photooxidized
polyethylene film.   Leventhal et al (23) separated only
stilbenequinone from vegetable oil heated at 180.5°C for 262 hours.
Warner et al (7,8) reported that TBHQ and BHA are oxidized to
tert-butylbenzoquinone (2TBBQ) at 190°C.   In addition, quinones are
subject to a wide variety of reactions of the 1,4-Michael addition
type as shown in Figure 6.   The reactions of TBBQ with various
nucleophilic substances (REH), such as water, carboxylic acids,
thiols, amines and alcohols, result in the substituted tert-
butylhydroquinone (SBHQ).   Apparently the SBHQ must be an excellent
antioxidant, since it is a sterically hindered hydroquinone with a
lower oxidation potential than TBHQ, because of the ring
substitution of the atom, E, with one or more unshared electron
pairs.   This may explain why TBHQ yields a fried product with a
longer shelf-life than the identical product fried in antioxidant-
free oil,   even though TBHQ is very unstable in heated fat (7,8,24).
Warner et al have also proposed that TBBQ as an initial product has
only a transitory existence and would undergo extensive
decomposition in a frying medium because hydroxyquinone (where REH =
H$_2$0) are unstable compounds.
Let us consider three major phenolic antioxidants: (a) tert-
butyl-hydroquinone; (TBHQ) (b) tert-butyl-hydroxytoluene; and tert-
butyl hydroxyanisol.

(a)   Tert-Butyl-Hydroquinone.   Volatility of TBHQ and TBBQ at 175°C
was measured by weight changes.   This is shown in Figure 7.   As
heating progressed, TBHQ peak was decreased, apparently in contrast
to the increased TBBQ.TBBQ was evaporated around 20% within 1 hour,
50% within 3 hours, and 99% within 20 hours, while TBHQ was
evaporated 50% at over 20 hours heating and around 10% was left even
after 70 hours heating.   This much higher volatility of TBBQ is
thought to be closely related to the antioxidant effectiveness of
TBHQ.
Thermolized products of TBHQ was detected G.C./M.S.   TBBQ was
detected as the major degradation compound and many dimerized
compounds with ether linkages were detected in which these products
may alter the typical TBHQ action in the food system.   Thermalized
products were dimerized as TBHQ itself.   (3-tert-butyl-4-hydroxy)-
phenyl acetate was really a minor peak in this sample.   When this
sample was heated 50 hours or more, it became a major peak.   The
derivation of (3-tert-butyl-4-hydroxy)-phenyl acetate is difficult
to visualize.
The most important carry-through effect of TBHQ must be
derived from the TBHQ and TBBQ recycling system.   This result could
be considered a 1.4 Michael addition type of reaction, which is
proposed by Warner et al (8) in that TBBQ was easy to be substituted
by nucleophiles at free ortho position, and is not favorable at high
temperatures.   This may explain why TBHQ shows higher antioxidant
effectiveness in the system during heat treatment.

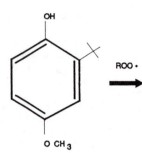
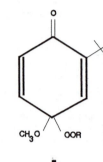
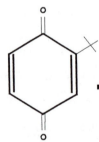
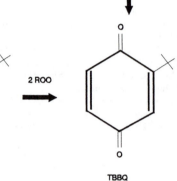
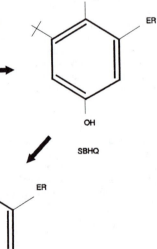

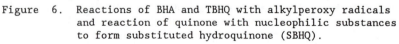

Figure 6. Reactions of BHA and TBHQ with alkylperoxy radicals and reaction of quinone with nucleophilic substances to form substituted hydroquinone (SBHQ).

Considerable amount of TBBQ was converted into TBHQ. However, we could not detect methoxy group substituted TBHQ. Other thermolized products could not be understood. But from these results, the TBHQ and TBBQ recycling system seems to make carry-through effect. The compound esterified with acetic acid was the major peak. In the case of fatty acid, the esterified tail is much longer than this. Therefore, it is believed that a large amount of phenolic antioxidant reacts with the carboxyl group of fatty acid which may lower the antioxidant activity.

(b) Tert-Butyl-Hydroxytoluene. Thermolized products of BHT in the absence of hydroperoxy or alkylperoxy radicals are shown in Figure 8. The identification of 3,3',5,5'-tetra-bis-(tert-butyl)-4,4'-dihydroxyl-1,2-diphenylethane as a major product under conditions which do not react with hydroperoxy radicals suggests that this dimer formation could be the preferred pathway of BHT thermolysis. This can be inferred that a phenoxy radical resulting from thermolysis will be dimerized.

The decay mechanism of the phenoxy radical has been the subject of much debate. Stebbins et al (25) advocated first order kinetics, supporting rearrangement of phenoxy radical to the hydroxybenzyl free radical as shown in Figure 8. The subsequent dimerization of the hydroxybenzyl free radical leads to the same product as the dimer identified in our laboratory. However, Bauer et al (26) obtained second order kinetics, supporting the disproportionation of phenoxy radical to the parent phenol (BHT) and quinone methide Figure 9.

Further reaction of quinone methide results in the final products of 3,3',5,5'-tetra-bis-(tert-butyl)-4,4'-dihydroxyl-1,2 diphenyl ethane and its oxidized form, stilbenequinone. In the thermolysis mechanism, the first order kinetics of dimer formation appears to be more plausible, since quinone methide was not determined and, furthermore, two forms of dimer should be of equal quantity if the thermolysis mechanism follows second order kinetics.

Reaction products of BHT may depend on the presence of hydroperoxy radicals, and on the temperature and the particular condition of reaction. Leventhal et al (23) isolated stilbenequinone from vegetable oil containing BHT at 190°C for 262 hours which means that dimerization, as well as 4-alkyl peroxy cyclohexadienone (APCD) formation still occurred even though hydroperoxy radical is present in the system. Dimerization and APCD formation are believed to be a competitive relation depending on reaction condition. The products are thought to be formed by several different types of reaction.

First, C-O bond scission at C-4 position contributes to BHT itself as a major product whose result is consistent with observation of Warner et al (8). Regeneration of parent phenol (BHT) might be a reason for carry-through effect.

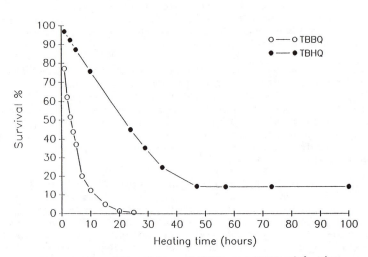

Figure 7. Volatility of TBHQ and TBBQ with time.

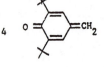

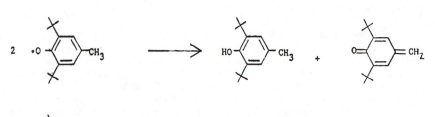

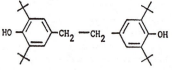

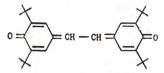

Figure 8. Dimerization of BHT by second order kinetics.

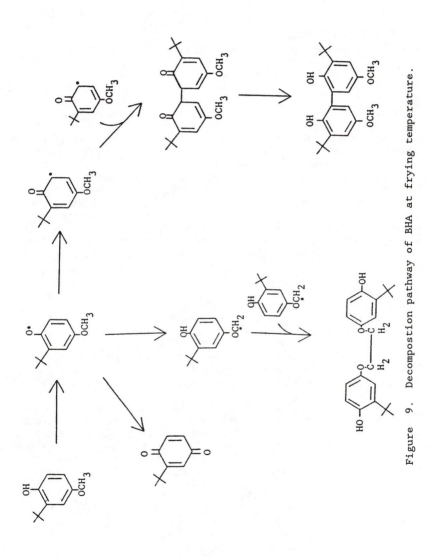

Figure 9. Decompostion pathway of BHA at frying temperature.

Second, 2,6-di-tert-butyl-benzoquinone (DBBQ) arose from C-C
bond scission at C-4 position. Also, trace amounts of 2,6-di-tert-
butyl-hydroquinone (DBHQ), reduced form of DBBQ, were detected.
Third, trace amounts of quinone methide (QM) were formed
through entire oxidation (C-O bond scission at C-4 and
dehydrogenation). QM may react easily with other compounds, such as
water, because of its relatively higher reactivity. This may
account for the 2,6-di-tert-butyl-4-hydroxymethol-phenol (DBHP)
which was detected as an intermediate decomposition product of HBHT.
Fourth, intermediate amounts of 3,5-di-tert-butyl-4-
hydroxybenzaldehyde (DBHB) were detected which suggests that this
phenol oxidized moderately to an aldehyde-type derivative rather
than to an acid as might be expected (27).
Fifth, 2,6-di-tert-butyl-4-hydroxy-4-methylcyclohexa-2,5-
dienone (MHCD) arose from direct hydrogen abstraction. Also, a
considerable amount of 3,3',5,5'-terra-bis-(tert-butyl)-4,4'-
dihydroxyl-1,2-diphenylethane, which must result from either first
order kinetics or second order kinetics as previously described, was
detected. Unidentified compounds of M.W. 414, 442, 358 and 454,
believed to be dimers, were detected.
When BHT and acetic acid were heated at equimolar
concentrations (1M), derivatives, esterified with acetic acids as
might be expected, were not determined. This was considered because
phenols having tertiary butyl groups in the 2 and 6 positions
exhibit steric hindrance and also do not show the usual chemical
properties of phenol, which means that these phenols do not form
derivatives, such as the acetate, by the method usually applicable
to phenol.

Tert-Butyl-Hydroxyanisol

Thermolized products of BHA in the absence of hydroperoxy or
alkylperoxy radicals are shown in Figure 10. 3,3'-bis-(1,1-
dimethylethyl)-5,5'-dimethoxy-1,1'-biphenyl-2,2'-diol (DHDP), and
its isomer (ethylene glycol type) were detected as major products.
Trace amounts of TBBQ were identified. The phenolic antioxidant
having a free ortho position is able to dimerize as shown in Figure
19.
The formation of TBBQ from BHA could arise from direct O-C
scission at methoxy groups in the para position and direct reaction
with molecular oxygen rather than pathways previously described due
to absence of peroxy free radicals. If peroxy radicals were present
in the system dimerization to DHDP and its isomers and adduct
formation with peroxy radical should be a competitive relation.
Even though BHA has free ortho position, dimerization through
mechanism of free radical formation at methoxy carbon atom is
analogous to dimerization of BHT. It is possibly inferred that two
dimers, DHDP and its ethylene glycol type of isomer, can form adduct
with peroxy free radicals at each para position.
When BHA and acetic acid were heated at equimolar
concentrations, derivatives esterified with acetic acid were
identified as major products (Figure 20). This is in contrast with
the result of BHT which suggests that absence of t-butyl group at 6
position leads to less sterid hindrance than that of BHT having two

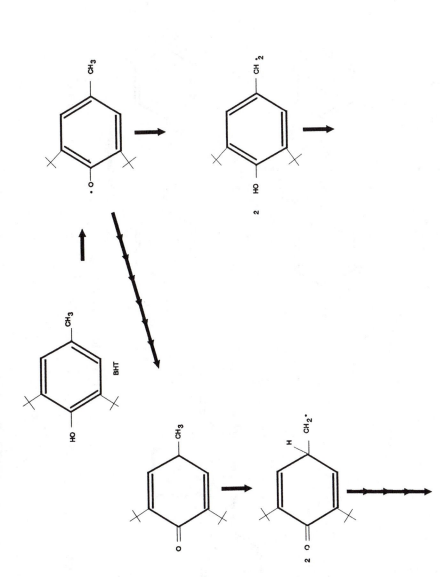

Figure 10. Dimerization of BHT by first order kinetics. *Continued on next page*

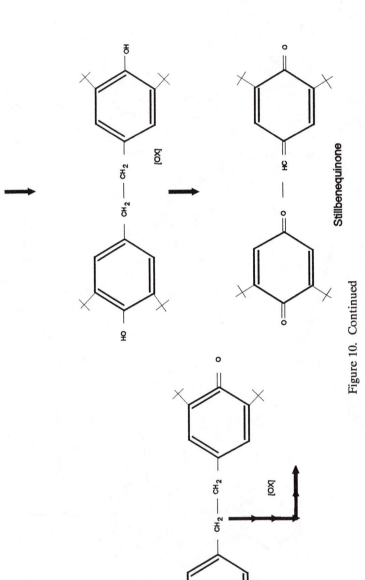

Figure 10. Continued

t-butyl groups at 2 and 6 positions. As described earlier, TBHQ forms esterified derivative with carboxyl group which is consistent with the result of BHA due to structural similarity (BHA and TBHQ have free ortho position).

Antioxidant Effectiveness

Antioxidant effectiveness at frying temperature was determined using a mixture of palmitic acid (16:0) and linoleic acid (18:2) by gas chromatography. The loss of unsaturated fatty acids was characteristic of oxidative deterioration. The results were expressed as the peak area ratio of linoleic acid (18:2) relative to total of linoleic acid (18:2) and palmitic acid (16:0). Phenolic antioxidants are much less effective at frying temperature than at normal storage temperatures. However, the antioxidants obviously showed a marked protective effect. BHT had a relatively lower stabilizing effect on polyunsaturated fatty acid than the others. TBHQ showed the most effective antioxidant action after 40 hours heating but BHA was more effective until around 40 hours heating. Thus, as the reaction progresses, derivatives (such as DBBQ, MHCD, DBHB, DBHP, and dimers from BHT) have low antioxidant potency. Derivatives formed from BHA were slightly more effective in lipid systems. However, TBHQ may produce derivatives having higher antioxidant potency, such as substituted tert-butylhydroquinone (SBHQ), with nucleophiles present in systems postulated by Warner et al, (7,8) or TBHQ and TBBQ could be recycled in the system as described earlier.

Loss of antioxidants during frying has been regarded as a factor contributing to the loss of antioxidant potency. Many factors, such as water boiled out of system, may accelerate volatilization of antioxidant. Moreover, volatilities of antioxidant molecules themselves may play a more important role for volatilization of antioxidant, which means that compounds with higher polarity have lower volatility regardless of their complicated interaction with lipids. BHT was formed to have the highest volatility due to lower polarity, which is consistent to observations of Warner et al (8) and Peled et al (28). Even though BHA was the least volatile until heated around 60 hours and volatilized around 94% at 100 hours heating, the volatilization of TBHQ was slower than that of BHA until 60 hours. However, final residue of TBHQ at 100 hours heating was 12% higher than that of BHA. The higher volatility of BHT could be overcome by incorporating the more polar groups in the phenol ring.

Literature Cited

1. Scott, G. *Atmospheric oxidation and antioxidants.* Elsevier Publishing Company, Amsterdam, London, New York. 1965
2. Prokorny, J. *In Autoxidation of Unsaturated Lipids.* H.W.S. Chan, ed. Academia Press: New York 1987, pp141-206
3. Freeman, I.P.; Padley, F.H.; and Sheppard, W.L. *J. Am. Oil Chem. Soc.* 1973 50;101
4. Augustin, M.A.; Berry, S.K. *J. Am. Oil Chem. Soc.* 1984, 61;873
5. Stuckey, B.N. *In Handbook of Food Additives,* T.E. Furia, ed. CRC Press, Inc.: Cleveland. 1972 pp. 139-150

6.   Peled, M.; Gutfinger, T.; Letar, A. *J. Am. Oil Chem. Soc.*
     1975 *26*;1655
7.   Warner, C.R.; Brumley, W.C.; Daniel, D.H.; Joe, F.L.; and
     Fazio, T. *Fd. Chem. Toxic.* 1986 *24*;1015.
8.   Warner, C.R.; Daniel, D.H.; Lin, F.S.D.; Joe, F.L.; Fazio, T.
     J. Agric. Food Chem. 1986 *34*;1.
9.   Hamama, A.A.; Nawar, W.W. *J. Agric. Food Chem.* 1991 *39*, 1063
10. Labuza, T.P. *CRC Critical Review in Food Technology.* 1971 *2*,
     pp. 355-405
11. Sherwin, E.R. *J. Am. Oil Chem. Soc.* 1976 *53*;430
12. Sherwin, E.R. *J. Am. Oil Chem. Soc.* 1978 *55*;809
13. Ingold, K.U. *J. Phys. Chem.* 1960 *64*;1636
14. Howard, J.A.; Yamada, T. *J. Am. Chem. Soc.* 1981 *103*;7102
15. Hodgeman, D.K.C. *J. Macromol. Sci. Chem.* 1980 *2*;173
16. Koch, J. Augew. *Macromol. Chem.* 1971 *20*;7
17. Buben, I.; Pospisil, J. *Colln. Chem. Commun.* 1975, *40*, 927
18. Rosenwald, R.H.; Chenicek, J.A. *J. Am. Oil Chem. Soc.* 1951
     5;185
19. Maga, J.A.; Monte, W.C. *Lebensm-Wiss. Technol.* 1977 *10*;102
20. Kurechi, T.; Kato, T. *Chem. Pharm. Bull.* 1983 *81*, 1982
21. Costcove, S.L.; Water, A.W. *J. Chem. Soc.* (London) 1951, 388
22. Daun, H.; Gilbert, S.G.; Giacin, J. *J. Am. Oil Chem. Soc.*
     1974 *51*;404
23. Leventhal, B.; Daun, H.; Gilbert, S.G. *J. Fd. Sci.* 1976
     41;467
24. Lin, F.S.; Warner, C.R. Fazio, T. *J. Am. Oil Chem. Soc.* 1981
     58;789
25. Stebbins, R. and Sicilio, H. Tetrahedron 1970 *26*:287.
26. Bauer, R.H.; Coppinger, G.M. *Tetrahedron* 1963 *19*;1201
27. Kim, M.C.; Pratt, D.E. *J. Food Sci.* 1990 *55*;848
28. Peled, M.; Gutfinger, T.; Letan, A. *J. Sci. Fd. Agric.* 1975
     26;1655

RECEIVED February 18, 1992

FLAVONOIDS AND CANCER PREVENTION

Chapter 16

Molecular Characterization of Quercetin and Quercetin Glycosides in *Allium* Vegetables

Their Effects on Malignant Cell Transformation

Terrance Leighton[1], Charles Ginther[1], Larry Fluss[1], William K. Harter[2], Jose Cansado[2], and Vicente Notario[2]

[1]Department of Biochemistry and Molecular Biology, University of California, Berkeley, CA 94720
[2]Department of Radiation Medicine, Georgetown University Medical Center, Washington, DC 20007–2144

Flavonol levels in the edible portions of *Allium* vegetables (leeks, shallots, green onions, garlic and onions) range from <0.03 to >1 g/kg of vegetable. Shallots contained uniformly high concentration of total flavonols, >800 mg/kg in each of five independent samples. However, onions varied widely in the amounts of flavonoids they contained. White onions contained no detectable flavonols, but 20 cultivars of yellow and red onions contained between 60 mg/kg and >1000 mg/kg. Individual samples of leeks, garlic, and scallions contained undetectable levels of flavonols.
The primary flavonols present in shallots were quercetin 4'-glucoside and quercetin aglycone; flavonols identified in onions were three quercetin diglucosides, quercetin 4'-glucoside, quercetin aglycone, and, in some cases, isorhamnetin monoglycoside or kaempferol monoglycoside. The relative concentrations of the various onion flavonols varied significantly in different onion cultivars. Treatment of vegetable flavonol extracts with human gut bacterial glucosidases (fecalase) efficiently deglycosylated the quercetin glucosides with concomitant release of free quercetin.
Quercetin selectively inhibits the growth of transformed-tumorigenic cells (*ras*/3T3 and H35) and prevents the neoplastic transformation of NIH/3T3 cells with the oncogene H-*ras*.

Flavonols are widely distributed in edible plants, primarily in the form of flavonol glycosides (*1*). It has been estimated that humans consuming high fruit and vegetable content diets ingest up to 1 g of these compounds daily (*2*). Glycosidases are found in both saliva and in intestinal tract bacteria that catalyze the removal of the sugar moieties from a wide array of flavonol glycosides, yielding the cognate aglycones (*3-5*). Among the most biologically active and common dietary flavonols is quercetin (3,3',4',5,7-pentahydroxyflavone).
Recent investigations of the potential effects of quercetin on human health began following the identification of quercetin as a mutagen in the Ames-test (*6*). The glycosides of quercetin were also found to be mutatgenic in the Ames-test, but only after activation by removal of the sugar moiety to release

0097–6156/92/0507–0220$06.00/0

free quercetin (*7*). These results were of general interest because compounds that are mutagens in the Ames-test are frequently also carcinogens.

We recently suggested that the mutagenic effect of quercetin could be specific to the nature of the Ames-test and other *in vitro* test systems (*8*). In the presence of molecular oxygen and redox metals, such as iron and copper, quercetin can autooxidize producing hydrogen peroxide and superoxide (*9*) which can react by non-enzymatic Haber-Weiss reactions to form hydroxyl radicals, a highly reactive and mutagenic species. We have proposed that *in vitro* mutagenicity tests, like the Ames-test, provide conditions and components required for generation of hydroxyl radicals, but that such conditions and components usually do not exist under normal physiological conditions.

Thus it is not surprising that although quercetin is mutagenic in several *in vitro* test systems, it does not appear to be carcinogenic. It is not reproducibly carcinogenic in rats, mice or hamsters (*10-14*). Although some reports suggest that quercetin induces cancer in rats (*15-16*), these observations have not been confirmed. It does not possess cancer initiating (*17*) or promoting activity in two-stage skin carcinogenicity assays (*13,18*). It does not act as a tumor promoter in rat urinary bladder (*19*) or liver carcinogenesis assays (*20*), nor does it induce unscheduled DNA repair in the Williams' hepatocyte system (*21*). Quercetin also does not exhibit teratogenic activity in rats or mice at 400 mg/kg (1 mmole/kg) (*22*).

To the contrary, quercetin has been reported to exhibit several types of anticarcinogenic activities:

Inactivation of carcinogens by quercetin. Quercetin significantly reduces the carcinogenic activity of several cooked food mutagens including bay-region diol epoxides of benzo[a]pyrene, and heterocyclic amines (*23-25*). These carcinogens require activation by cytochrome P-450 dependent mixed function oxidases; quercetin inhibits these oxidases *in vitro* (*26*). Quercetin also inhibits the binding of polycyclic aromatic hydrocarbons to DNA *in vitro* (*27*), and in epidermal and lung tissue of SENCAR rats (*28-29*). Of particular importance in defining the *in vivo* correlation between dietary quercetin inputs and inhibition of carcinogenesis is the report that quercetin inhibits the induction of mammary cancer by 7,12-dimethylbenz(a)anthracene and N-nitrosomethylurea in rats fed quercetin containing diets (*30*).

Tumor promoting compounds which are also inhibited by quercetin include 12-O-tetradecanoylphorbol-13-acetate (TPA) a phorbol ester (*31-33*), and aflatoxin B1 (*34*). Quercetin also inhibits the cytotoxic effects of T-2 mycotoxins (*35*).

Inhibition of cancer associated enzyme activities. Quercetin directly inhibits enzymatic activities associated with several types of tumor cells *in vitro* and *in vivo* including the calcium and phospholipid-dependent protein kinase (protein kinase C) (*36-40*); TPA-induced lipoxygenase and ornithine decarboxylase in mouse epidermal tissue (*41*); the high aerobic glycolytic enzyme levels of tumor cells (*42*); the activity of the oncogene pp60 v-src, and other tyrosine kinases (*43*); cyclo-oxygenases and 15-lipoxygenase (*44*); cyclic GMP phosphodiesterase (*45*); and cytochrome P-450/P-448 monooxygenases (*26*). Quercetin also inhibits cigarette smoke inhalation toxicity (*46*).

Synergistic enhancement of known anti-proliferative agents. Quercetin synergistically enhances the antiproliferative activity of the anticancer agents cis-diamminedichloroplatinum(II) (cis-DDP) (cultured Walker rat sarcoma cells and Ehrlich cells), nitrogen mustard (*39*), and busulphan (*47*) in human tumor cell culture systems. At 100 μM concentrations of quercetin, the antiproliferative activity of cis-DDP and nitrogen mustard were enhanced 5-10 fold. Since toxicity of these antiproliferative agents is high at therapeutic concentrations, their synergism with quercetin may be of clinical importance.

Other quercetin mediated effects. Quercetin is reported to inhibit allergic and inflammation responses of the immune system (48-52), to inhibit the growth of many bacteria and fungi (53-54), and to be a vasoprotective and antithrombotic agent (55).

Given these effects on human health, there is increasing interest in foods and beverages that contain quercetin. *Allium* vegetables contain high levels of quercetin and quercetin glycosides (1,2). We report here that these vegetables provide a useful model system to investigate the genetic and geographical factors which determine the quantitative levels and individual species of plant flavonols.

Aglycone flavonols have greater pharmacological activity than glycosylated species (3-5). However, flavonol glycosides are the most abundant species in edible plants (1-5). For this reason we have investigated the *in vitro* molecular processing of onion flavonol glycosides by human gut bacterial enzyme systems.

The role of quercetin or other compounds, in dietary anticarcinogenesis is uncertain unless a specific molecular target interaction can be established which is of relevance to the cancer-generating processes. We report here the development of a DNA transfection system which allows the introduction of a single activated human oncogene (H-*ras*) into a "normal" cell line (NIH/3T3) leading to the malignant transfromation of these cells. We show that quercetin is capable of inhibiting this malignant transformation model when present in the cell culture medium at μM concentrations.

Materials and methods

Flavonol and vegetable sources. Cyanidin-chloride, fisitin, myrecetin, and quercetrin were generously provided by Dr. McGregor of the U.S.D.A. (Albany, CA). Other flavonol standards were obtained from commercial sources. Apignin, catechin and malvidin-3,5-diglucoside (Sigma), quercetin-3,5-rutinoside (Pfaltz and Bauer) and quercetin (Kodak and Sigma). Twenty cultivars of onions grown in different regions of the United States and Mexico were provided by Rio Colorado Seeds Ltd (Table I). Other *Allium* vegetables were purchased locally.

Flavonoid Extraction. 250-1000 g of raw vegetable with inedible portions removed (the removed portion amounted to 10-15% of weight), were chopped, mixed with 1 ml butanol containing 0.75% glacial acetic acid/g of tissue, and blended for 2 min, incubated overnight at 4°C, and filtered through Whatman #1 filter paper. The filtrate was taken to dryness under vacuum at 65°C. The dried material was resuspended in 2.5-10 ml DMSO, and stored in foil covered screw-capped tubes. Cooked (boiled or fried) onions were extracted using an identical method. Whole onion skins (10 g) were extracted with 250 ml butanol containing 0.75% acetic acid. The butanol fraction was dried under vacuum at 65°C, and the residue was resuspended in 2.5-10 ml DMSO.

Column chromatography on Sephadex LH-20. Flavonol DMSO extracts (1 ml) were mixed with 1 ml methanol containing 7.5% glacial acetic acid and applied to a Sephadex LH-20 column (2.5 x 65 cm). Fractions were eluted with a methanol/7.5% glacial acetic acid mobile phase at 1 ml/min, and 200 fractions of 3.5 ml were collected. The absorbance of the fractions at 365 nm was

Table I. Estimated Flavonoid Content of Onions and Other *Allium* Vegetables

Vegetable Variety (Origin)	Flavonoid Content (A_{365} units/g)
Yellow onion	
Utah Hard Yellow Globe (Bakersfield, CA)	34.4
Yellow segregant from white dehydrator (Bakersfield, CA)	15.3
Rio Ringo (Mexico) onion #1	13.5
onion #2	11.5
onion #3	12.6
RCS 1506 (Rio Grande Valley, TX)	12.3
Rio Ringo (Imperial Valley, CA)	12.2
RCS1542 (Rio Grande Valley, TX)	11.8
Rio Ringo (Rio Grande Valley, TX)	10.6
RCS 1502-1 (Mexico)	10.3
RCS 1507 (Mexico)	9.6
Rio Enrique (Mexico) onion #1	9.2
onion #2	9.3
onion #3	8.7
RCS 1506 (Mexico)	9.3
Rio Honda (Imperial Valley, CA)	9.1
Yellow Sweet Spanish (Bakersfield, CA)	8.7
Texas Early Gruno 502 (El Centro, CA)	8.5
Yellow Bermuda #1 (El Centro, CA)	8.2
Yellow Bermuda #2	8.1
Rio Bravo (Rio Grande Valley, TX)	7.7
RCS 1502-1 (Rio Grande Valley, TX)	7.6
Rio Enrique (Rio Grande Valley, TX)	6.7
RCS 1542 (Mexico)	6.2
Ringer Select (Mexico) onion #1	7.5
onion #2	5.8
onion #3	6.2
Sweet Vidalia (Vidalia, GA)	6.1
RCS 1553 (Rio Grande Valley, TX)	5.8
Rio Estrella (Imperial Valley, CA)	3.5
Deming (New Mexico)	2.1
Red onion	
California Early Red (Bakersfield)	22.9
Southport Red (Hollister)	9.6
White onion	
White #1	<0.1[a]
White #2	<0.1
White boiling	<0.1
Shallot	
Shallot #1	39.6
Shallot #2	27.0
Garlic	<2.9[b]
Leeks	<0.3[b]
Scallions (green onions)	<1.2[b]

[a]White onions contained no detectable absorbance at 365 nm. The limit of detection is 0.1 A_{360} units/g of vegetable.

[b]Extracts of these vegetables contain complex mixtures of non-flavonoid components that absorb at 365 nm. Complete separation of these interfearing components was not possible using LH-20 column chromatography. The measurments presented reflect absorbance by both flavonoids and non-flavonoid contaminants, and thus represent upper limits of relative flavonoid content.

measured in a Beckman DU Spectrophotometer. Fractions from each peak were pooled, dried under vacuum at 67°C, and stored in the dark at -20°C.

XAD-2 column purification of flavonoids for HPLC. 2-5 ml of extract was passed over an XAD-2 column, and then dried with nitrogen gas. The flavonoids were eluted with 10 ml of acetone. The eluted material was dried in a heating block at 67°C under a stream of nitrogen. The dried material was resuspended in 0.2-0.5 ml of DMSO. We had previously demonstrated that XAD-2 purification procedures yielded greater than 80% recovery of quercetin and rutin (8).

UV/visible light spectra. UV-visible spectra (240-440 nm) of samples in methanol/7.5% acetic acid were determined using a Beckman DU-7 spectrophotometer.

Thin layer chromatography of flavonol extracts. Samples of flavonol extracts (20-50 μg material in 5-20 μl) were spotted onto cellulose thin-layer plates (Macherey-Nagel Polygram Cell 300 UV254, 20 x 20 cm x 0.1 mm) and developed with one of the following solvents: (1) 15% acetic acid; (2) water; (3) phenol:water (3:1); (4) butanol:acetic acid:water (4:1:5, top layer); and (5) acetic acid:methanol:water: acetone (25:15:50:5). Compounds were detected under long- and short-wavelength UV light. Their relative mobilities and fluorescence under UV light were compared to reference compounds and to published data (56).

HPLC analysis of flavonols. The HPLC apparatus consisted of a Beckman 421 controller, Beckman 160 absorbency detector (254 nm), Beckman 112 solvent delivery pumps, Beckman mixer (340 organizer with an injection attachment), and a Hewlett Packard 3390A integrator. Ultrasphere ODS or Ultrasphere IP (5 μdp, 4.6 mm ID x15 cm) columns have been used with similar results. Solvents were degassed under vacuum, in a sonicator bath prior to use. For isocratic quantitation, the solvent mixture was 48% methanol, 44.5% water, and 7.5% acetic acid. Samples were processed by absorption to, and elution from XAD-2 resin columns. Typically 10-20 μl of the concentrated XAD-2 extract, dissolved in DMSO, was injected. The solvent flow rate was set at 1 ml/min and the pressure at the pump at 1.60 KPSI. The method was standardized with pure quercetin and rutin reference standards. The standard deviation for measurements was +4%. The detection limit of the procedure was in the range of 0.1 mg quercetin/l. The high resolution HPLC gradient separation system employed a mixture of two solvents: (A) 90% acetonitrile, 10% glacial acetic acid, and (B) 7.5% glacial acetic acid in water. Following injection of 10-20 μl of sample into the column, the percentage of solvent B decreased from 100% to 10% over a 60 min separation period (57). The HPLC apparatus was attached to a LKB 2140 Diode Array Rapid Spectral Detector, and a Hewlett Packard 3390A Integrator. UV-vis spectra were recorded and areas under curves quantitated by this equipment.

Mass spectroscopy of flavonols. Samples from LH-20 chromatography were analyzed by electron ionization mass spectroscopy. 3-10 mg of dried compound were resuspended in 0.5 ml methanol, and analyzed by the U.C. Berkeley Mass Spectroscopy Laboratory.

Fecalase preparation. We have modified the method of Ames to produce high specific activity human fecal flavonol glycosidase enzyme preparations (fecalase). Feces was suspended in breakage buffer (50 mM phosphate buffer, pH 7.4, and 2 mM dithiothreitol) (buffer:feces 2:1 v/v). The slurry was sonicated until 95% of the fecal bacteria were disrupted as determined by phase contrast microscopy. The suspension was then centrifuged at 15,000 x g for 20 minutes. The supernatant fraction was withdrawn and brought to 80% saturation with ammonium sulfate, maintained at pH 7.0. The solution was again centrifuged (18,000 x g for 30 minutes) and the pellet was resuspended in breakage buffer to a protein concentration of 25 mg/ml. Following extensive dialysis against breakage buffer, the solution was sterilized by passage through a 0.45 micron low protein binding membrane filter. The sterile fecalase solutions could be stored at -80°C for several months. The specific activity of the fecalase preparations was assayed by measuring cleavage of the synthetic nitrophenyl-glycoside substrate o-nitrophenyl β-D-glucoside (*58*).

Cell culture and quercetin treatments. The following cell lines were used: mouse NIH/3T3, primary rat embryo fibroblasts (REF), H35 rat hepatoma, and c-H-*ras*-1 transfected NIH 3T3 cells. All cell lines were cultured in DMEM medium supplemented with 10% fetal calf serum and antibiotics. Quercetin was dissolved in ethanol at 3.34 mg/ml, and varying amounts (up to 35 μg/ml) were added to the cells at a density of 2.5 x 10^5 cells per 35 mm dish. Every three days the cells received fresh medium containing the appropriate concentration of quercetin.

Cell growth, toxicity and transfection assays. Viable cell counts were used to determine the growth curves of the cell lines under study. LD_{50} determination were carried out by clonogenic assays performed in the presence of increasing concentrations of quercetin. These experiments were performed twice, each time in triplicate. Transfection were conducted by the calcium phosphate/DNA precipitation method (*59*) as modified by Notario (*60*).

Results

The concentration of flavonols in *Allium* vegetables. Butanol phase partitioning was found to efficiently extract flavonols from *Allium* vegetables. Alternative methods of extraction employing ethyl acetate, methanol or ethanol yielded the same or lower levels of recovery as determined by absorbance at 365 nm (data not shown). The relative concentration of flavonols in the butanol extracts could be estimated from the absorbance of the extracts at 365 nm (Table I).   Individual flavonols absorb characteristically at different wavelength peaks between 345-375 nm, and possess varying coefficients of extinction (ξ). However, all of them absorb to some extent at 365 nm. This specific wavelength corresponded to the maximal absorbance value in extracts from onion and scallions (reflecting the high levels of quercetin 4'-glucoside, which absorbs maximally at 365 nm). Thus, the measurement of absorbance at 365 nm provides a useful semi-quantitative estimation of total flavonoid content. Fortunately, few other compounds in yellow onions absorb significantly at this wavelength. However, red onions contain high levels of related anthrocyanins that contribute to absorbance at 365 nm, and yield flavonol estimates that are artificially high. The garlic and scallion samples examined also contained levels of non-flavonol compounds which absorbed at 365 nm.

 White onions contained undetectable levels of flavonols. In yellow and red onions, the total flavonol concentration differed significantly between onion types, ranging between a low of 2.1 A_{365} units/g of Deming onions from New Mexico to a high of 34.4 A_{365} units/g of Utah Hard Yellow Globe onions from Bakersfield, a variation of over 16-fold. Shallots contained flavonol levels equivalent to the highest levels found in onions. Other *Allium* vegetables contained significantly lower concentrations (Table I). The flavonol contents of individual onions of the same cultivar, grown in the same area, and from the same growing season, were compared for Rio Ringo, Rio Enrique, and Ringer Select onions grown in Mexico (cultivars selected for their high, medium and low concentrations of total flavonols). The Rio Ringo samples averaged 12.5 A_{365}/g of onion with a standard deviation among individual onions of 1.0 A_{365}/g. Similarly the Rio Enrique onions averaged 9.2 A_{365}/g with a standard deviation of 0.3, and the Ringer Select averaged 6.5 A_{365} with a standard deviation of 0.35 (Table I).

Samples of Rio Ringo, Rio Enrique, RCS 1542 and RCS 1502-1 were obtained from more than one region. Rio Ringo onions grown in Mexico, the Imperial Valley of California, or the Rio Grande Valley of Texas were all relatively high in total flavonol concentration, averaging 12.1 A_{365}/g with a standard deviation of 2.25 A_{365}/g. Rio Enrique and RCS 1502-1 onions grown in Mexico and the Rio Grande Valley produced medium levels of flavonols when grown in both areas, varying by 30-40%. Only the RCS 1542 grown in Mexico and Rio Grande differed by as much as 2-fold (Table I). Thus, while both site of growth and the genetic nature of the cultivar affect the content of flavonols in onions, genetics characteristics appear predominant in determining the observed 16-fold variation in the flavonol content of onions. Quantitation of total and individual flavonol compounds in butanol extracts was also derived from HPLC and LH-20 column analyses. These calculated concentrations were consistent with the estimates based on absorbance data.

Identification of the flavonols in onions and shallots. Representative LH-20 column elution profiles for shallots and several onion cultivars examined are presented in Figures 1 and 2. The peaks from the LH-20 column were analyzed using UV-vis absorption spectra, mass spectra, and mobility and appearance in thin-layer chromatography. Greater than 80% of the flavonols in most onion cultivars are present as quercetin diglycosides, monoglycoside, and quercetin aglycone.

Three diglycosides were identified by HPLC and TLC analysis. All three-compounds had a molecular weight of 626 as determined by mass spectroscopy. Thin-layer chromatography identified two diglycosides having the mobility and chemical characteristics of quercetin 4',7-diglucoside and quercetin 3,7-diglucoside (*56*). These and quercetin 3,4'-diglucoside have been reported as being present in onions (*1,61*). Only one quercetin monoglycoside was identified (although there were monoglycosides of other flavonols). By mass spectroscopy its molecular weight was 464. In thin-layer chromatograms, the compound comigrated with quercetin 4'-glucoside. This compound has been reported in the literature as a major component of onions (*61*). Quercetin was identified by TLC, HPLC, and UV-vis spectral analysis. It has also been identified in onions previously (*61*). In addition, two other flavonols were sometimes present in significant amounts, and they were tentatively identified as an isorhamnetin monoglycoside and a kaempferol monoglycoside, by mass spectroscopy, TLC, and HPLC.

Following identification of the primary flavonols in onions and shallots, it was possible to calculate the concentration of individual and total flavonols from HPLC or LH-20 separations. The concentrations of specific quercetin compounds differed significantly among onion varieties (Table II, Figures 2). In all onion samples, quercetin 4'-glucoside was the major quercetin containing compound present, but its concentration ranged from less than 40% to more than 80% of the total quercetin containing compounds by weight, ranging between 30-700 mg/kg. Free quercetin concentrations ranged between 1-76

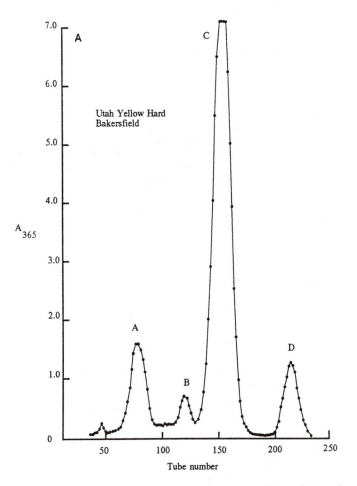

Figure 1. Representative LH-20 column elution profiles of A_{365} absorbing compounds from selected onion cultivars. (A) Utah Yellow Hard grown in Bakersfield , extract from 0.5 g of onion tissue. This sample produced the highest levels of flavonols. The peaks were identified as quercetin diglyco-sides [A], isorhamnetin glucoside [B], quercetin 4'-monoglucoside [C], and quercetin [D]. (B) Deming onion from New Mexico, extract from 1 g onion tissue. This strain produced the lowest level of flavonols. (C-E) Onions of intermediate quercetin content, extract from 1 g of tissue. The shoulder on the quercetin 4'-glucoside peak is kaempferol-glucoside.

Continued on next page

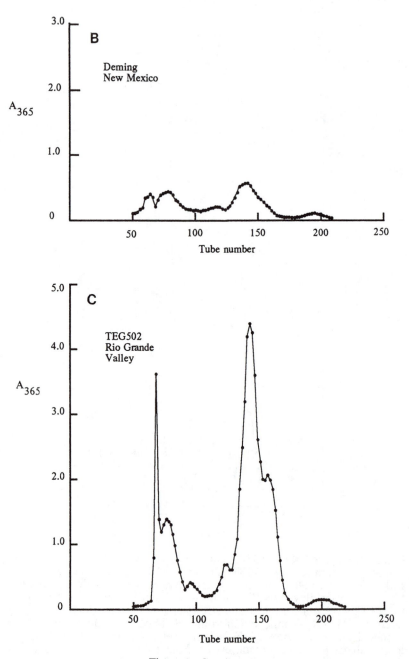

Figure 1. Continued

Continued on next page

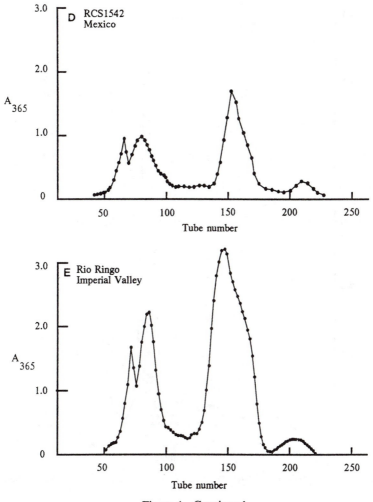

Figure 1. Continued

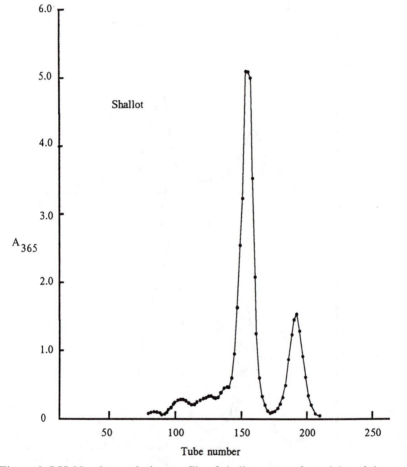

Figure 2. LH-20 column elution profile of shallot extract from 0.3 g of tissue.

Table II. Quercetin and Quercetin Glycoside Content of Yellow and Red Onion
Varieties and Shallots

Vegetable Variety (Area)[b]	Quercetin Glycosides Diglucoside	Quercetin Glycosides Monoglucoside	Quercetin
Yellow onion			
Utah Hard Yellow Globe (B)	319	701	76
Yellow segregant from white dehydrator (B)	269	264	9
Rio Ringo (M)	248	230	2
RCS1506 (RG)	166	144[a]	10
Rio Ringo (E)	183	158[a]	10
RCS1542 (RG)	185	200	9
Rio Ringo (RG)	145	114[a]	2
RCS1502-1 (M)	76	111[a]	2
RCS1507 (M)	96	100[a]	1
Rio Enrique(M)	145	179	2
RCS1506 (M)	78	123	1
Rio Honda (E)	150	140[a]	7
Yellow Sweet Spanish (B)	94	162	3
Texas Early Gruno 502 (E)	88	190	3
Yellow Bermuda #1 (E)	100	158	9
Yellow Bermuda #2	38	195	9
Rio Bravo (RG)	126	115	3
RCS1502-1 (RG)	135	144	4
Rio Enrique (RG)	96	83	4
RCS1542 (M)	82	50	1
Ringer Select (M)	43	96	7
Sweet Vidalia (V)	71	68	
RCS1553 (RG)	93	107	5
Rio Estrella (E)	64	51	1
Deming (NM)	26	30	1
Red onion			
California Early Red	207	450	42
Southport Red	100	65	3
Shallot			
Shallot #1	100	890	120
Shallot #2	94	589	97

[a] Contained significant levels of kaempferol glycoside
[b] Grown in Bakersfield, CA (B), El Centro, CA (E), Rio Grande Valley, TX (RG), Vidalia, GA (V), New Mexico (NM), or Mexico (M)

mg/kg of onion. Neither boiling nor frying of onions affected the relative concentrations of the flavonols. The brown outer skins of onions and shallots contained significantly higher levels of flavonols than the edible portion of the vegetable, 2000-10,000 mg/kg of skins (data not shown). Two major flavonol compounds are present in the skins, quercetin and quercetin 4'-glucoside. The quercetin compounds of shallot differed from onions in that they contain relatively lower amounts of quercetin diglycosides. Quercetin 4'-glucoside and free quercetin constitute 85-90% of the flavonol content of shallots (Table II; Fig. 1).

Fecalase treatment of onion flavonol extracts. Human fecal bacterial glucosides (fecalase) significantly altered the flavonol elution profiles from LH-20 columns and HPLC (Figure 3). Quercetin glucosides and diglucosides were efficiently cleaved to the corresponding aglycone quercetin.

Toxicity and cell growth effects of quercetin. Clonogenic assays were used to establish the LD_{50} for quercetin with normal (NIH/3T3 and REF) and transformed/tumorigenic (*ras*/3T3 and H35) cells. Quercetin showed more toxicity to transformed cells than to normal cells. The LD_{50} values obtained were 6.1-7.8 μg/ml (REF) and 3.5-4.1 μg/ml (NIH/3T3) for normal cells versus 2.4-3.05 μg/ml (H35) and 1.0-1.3 μg/ml(*ras*/3T3) for transformed cells. In agreement with these data quercetin also demonstrated a selective inhibitory effect on the growth of transformed cells. As shown in Table III, at 4 μg/ml of quercetin inhibited the growth of NIH/3T3 cells by only 23%, whereas the growth of *ras*/3T3 was inhibited 58% and that of H35 cells was inhibited by 44%; at 8 μg.ml, the growth of *ras*/3T3 cells was inhibited by 79-88% while that of normal NIH/3T3 cells was inhibited by 40-43%.

Quercetin effects on neoplastic transformation of NIH/3T3 cells. Quercetin was added at the LD_{25} to NIH/3T3 at different times during the transfection protocol: during cell growth prior to the actual transfection, during the DNA precipitation and uptake into the cells, during growth for foci formation after transfection, or during the entire process. Controls without quercetin were kept for each of the times of addition. When a plasmid (pT24) carrying the human activated oncogene H-*ras* was used to transform NIH/3T3 cells, it was found that, regardless of the timing and pattern of addition, quercetin nearly completely blocked the appearance of transformed foci: 49 and 46 foci were obtained in the two independent experiments with untreated cells, whereas only 2 and 4 foci were found with quercetin-treated cells. Quercetin addition exclusively during DNA precipitation and uptake did not block the appearance of transformed foci, although it reduced slightly the transfection efficiency by 10-15%. The inhibitory effect of quercetin on the neoplastic transformation of NIH/3T3 cells was reversed by removal of quercetin from the medium two weeks after the original transfection; 50-55% of the foci (relative to the numbers obtained in the untreated cells) grew within 15 days in quercetin-free medium. The cells from foci obtained from untreated cultures and those from "reverted" cultures were tumorigenic in nude mice with identical lag periods and tumor growth rates.

Discussion

Epidemiological analysis suggests that the etiology of human cancer is significantly impacted by factors such as diet and "lifestyle." Doll and Peto (*63*) have suggested that 35-40% of all cancer mortality in the United States could be attributed to dietary factors. It is clear that dietary composition, beverage consumption, cigarette smoking, etc. are carcinogenic modifying factors largely

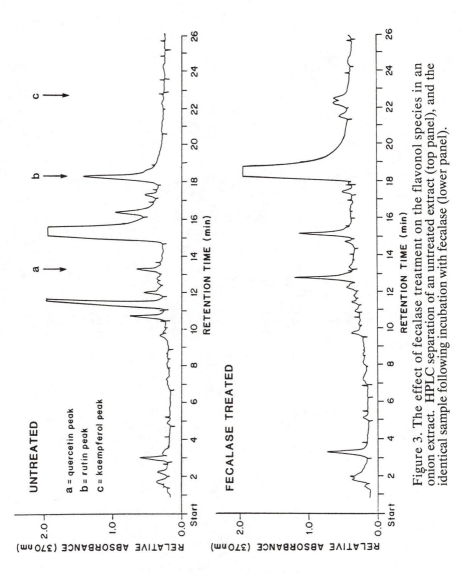

Figure 3. The effect of fecalase treatment on the flavonol species in an onion extract. HPLC separation of an untreated extract (top panel), and the identical sample following incubation with fecalase (lower panel).

Table III. Inhibitory Effect of Quercetin on the Growth of Normal and
Transformed Cells in Culture

Cell Type	Quercetin Concentration		
	4 μg/ml	8 μg/ml	12 μg/ml
NIH/3T3	23-28	40-43	60-62
ras/3T3	58-62	79-88	81-90
H35	44-52	54-62	56-61

Results are indicated as percentage inhibition relative to cultures which were
not exposed to quercetin, and show the range of effect in three independent
experiments.

within the control of each individual. The study of dietary factors which could interfere with the initiation or progression phases of carcinogenesis offers promising prospects for "chemoprevention."

It is known that diets which are high in fruits and green/yellow vegetables, and which are low in animal fat are low cancer-risk; whereas diets which are high in animal fat and low in fruit and green/yellow vegetables are high cancer-risk (*64-65*). While several plant derived compounds (vitamin C, carotenes, etc.) have received considerable attention as possible dietary anticarcinogens, it would seem that quercetin and its cognate glycosides are worthy of equally serious attention.

Our research has focused on the origin, distribution and molecular processing of quercetin, a plant flavonol that is present in many fruits and vegetables, and which has been shown to block the action of a variety of endogenous and exogenous cancer initiators or promoters, both *in vivo* and *in vitro*. It is well established that yellow and red onions contain high levels of quercetin and quercetin glycosides relative to other vegetables (1-2). However, the variation in total flavonol concentration which we have observed among different onion cultivars is unexpectedly wide, ranging over 15-fold. Genetic determinants are clearly of substantial importance in setting the cultivar-specific quercetin content. Rio Ringo onions grown in three different localities, for example, were consistently among the highest flavonol producers. Previous studies suggest that flavonols are frequently produced at increased levels in times of environmental stress, such as insect or fungal attack, or water limitation (*61*). Perhaps it is not coincidental that the samples containing the highest quercetin and quercetin glycoside concentrations were grown near Bakersfield, California.

Onions are unusual in the complexity of their flavonol components. The major flavonols are all derivatives of quercetin, either as the free aglycone, quercetin 4'-glucoside, or one of three diglucosides. Quercetin 4'-glucoside is generally the major component, but its relative concentration varies significantly from 40-80% of the total flavonol content in different varieties. It is known that storage of onions increases the relative concentrations of quercetin diglucosides, probably at the expense of quercetin 4'-glucoside (*61*).

Although the majority of quercetin is present in onions as pharmacologically less active glycoside species (*1-5*), glycosidases produced by the human intestinal flora ("fecalases") are capable of "activating" a wide array of these flavonol glycosides to yield the cognate aglycone quercetin. The substantial concentrations of flavonol glycosidases produced by the human gut flora (*3-5*) suggests that flavonol glycosides are indeed present at significant levels in the human alimentary tract and that the liberated glucoside moieties could represent important secondary carbon and energy sources to these organisms.

We have shown that quercetin inhibits the growth of NIH/3T3 cells containing the activated human oncogene H-*ras*, and significantly inhibits transformation of these cells by H-*ras*. These results provide evidence that quercetin can effectively alter the malignant transformation process caused by a known oncogene. The oncogene transfection model provides a powerful tool for directly investigating the molecular mode of action of dietary anticarcinogens. The ability to induce malignant transformation by the introduction of a single cloned oncogene obviates many limitations of earlier models which have utilized "established" tumor cell lines containing multiple and generally unknown genetic alterations (*39,47*).

In view of the possible human health benefits which could result from ingestion of quercetin, and recent epidemiological evidence that *Allium* vegetable consumption reduces the risk of some cancers (*62*), it is of interest to note (1) the high levels of quercetin compounds in shallots and many onion varieties, (2) the large variation in quercetin compounds among various onion cultivars,

(3) the low levels of quercetin compounds in other *Allium* vegetables, (4) the generation of free quercetin from onion quercetin glycosides via the activity of human intestinal bacterial enzyme preparations (fecalase), and (5) the effect of quercetin in slowing the growth of cells containing an activated human oncogene, and inhibiting the transformation of normal cells by this oncogene.

We would like to draw the readers attention to the use of the experimental tools developed during the course of these studies in providing a "molecular trace" which tracks the pathway of dietary anticarcinogen formation, bioconversion and target interaction. More detailed and integrated molecular and cellular biological data, such as those presented here, will be essential to assessing the role of dietary factors in the etiology of human carcinogenesis.

Acknowledgements

We thank Dr. McGregor of the U.S.D.A. (Albany, CA) for generously providing the following flavonols: cyanidin-chloride, fisitin, myrecetin, and quercetrin. We are very grateful to Rio Colorado Seeds for providing samples of genetically defined onion cultivars grown in several geographical locations. This reseach was supported by grants from the National Cancer Institute and the Cancer Research Coordinating Committee.

References

1. Kuhnau, J. *World Rev. Nutr. Dig.* **1976,** *24,* 117
2. Brown, J.P. *Mutation Res.* **1980,** *75,* 243.
3. MacDonald, I.A.; Bussard, R.G.; Hutchinson, D.M.; Holdeman, L.V. *Appl. Environ. Microbiol.* **1984,** *47,* 350.
4. MacDonald, I.A.; Mader, J.A.; Bussard, R.G. *Mutation Res.* **1983,** *122,* 95.
5. Tamara, G.; Gold, C.; Ferro-Luzzi, A.; Ames, B.N. *Proc. Natl. Acad. Sci. (U.S.A.)* **1980,** *77,* 4961.
6. Bjeldanes, L.S.; Chang, G.W. *Science* **1977,** *197,* 577.
7. Brown, J.P.; Dietrich, P.S. *Mutation Res.* **1979,** *66,* 9.
8. Nguyen, T.; Fluss, L.; Ginther, C.; Leighton, T. *J. Wine Res.* **1990,** *1,* 25.
9. Hodnick, W.F.; Kalyanaraman, B.; Pristsos, C.A.; Pardini, R.S. *Basic Life Science* **1988,** *49,* 149.
10. Ito, N.; Hagiwara, A.; Tamano, S.; Kagawa, M.; Shibata, M.-A.; Kurata, Y.; Fukushima, S. *Jpn. J. Cancer Res.* **1989,** *80,* 317.
11. Saito, D.D.; Shirai, A.; Matsushima, T.; Sugimura. T.; Hirono, I. *Carcinogen. Mutagen.* **1980** *1,* 213.
12. Stoewsand, G.S.; Anderson, J.L.; Boyd, J.N.; Hrazdine, G.; Babish, J.G.; Walsh, K.M.; Losco, P. *J. Toxicol. Environ. Health* **1984,** *14,* 105.
13. Morino, K.; Matsushima, N.; Kawachi, T.; Ohgaki, H.; Sugimura, T.; Hirono, I. *Carcinogenesis* **1982,** *3,* 93.
14. Hirono, I.; Ueno, I.; Hosaka, S.; Takanashi, H.; Matsushima, T.; Sugimura, T.; Natori, S. *Cancer Lett.* **1981,** *13,* 15.
15. Pamukcu, A.M.; Yalciner, S.; Hatcher, J.F.; Bryan, G.T. *Cancer Res.* **1984,** *40,* 3468.
16. Ertuck, E.; Hatcher, J.F.; Nunoya, T.; Pamukcu, A.M.; Bryan, G.T. *Ann. Meet Am. Cancer Res.* **1984,** *25,* 95.
17. Sato, H.; Takahashi, M.; Furukawa, F.; Miyakawa, Y.; Hasegawa, R.; Toyoda, K.; Hayashi, Y. *Cancer Lett.* **1987,** *38,* 49.
18. Van Duuren, B.L.; Sivak, A.; Langseth, L.; Goldschmidt, B.M.; Segal, A. *NCI Monograph* **1968,** *28,* 173.
19. Hirose, M.; Fukushima, S.; Sakata, T.; Inui, M.; Ito, N. *Cancer Lett.* **1983,** *21,* 23.

20. Kato, K.; Mori, H.; Tanaka, T.; Fujii, M.; Kawai, T.; Nishikawa, A.; Taka hashi, M.; Hirono, I. *Ecotoxicity and Environmental Safety* **1985**, *10*, 63.
21. Williams, G.M. *Genetic Toxicology of the Diet*; Knudsen, I., Ed.; Liss: New York, NY, 1986, pp 73.
22. Willhite, C.C. *Fd. Chem. Toxic.* **1982**, *20*, 75.
23. Huang, M.-T.; Wood, A.W.; Newmark, H.L.; Sayer, J.M.; Yagi, H.; Jerina, D.M.; Conney, A. H.*Carcinogenesis* **1983**, *4*, 1631.
24. Alldrick, A.J.; Flynn, J.; Rowland I.R. *Mutation Res.* **1986**, 163, 225.
25. Khan, W.A.; Wang, Z.Y.; Athar, M.; Bickers, D.R.; Mukhtar, H. *Cancer Letters* **1988**, *42*, 7.
26. Sousa, R.L.; Marletta, M.A. *Arch. Biochem. Biophys.* **1985**, *240*, 345.
27. Shah, G.M.; Bhattacharya, R.K. *Chem.-Biol. Interactions* **1986**, *59*, 1.
28. Das, M.; Khan, W.A.; Asokan, P.; Bickers, D.R.; Mukhtar, H. *Cancer Res.* **1987**, *47*, 767.
29. Das, M.; Mukhtar, H.; Bik, D.; Bickers, D. *Cancer Res.* **1987**, *47*, 760.
30. Verma, A.K.; Johnson, J.A.; Gould, M.N.; Tanner. M.A. *Cancer Res.* **1988**, *48*, 5754.
31. Nishino, H.; Nishino, A.; Iwashima, A.; Tanaka, K.; Matsuura, T. *Oncology* **1984**, *41*, 120.
32. Tanaka, K.; Ono, T.; Umeda, M. *Jpn. J. Cancer Res. (Gann)* **1987**, *78*, 819.
33. Kato, R.; Nakadate, T.; Yamamoto, S.; Sugimura, T. *Carcinogenesis* **1985**, *4*, 1301.
34. Bhattacharya, B.K.; Firozi, P.F. *Cancer Lett.* **1988**, *39*, 85.
35. Markham, R.J.F.; Erhardt, N.P.; Dininno, V.L.; Penman, D.; Bhatti, A.R. *J. Gen. Microbiol.* **1987**, *133*, 1589.
36. Gschwendt, M.; Horn, F.; Kittstein, W.; Furstenberger, G.; Besenfelder, E.; Marks, F. *Biochem. Biophys. Res. Commun.* **1984**, *124*, 63.
37. Gschwendt, M.; Horn, F.; Kittstein, W.; Marks, F. *Biochem. Biophys. Res. Commun.* **1983**, *117*, 444.
38. Mookerjee, B.K.; Lipps, H.A.; Middleton, E. Jr. *J. Immunopharmacology* **1981**, *8*, 371.
39. Hofmann, J.; Doppler, W.; Jakob, A.; Maly, K.; Posch, L.; Uberall, F.; Grunicke, H. *Int. J. Cancer* **1988**, *42*, 382.
40. Horn, F.; Gschwendt, M.; Marks, F. *Europ. J. Biochem.* **1985**, *148*, 533.
41. Kato, R.; Nakadate, T.; Yamamoto, S.; Sugimura, T. *Carcinogenesis* **1985**, *4*, 1301.
42. Suolina, E.M.; Buchsbaum, R.N.; Racker, E. *Cancer Res.* **1975**, *35*, 1825.
43. Graziani, Y. In *Plant flavonoids in biology and medicine: biochemical pharmaecological and structural-activity relationships;* Cody, V; Middleton, E. Jr.;J.F. Harborne, J.F., Eds.; Liss: New York, NY, 1986.
44. Kingston, W.P. *Br. J. Pharmacol.* **1983**, *80*, 515.
45. Ruckstuhl, M.; Beretz, A.; Anton, R.; Landry, Y. *Biochem. Pharmacol.* **1979**, *25*, 535.
46. Harada, T.; Maita, K.; Odanaka, Y.; Shirasu, Y. *Jpn. J. Vet. Sci.* **1984**, *46*, 527.
47. Hoffman, R.; Graham, L.; Newlands, E.S. *Br. J. Cancer* **1989**, *59*, 347.
48. Mookerjee, B.K.; Lee, T.-P.; Lippes, H.A;Middleton, E., Jr. *J. Immunophar macology* **1986**, *8*, 371.
49. Schwartz, A.; Sutton, S.L.; Middleton, E. *Immunopharmacology* **1982**, *4*, 125.
50. Middleton, E. Jr.; Drzewiecki, G.; Krishnarao, D. *J. Immunol.* **1981**, *172*, 504.
51. Middleton, E. Jr.; Drzewiecki, G.*Biochem. Pharmacol.* **1982**, *31*, 1449.
52. Middleton, E. Jr.; Drzewiecki, G. *Biochem. Pharmacol.* **1984**, *33*, 3333.
53. El-Gammal, A.A.; Mansour, R.M.A. *Zentralbl. Mikrobiol.* **1986**, *151*, 561.
54. McClure, J.W. In *Plant flavonoids in biology and medicine: biochemical pharmaecological and structural-activity relationships.* Cody, V; Middleton, E. Jr.;J.F. Harborne, J.F., Eds.; Liss: New York, NY, 1986.

55. Gryglewski, R.J.; Korbut, R.; Robak, J.; Swies, J. *Biochem. Pharmacol.* **1987,** *36,* 317.
56. Markham, M.R. In *Methods in Plant Biochemistry;* Academic Press: New York, NY, 1989, Vol. 1, p 197.
57. Vande Casteele, K.; Geiger, H.; Van Sumere, C.F. *J. Chromatogr.* **1982,** *240,* 81.
58. Macdonald, I.A.; Mader, J.A. In *Carcinogens and Mutagens in the Environ ment;* Academic Press: New York, NY, 1981, Vol. 2, p 41.
59. Wigler, M.; Silverein, S.; Lee, L.S.; Pellicer, A.; Cheng, Y.; Axel, R. *Cell* **1977,** *11,* 223.
60. Notario, V.; Castro, R.; Flessate, D.M.; Doniger, J.; DiPaolo, J.A. *Oncogene* **1990,** *5,* 1424.
61. Herrmann, K. *J. Fd. Technol.* **1976,** *11,* 433.
62. You, W.-C.; Blot, W.J.; Chang, Y.-S.; Ershow, A.; Yang, Z.T.; An, Q.; Henderson, B.E.; Fraumeni, J.F. Jr.; Wang, T.-G. *J. Natl. Cancer Inst.* **1989,** *81,* 162.
63. Doll, R.; Peto, R. *J. Natl. Cancer Inst.* **1981,** *66,* 1192.
64. Ames, B.N. In *Genetic Toxicity of the Diet* Knudsen, I., Ed.; Liss, Inc., New York, NY, 1986, p 3.
65. Hirayama, T. *Toxicology of the Diet"* Knudsen, I., Ed.; Liss, Inc., New York, NY, 1986, p 299.

RECEIVED December 17, 1991

Chapter 17

Effect of Flavonoids on Mutagenicity and Bioavailability of Xenobiotics in Foods

B. Stavric[1], T. I. Matula[2], R. Klassen[1], R. H. Downie[2], and R. J. Wood[2]

[1]Bureau of Chemical Safety and [2]Bureau of Drug Research, Health Protection Branch, Health and Welfare Canada, Tunney's Pasture, Ottawa, Ontario K1A 0L2, Canada

The role of dietary components, both natural and man-made, in the etiology of human cancers has been well established. There is also epidemiological evidence suggesting that certain food components may have a protective effect against cancer. It appears that plant polyphenols, and flavonoids in particular, are one such group of compounds. To investigate the role of polyphenols (quercetin, ellagic acid, chlorogenic acid) on the mutagenicity and bioavailability of food carcinogens {benzo(a)pyrene [B(a)P] and 2-amino-3-methylimidazo[4,5-f] quinoline [IQ]}, we used an *in vitro* mutagenicity assay and several *in vivo* systems (bile cannulated rat, host-mediated assay and uptake of B(a)P in the mouse). It appears that some flavonoids interact with carcinogens in the gastrointestinal tract, thereby reducing their bioavailability. Overall, our data suggest that certain polyphenols may play a dual protective role in carcinogenesis by reducing bioavailability of carcinogens, and by interfering with their bio-transformation in the liver.

Overwhelming evidence from various epidemiological studies suggest that dietary factors greatly contribute to the overall cancer risk (*1,2*). Recently, additional data also indicate that some substances in foods possess the ability to inhibit carcinogenesis (*3*). However, no firm conclusions regarding the mechanism of prevention/ reduction of cancer risk by dietary modifications have been reached so far (*1,4,5*).

In recent decades, certain compounds present in human diet, such as traces of some agricultural chemicals or impurities in food additives, bacterial toxins, or toxins formed during the preparation of foods, have received much attention of scientists and the general population (6,8). In addition, the diet contains many naturally occurring non-nutritional substances, which could have beneficial or adverse effects on human health.   Recently, more intensive research has been conducted to identify these naturally occurring chemicals (3).

Numerous studies have been conducted to prove a correlation between the enhancement or inhibition of human cancers with diet (9). Many investigators are attempting to identify components in foods which could either enhance or reduce the developments of tumors (1,10). This report is an attempt to determine the mode of action of certain food ingredients with specific ability to reduce the risk of cancers. The list of regular food ingredients found to exhibit either antimutagenic or anticarcinogenic properties is quite impressive and new compounds are continuously added to that list (3,11).   It appears that a number of polyphenols generally, and some flavonoids in particular, are one such group of food compounds, although the mechanism of their protective effects is still not well understood (5,12).   Some of these polyphenols are capable of inducing mutagenic effects, such as gene mutations and chromosome aberrations, in a great variety of test systems (5,13). However, the same compounds can also suppress the genotoxic activity of numerous carcinogenic compounds in both in vitro and in vivo test systems. Because of this dual effect of dietary polyphenols, the issue as to the extent to which dietary polyphenols present a health hazard or beneficial effect on human health (14) remains unresolved.

Since relatively little attention has been devoted to the effect of polyphenols on the bioavailability of some food xenobiotics, we herein present preliminary results of such experiments. This is a summarized report from several in vitro and in vivo studies, in which B(a)P and IQ were used as xenobiotics and quercetin, ellagic acid and chlorogenic acid as representative of food polyphenols.

In Vitro Mutagenicity Studies

Since B(a)P and IQ are mutagens and carcinogens, all required safety procedures have been strictly applied when working with these chemicals.

The purpose of these studies was to investigate the mutagenic response of selected food mutagens when tested in combination with three plant polyphenols: quercetin, ellagic acid and chlorogenic acid. For these investigations the standard Ames Salmonella assay, using TA 98 or TA 100 strains of Salmonella, with a rat S9 activation system was employed (15). In these tests the following food carcinogens were tested: benzo(a)pyrene [B(a)P], aflatoxin B_1, and several amino-imidazoazaarenes [AIA]: 2-amino-3-methylimidazo[4,5-f]quinoline [IQ]; 2-amino-3,4-dimethylimidazo[4,5-f]quinoline [MeIQ]; 3-amino-1,4-dimethyl-5H-pyrido[4,3-b]-quinoline[Trp-P-1]and3-amino-1-methyl-5H-pyrido[4,3-b]-indole [Trp-P-2]. B(a)P is a polycyclic aromatic hydrocarbon found in human diet, and is produced during the preparation of foods (*16*). The AIAs are food mutagens produced by heating meat-containing foods (*17,18*).

Figure 1 illustrates the inhibitory effect of quercetin, ellagic acid and chlorogenic acid on the *in vitro* mutagenicity of B(a)P. All three polyphenols inhibited the mutagenicity of B(a)P. The most pronounced inhibition was obtained with quercetin, although this compound by itself was mutagenic in the same system. Quercetin also inhibited the mutagenicity of aflatoxin B_1, IQ, MeIQ, Trp-P-1 and Trp-P-2. Ellagic acid, a non-mutagenic substance, inhibited the mutagenicity of B(a)P, IQ and MeIQ, but had no effect or increased the mutagenicity of Trp-P-1 and Trp-P-2 (*19*). Figure 2 illustrates the increase of mutagenicity of Trp-P-1 by ellagic acid in a dose-response manner.

In Vivo Studies

The main objective of the *in vivo* studies was to compare the effect of selected flavonoids on the bioavailability of B(a)P or IQ. For this purpose three different types of studies with rats and mice were used: (a) bile-cannulated rats, to measure the percentage of radioactivity excreted in the bile (*20*); (b) the intrasanguineous host-mediated assay with mice, to evaluate *in vivo* mutagenicity (*21,22*) and (c) measurements of the uptake of (^{14}C)-B(a)P in mice, in the presence of polyphenols (*23*).

Studies with Bile-cannulated Rats. Rat chow was introduced by gavage and followed immediately by an oral dose of ^{14}C-labelled B(a)P or IQ (200,000 c.p.m./dose) into over-night fasted groups of 4 bile-cannulated rats per dose/treatment (controls). A parallel group of rats was concomitantly treated with identical rat chow blended with 2% of either quercetin, ellagic acid or chlorogenic acid, followed with an oral dose of radioactive B(a)P or IQ. Bile was collected at two-hour intervals for the first 14 hours. Each sample was weighed to determine the amount, and an aliquot was used to determine the radioactivity. Aquasol was used as scintillation cocktail and the radioactivity measured on the Beckman scintillation counter. The results were expressed as a percentage of the given radioactivity recovered in the bile for the first 14 hours.

Tests with B(a)P. Since the major elimination route of B(a)P and its metabolites in mammals is through bile (*24*), it was expected that if a flavonoid reacted with B(a)P and/or prevented its absorption from the intestine, then the number of ^{14}C counts in the bile, whether it comes from the parent B(a)P or its metabolites, should be reduced in

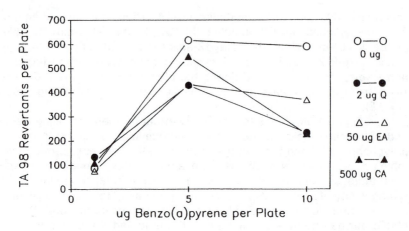

Fig. 1.   Effect of quercetin [Q], ellagic acid [EA] and chlorogenic acid [CA] on the *in vitro* mutagenicity of benzo(a)pyrene

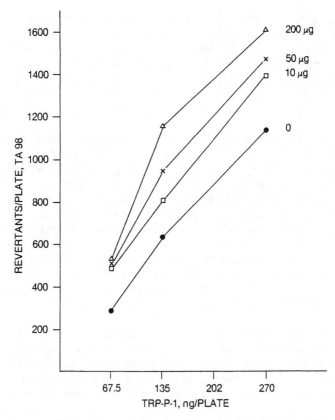

Fig. 2.   The effect of ellagic acid on mutagenicity of Trp-P-1

comparison with the controls not receiving the flavonoid. The results (Table I) indicate that all three polyphenols reduced the bioavailability of B(a)P from the intestine. Quercetin, chlorogenic acid and ellagic acid reduced the absorption of B(a)P by 21, 16 and 59 percent, respectively.

Tests with IQ. IQ is known to be rapidly absorbed from the gastrointestinal tract. It forms a number of metabolites, many of which are excreted via bile in the feces (*25*). Therefore, by measuring the radioactivity of the bile samples, it was possible to compare the absorption rate of IQ from the intestine under different experimental conditions. At this time only ellagic acid has been tested. In two separate experiments, no difference was found in the number of ^{14}C counts of the bile, between the experimental groups dosed with food, with or without 2% ellagic acid added. This would suggest no significant influence of ellagic acid on the absorption of IQ.

The Intrasanguineous Host-mediated Assay. For the intrasanguineous host-mediated mutation assays (*21*), overnight fasted male mice were treated orally by gavage with either B(a)P and IQ alone (controls) or in combination with one of the following polyphenols: quercetin, chlorogenic acid and ellagic acid. A concentrated suspension of *Salmonella typhimurium* (strain TA 98) in saline, was injected via the tail vein. For the tests with B(a)P, mice were injected with Salmonella one hour after the dosing. Mice treated with IQ received their dose of Salmonella immediately after the oral dosing. After one hour, the mice were killed by cervical dislocation, their liver excised, weighed and used to recover bacteria from the homogenates for the mutation assay (Ames test). After 48 hr incubation at 37°C the mutant colonies were counted. The results were expressed as the "Response Ratio", which indicates the ratio of the mutagenic response of treated vs. control groups.
        The aim of these experiments was to investigate the effects of the above mentioned polyphenols on the *in vivo* mutagenic activity of B(a)P and IQ. It was postulated that if the polyphenols interact with B(a)P or IQ in the gastrointestinal tract, then a probable reduction of their bioavailability will occur, which should also be reflected in reduced *in vivo* mutagenicity. However, a reduction in the mutagenic activity of liver homogenates could also indicate an inhibition of liver enzymes (*3*) responsible for biotransformation of B(a)P or IQ into mutagenic species, by polyphenols or their metabolites.

Test with B(a)P. Table II gives the results of polyphenols tested on the mutagenic activity of B(a)P. Quercetin, chlorogenic acid and ellagic acid all reduced the mutation rate of B(a)P by 52, 72 and 66 percent, respectively.

Table I. Effect of Polyphenols on the Recovery of the ^{14}C-radioactivity in the Bile* After Oral Dosing of (^{14}C)-Benzo(a)pyrene** [B(a)P]

Added into the rat chow	Average percentage of the recovery of the given dose	Uptake***
B(a)P** (control)	15.7 ± 4.3	1.00
B(a)P** + Quercetin****	12.4 ± 5.4	0.79
B(a)P** + Chlorogenic acid****	13.2 ± 4.3	0.84
B(a)P** + Ellagic acid****	6.5 ± 3.1	0.41

*   Bile collected for the first 14 hours
**   Each rat received a dose of 200,000 cpm (1.9 μg/kg) of ^{14}C-B(a)P
***   cpm of B(a)P alone expressed as 1.00
****   dose: 2% in the rat chow

Table II. Effect of Flavonoids on the Host-Mediated Mutagenic Activity of Benzo(a)pyrene [B(a)P]

Treatment	No. of animals	Revertants/ plate ± S.E.	Response Ratio, T/C*
Control (vehicle)	12	26.7 ± 3.4	1.0
B(a)P**	15	104.6 ± 11.1	3.91
B(a)P** + Quercetin***	22	50.0 ± 3.1	1.87
B(a)P**	10	172.5 ± 18.9	6.46
B(a)P** + Chlorogenic acid***	12	48.9 ± 5.1	1.83
B(a)P**	11	137.9 ± 20.0	5.16
B(a)P** + Elagic acid***	11	47.3 ± 4.8	1.77

*   Ratio of number of revertants (treated/control)
**   Dose: 200 mg/kg
***   Dose: 1000 mg/kg

SOURCE: Reference 19. Copyright 1990, National University of Singapore.

Tests with IQ. No differences were found in the mutagenic potencies between the livers of mice treated with IQ alone, or in combination with either quercetin or ellagic acid (Table III).

The Uptake of (^{14}C)-B(a)P in the Mouse. Since the *in vivo* mutagenicity of B(a)P, but not of IQ was reduced by all three polyphenols, the possible interaction of flavonoids and B(a)P in the intestinal tract was investigated. For these tests, male mice (8-9 per group), fasted overnight, were orally dosed with (^{14}C)-B(a)P alone (100,000 c.p.m.) (controls) or as mixtures of B(a)P and quercetin, chlorogenic acid or ellagic acid. After 3 hours, mice were killed by cervical dislocation, livers excised, weighed and solubilized in Soluene. Aliquots were added to scintillation cocktail and counted in a Beckman scintillation counter. The number of (^{14}C) counts per milligram of liver tissue was calculated and the results expressed as the percentage of the recovered c.p.m. in comparison to the controls.

The results (Table IV) indicate that all three tested polyphenols, namely, quercetin, chlorogenic acid and ellagic acid reduced the uptake of B(a)P by 17, 29 and 26 percent, respectively, in comparison with the uptake of B(a)P if given alone.

Discussion

The results obtained in these studies indicate that plant polyphenols can reduce the *in vitro* as well as the *in vivo* mutagenesis of B(a)P when tested polyphenols are administered together with the carcinogens. The *in vitro* effects appear to be exerted via modification of the activities of liver enzymes responsible for biotransformation of B(a)P. In contrast, the *in vivo* experiments showing reduced uptake of B(a)P from the gastrointestinal tract, and the *in vivo* inhibition of mutagenicity of B(a)P by quercetin, ellagic acid and chlorogenic acid, point to the ability of these polyphenols to modify, possibly by physico-chemical interactions, the absorption of B(a)P from the gastrointestinal tract. Theoretically, polyphenols could also act by other means, such as by inducing phase II enzymes that are responsible for detoxification processes, and by modifying the activity of phase I enzymes, that produce mutagenic metabolites. Still another possibility is that plant phenolics may block carcinogenic electrophiles in the cell itself, thus preventing DNA alterations (*26*).

For those possibilities to occur, however, the presence of flavonoids, at certain concentrations, in the liver and target cells is required. However, most pharmacokinetic studies with polyphenols seem to indicate that these substances are only weakly absorbed, and are mainly metabolized by the microbial flora endogenous to the gastrointestinal tract. For instance, an oral dose of 4 grams of quercetin given to human volunteers failed to produce measurable levels of

Table III. Effect of Quercetin and Ellagic Acid on
the Host-Mediated Mutagenic Activity of IQ

Treatment	No of animals	Revertants/ plate ± S.E.	Response Ratio, T/C*
Control (vehicle)	3	14.0 ± 4.0	1.0
IQ**	9	962.4 ± 123.9	68.74
IQ** + Quercetin***	7	1023.8 ± 101.6	73.13
IQ** + Ellagic Acid***	4	1016.0 ± 120.9	72.57

*   Ratio of number of revertants (treated/control)
**   Dose: 2 mg/kg
*** Dose: 20 mg/kg

Table IV. Effect of Polyphenols on the Uptake of
^{14}C-Benzo(a)pyrene [B(a)P] in the Mouse†

Treatment	No of animals	cpm/mg liver		Uptake***
		mean	range	
B(a)P*	8	3.3	2.8-3.4	1.00
B(a)P* + Quercetin**	8	2.8	2.0-3.1	0.83
B(a)P* + Chlorogenic acid**	9	2.3	1.6-3.7	0.71
B(a)P* + Ellagic acid**	9	2.5	1.6-3.6	0.74

*   Dose: 0.01 mg/kg
**   Dose: 5.0 mg/kg
*** cpm of B(a)P alone expressed as 1.00
†   Adapted, with permission, from Ref. 19, 1990 National University of Singapore.

quercetin in plasma and urine (*27*). Efficient metabolism by intestinal flora and elimination of quercetin as such via feces has been confirmed (*28*). Therefore, only limited amounts of quercetin could be expected in the liver and other body tissues. The frequently reported inhibition of liver enzymes, which could play a role in the genesis of cancer by dietary flavonoids (*29*), were observed in *in vitro* studies. However, due to the limited absorption of flavonoids (especially quercetin) from the gastro-intestinal tract, the relatively high doses of flavonoids required for these inhibitions are not achieved in the liver (*30*). Therefore, in addition to the inhibitory effect on the liver enzymes, some other mechanism must be involved, by which the "beneficial" effects of flavonoids (e.g. quercetin) is manifested. From our data, it appears, that polyphenolic compounds exert their antimutagenic effect against B(a)P and perhaps other polycyclic aromatic hydrocarbons (PAHs) by forming adducts, thereby rendering them, at least partly, less absorbable (bioavailable). A similar mode of action was proposed for the reduction of the mutagenic (and carcinogenic) activity of B(a)P with ellagic acid, that was attributed to the formation of a covalent adduct between ellagic acid and diol metabolite of B(a)P (*31*).

The inhibitory effects of polyphenols on the mutagenicity of B(a)P has been confirmed by different laboratories. It appears also, that B(a)P is a fair representative for the biological activities of other PAHs in similar environment. In contrast to this, the effect of polyphenols on the mutagenicity of AIAs is not equal for all AIAs. Although they all are heterocyclic aromatic amines, their basic structures differ between carbolines, quinolines and quinoxalines (*32*). These different structures of the parent compounds may influence different biological (mutagenic) response of AIAs with various polyphenols (or other food constituents). For instance, while ellagic acid reduced the *in vitro* mutagenicity of IQ and MeIQ, it enhanced the mutagenicity of Trp-P-1 and Trp-P-2. This suggests that to obtain reliable data for the interactions between polyphenols and AIAs, each AIA should be tested separately. To this end it would be of interest to mention that it was recently reported that boiling-water extracts from plants used as Chinese drugs (Bupleuri Radix), enhanced the *in vitro* mutagenic activity of Trp-P-1, Trp-P-2 and B(a)P (*33*).

Additional chemical differences between B(a)P and IQ may further influence the different bioavailability of these two mutagens, as observed in our experiments.

Conclusions

Dietary polyphenols quercetin, ellagic acid and chlorogenic acid appear to exert their "beneficial" effect by reducing the uptake of the mutagen B(a)P from the intestinal tract. They do not appear to affect the *in vivo* activity of AIA type substances. Our results suggest that the *in vitro* test can not always provide a firm prediction for the *in vivo* results.

Literature Cited

1.   Alfin-Slater, R.B.; Kritchevsky, D. Eds., *Cancer and Nutrition*, Plenum Press, New York, 1991.
2.   Pariza, M.W.; Felton, J.S.; Aeschbacher, H.U.; Sato, S. Eds., *Mutagens and Carcinogens in the Diet*, Wiley-Liss, Inc., New York, 1990.
3.   Birt, D.F.; Bresnick, E. In *Cancer and Nutrition*, Alfin-Slater, R.B.; Kritchevsky, D., Eds., Plenum Press, New York, 1991, 221-260.
4.   NAS (National Academy of Sciences), *Diet, Nutrition and Cancer*, National Academy Press, Washington, D.C., 1982, chapter 18-11.
5.   Stich, H.F. *Mutat. Res.* 1991, *259*, 307-324.
6.   Mailman, R.B.; Sidden, J.A. In *Introduction to Environmental Toxicology*; Guthrie, F.E.; Perry, J.J. Eds; Elsevies, New York, 1980, 313-328.
7.   Grasso, P. In *Chemical Carcinogens*, Second Ed.; Searle, C.E. Ed., ACS Monograph *#182*; American Chemical Society, Washington, D.C., 1984, 1205-1239.
8.   Knudsen, I., Ed. *Genetic Toxicology of the Diet*; Progress in Clinical and Biological Research, vol. *206*; Alan R. Liss, Inc., New York, 1986.
9.   Higginson, J.; Sheridan, M.J. In *Cancer and Nutrition*, Alfin-Slater, R.B.; Kritchevsky, D. Eds., Plenum Press, New York,1991, 1-50.
10. Negri, E.; La Vecchia, C.; Franceshi, S.; D'Avanzo, B.; Parazzini, F. *Int. J. Cancer*, 1991, *48*, 350-354.
11. Fontham, E.T.H. *Intern. J. Epidemiol.* 1990, *19*, Suppl. 1, S32-S42.
12. Alldrick, A.J.; Flynn, J.; Rowland, I.R. *Mutat. Res.* 1986, *163*, 225-232.
13. MacGregor, J.T. In *Nutritional and Toxicological Aspects of Food Safety*, Friedman, M., Ed., Plenum Press, New York, 1984, 497-526.
14. Stavric,B.; Matula, T.I. In *Proc. XIVth Intern. Conference of Groupe Polyphenole*, St. Catharines, Ont. Canada, Aug. I6-I9, I988, 95-104.
15. Maron, D.M., Ames, B.N. *Mutat. Res.* 1983, *113*, 173-215.
16. Howard, J.W.; Fazio, T. *J. Assoc. Off. Anal. Chem.* 1980, *63*, 1077-1104.
17. Sugimura, T.; Wakabayashi, K. In *Mutagens and Diet*, Pariza,M.W.; Felton, J.S.; Aeschbacher, H. U.; Sato, S. Eds.,Wiley-Liss, Inc. New York, 1990, 1-18.

18. Felton, J.S.; Knize, M.G. In *Chemical Carcinogenesis and Mutagenesis I*, Cooper, C.S.; Grover, P.L. Eds., Springer-Verlag, Berlin, 1990, 471-502.
19. Stavric, B.; Matula, T.I.; Klassen, R.; Downie, R.H. In *Flavonoids in Biology and Medicine III: Current Issues in Flavonoid Research*, Das, N.P., Ed., National University of Singapore, 1990, 515-529.
20. Stavric, B.; Klassen, R. Third North American ISSX Meeting, San Diego, CA, October 21-25, 1990, abstr. 47.
21. Arni, P.; Mantel, T.; Deparade, E.; Muller, D. *Mutat. Res.* 1977, *45*, 291-307.
22. Howes, A.J.; Rowland, I.R., Lake, B.G.; Alldrick, A.J. *Mutat. Res.* 1989, *210*, 227-
23. Matula, T.I.; Stavric, B.; Downie, R.H.; Klassen, R. Third International Conference on Mechanism of Antimutagenesis and Anticarcinogenesis, Lucca, Italy, May 5-10, 1991. Abstr. 66.
24. Chipman, J.K.; Hirom, P.C.; Frost, G.S., Millburn, P. *Biochem. Pharmacol.* 1981, *30*, 937-944.
25. Turesky, R.J. In *Mutagens and Carcinogens in the Diet*, Pariza, M.W.; Aeshbacher, H.-U.; Felton, J.S., Sato, S., Eds., Wiley-Liss, Inc. New York, 1990, 39-53.
26. Newmark, H.L. *Can. J. Physiol. Pharmacol.* 1987, *65*, 461-466.
27. Gugler, R.; Leschik, M.; Dengler, H.J. *Europ. J. clin. Pharmacol.* 1975, *9*, 229-234.
28. Ueno, I.; Nakano, N.; Hirono, I. *Japan. J. Exp. Med.* 1983, *53*, 41-50.
29. Alcaraz, M.J.; Ferrandiz, M.L. *J. Ethnopharmacol.* 1987, *21*, 209-229.
30. Steele, C.M.; Lalies, M.; Ioannides, C. *Cancer Research*, 1985, *45*, 3573-3577.
31. Sayer, J.M. Yagi,H..; Wood, A.W.; Conney, A.H.; Jerina, D.M. *J.Amer. Chem. Soc.* 1982, *104*, 5562-5564.
32. Hargraves, W.A. In *Nutritional Toxicology*, Hathcock, J.N. Ed., Vol. *II*, Academic Press, Inc. Orlando, FA, 1987, 157-171.
33. Niikawa, M.; Sakai, Y., Ose, Y.; Sato, T., NagaseH.; Kito, H.; Sato, M; Mizuno, M. *Chem. Pharm. Bull.* 1990, *38*, 2035-2039.

RECEIVED December 2, 1991

Chapter 18

Modulation of Mouse Skin Carcinogenesis and Epidermal Phospholipid Biosynthesis by the Flavonol Quercetin

Ajit K. Verma

Department of Human Oncology, University of Wisconsin Clinical Cancer Center, Madison, WI 53792

Flavonoids are ubiquitous in plants and are an integral part of the human diet. Quercetin is found in the edible portion of most plants and has been demonstrated to be mutagenic in bacterial assay systems. Therefore, flavonoids may pose a carcinogenic risk to humans. We evaluated the carcinogenicity and the anticarcinogenicity of quercetin in detail in the two-stage mouse skin carcinogenesis model. Quercetin was not found to be carcinogenic but instead was found to inhibit both initiation with DMBA[1] and tumor promotion with TPA of mouse skin tumor formation in a dose-dependent manner. The number of papillomas/mouse was reduced by 43% ($p<0.05$) when 30 µmol of quercetin was applied topically in conjunction with DMBA. Application of 30 µmol of quercetin either 1 hr before or 1 hr after each twice weekly application of TPA to the skin inhibited the papillomas/mouse by 60-75%. The incidence of carcinoma formation was also reduced by quercetin treatment. In contrast, quercetin applied 24 hrs after TPA treatment affected neither the incidence nor the tumor multiplicity. Quercetin, when applied 1 hr before TPA treatment, inhibited TPA-induced ^{32}P-incorporation into phospholipids (e.g., phosphatidylcholine). The results summarized in this chapter indicate that: 1) quercetin is not a carcinogen *per se* but inhibits both the initiation and promotion of mouse skin tumor formation, and 2) interference with TPA-induced phospholipid biosynthesis may be one of the mechanisms underlying the antitumor promoting activity of quercetin.

Flavonoids are ubiquitous in plants and are an integral part of the human diet (*1*). The human dietary intake of flavonoids is estimated to be 1 gram per day, approximately 5% of which is quercetin (3,3',4',5,7-pentahydroxyflavone).

---

[1]The abbreviations used are: DMBA, 7,12-dimethylbenz[a]anthracene; TPA, 12-O-tetradecanoylphorbol-13-acetate; PC, phosphatidylcholine; PE, phosphatidylethanolamine; TLC, thin layer chromatography; PA, phosphatidic acid; PI, phosphatidylinositol; PS, phosphatidylserine; ODC, ornithine decarboxylase.

0097–6156/92/0507–0250$06.00/0
© 1992 American Chemical Society

Quercetin is found in the edible portion of most plants and is present in tea, coffee, citrus, berries, leafy vegetables, roots, tubers, legumes, and cereal grains (*1*).

A number of flavonoids have demonstrated mutagenicity and genotoxicity in *Salmonella typhimurium* (*2-5*). Among the compounds evaluated, quercetin was one of the most potent mutagens (*2,3,5*). The mutagenic effects appear to be strongly dependent upon the absence of excision repair capability (*2*). In assay systems proficient in excision repair, the mutagenic activity of the flavonoids was extremely weak or absent (*2*). Quercetin and related compounds have also been shown to induce specific locus mutations, sister chromatid exchange and chromosomal aberrations in Chinese hamster ovary cells (*6*). The significance of these lesions are yet to be determined, however, these observations do suggest that flavonoids may pose some carcinogenic risk to humans.

Pamukcu *et al.* (*7*) reported a high incidence (80%) of intestinal carcinomas and a low incidence (20%) of bladder carcinomas in Norwegian rats which were fed with a diet containing 0.1% quercetin. Ertürk *et al.* (*8*) have also documented, in abstract form, that quercetin significantly increased the incidence of hepatoma in Fisher 344 and Sprague-Dawley rats fed a 2% quercetin diet. These two reports are contrary to a number of other independent investigations which indicate that quercetin is not carcinogenic in laboratory animals (*9-17*). Furthermore, recent evidence suggests that quercetin and related flavonoids may actually be inhibitors of carcinogenesis (*18-27*).

Inhibitory effects of quercetin on mouse skin carcinogenesis have been determined, but the reports have been fragmentary and also lack the effect of dose and time of application of quercetin on the induction of mouse skin cancer (*15,18,27-30*). Thus, the purpose of the present investigation was to analyze, in a single study, whether quercetin: 1) is an initiator or promoter in the mouse skin system, 2) inhibits, when given either topically or in the diet, the induction of mouse skin tumors elicited by either the initiation-promotion or complete carcinogenesis protocol, 3) inhibits TPA-induced synthesis of epidermal phospholipids.

Materials and Methods

Materials. Virus-free female CD-1 mice 7 to 9 weeks of age were obtained from Charles River Breeding Laboratories, Inc., Wilmington, MA. TPA was purchased from LC Services Corporation, Woburn, MA. DMBA, quercetin dihydrate, and TLC sheets coated with silica gel (60Å) were purchased from Aldrich Chemical Company, Milwaukee, WI. AIN-76A purified diet without ethoxyquin was purchased from Tekland, Madison, WI. PA, PI, PC, PE and PS were purchased from Sigma Chemical Company, St. Louis, MO. DL-[1-^{14}C]-Ornithine (49.9 mCi/mmol) and ^{32}P-phosphoric acid were obtained from New England Nuclear, Boston, MA. Reagents were of analytical grade.

Tumor Induction Experiments. All mice were housed in plastic cages in light-, humidity-, and temperature-controlled rooms. Food and water were available *ad libitum*, and the animals were maintained on a 12 hr light and 12 hr dark daily cycle. The dorsal skins of the mice were shaved 3-4 days prior to experimentation and only those mice in the resting phase of the hair cycle were used in the experiment. Mouse skin tumors were induced either by the initiation-promotion protocol (31) or by the complete carcinogenesis regimen (32). In the initiation-promotion protocol, a single initiating dose of DMBA (200 nmol) in 0.2 ml acetone was applied to the shaved backs of the mice; two weeks later, 10 nmol of TPA in 0.2 ml of acetone was applied twice weekly for the duration of the experiment. In the complete carcinogen carcinogenesis regimen, 50 nmol of DMBA in 0.2 ml acetone was applied twice weekly. Quercetin was applied either to skin in acetone:ethanol (4:1) or given in the purified diet as described in each experiment. There were 24 mice per treatment. The incidence and tumor multiplicity were observed on a biweekly basis for 20 weeks at which time the experiment was either terminated or extended to 35 weeks for the determination of carcinoma incidence. The term carcinoma was used for lesions grossly invading the dermis or panniculus carnosus (31). Quercetin treatment did not affect body weight gains. The survival in each experiment was greater than 80%. Data was statistically analyzed using the student's t-test ($p < 0.05$).

Extraction and quantification of phospholipids from mouse epidermis. At appropriate times after treatment, mice were injected intraperitoneally with ^{32}P-phosphoric acid (100 μCi/0.2 ml saline/mouse) as described in the experiment. The mice were sacrificed by cervical dislocation, the skin was removed and then the epidermis was separated by a brief heat treatment (57°C for 30 seconds). The epidermal preparations from the 4 mice in each group were pooled, homogenized in ice-cold saline (1 ml saline/mouse skin), and the protein content of each sample was determined (33). An aliquot of each sample, representing an equivalent quantity of protein was used for the extraction of phospholipids with chloroform:methanol (1:2) containing 2% acetic acid (4 ml/1 ml of sample aliquot). The samples were dried under a stream of nitrogen and resuspended in 0.5 ml chloroform:methanol (95:5). Individual phospholipids were separated by the two-dimensional silica-gel TLC. Chloroform:methanol:ammonium hydroxide (65:35:5; v/v/v) was used to develop the plates in the first dimension and chloroform:acetone:methanol:acetic acid:water (10:4:2:2:1; v/v/v/v/v) for the second dimension (34). The individual phospholipids were identified by autoradiography. Standard phospholipids were cochromatographed with the radioactive samples, and visualized by exposure to iodine vapor or molybdenum spraying (35).

Results

Mouse skin carcinogenicity of Quercetin. To determine whether quercetin possessed the properties of either a mouse skin tumor initiator or a promoter, a dose of 30 μmol of quercetin replaced initiation with DMBA or promotion with TPA. As shown in Figure 1, quercetin is neither a tumor initiator nor a promotor when compared to the respective controls. Mice initiated with 30 μmol quercetin and promoted twice weekly with 10 nmol of TPA had 0.08 ± 0.06 pa/mouse and

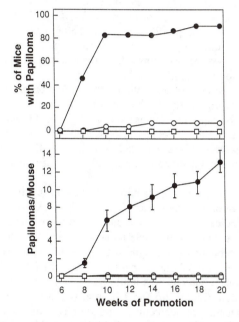

Figure 1.   Mouse skin tumor initiating and promoting activity of quercetin.
Female CD-1 mice were initiated by application of 0.2 μmol
DMBA or 30 μmol quercetin to their shaved backs; 2 weeks
later, 10 nmol of TPA or 30 μmol quercetin was applied to skin
twice a week for the entire duration of the experiment. (●),
initiation with DMBA and promotion with TPA; (○), initiation
with quercetin and promotion with TPA; (□), initiation with
DMBA and promotion with quercetin. The individual values for
the average number of papillomas/mouse represents the mean ±
S.E. of 24 mice.

only 8% of mice bore papillomas at 20 weeks of promotion. Both tumor incidence and multiplicity after initiation or promotion with quercetin were not significantly (p>0.1) different from their appropriate controls (Figure 1). Similarly, Sato et al. (36) have shown that quercetin failed to initiate mouse skin tumor formation even after twice weekly applications of quercetin (total dose 100 mg) for 5 weeks.

Effects of quercetin on the stages of mouse skin carcinogenesis. The effect of quercetin treatment on the initiation of tumors with DMBA is illustrated in Figure 2a. Quercetin at a dose of 30 μmol, applied 1 hr before and 1 hr after the DMBA application, did not significantly affect the percent of mice with tumors; however, the tumor multiplicity was reduced throughout the experiment and the inhibition at 20 weeks of promotion was about 43% (p<0.05).

The effect of the dose of quercetin on the promotion of mouse skin tumor formation is shown in Figure 2b. In the control group the incidence was 92% and the average number of papillomas/mouse was 13.2 ± 1.3. Quercetin, at doses of 10 and 30 μmol, reduced the incidence by 58% and 71%, respectively; the number of papillomas/mouse was inhibited by 27% (p<0.05) at 1 μmol, 62% at 10 μmol and 74% at 30 μmol quercetin dose by 20 weeks of promotion treatment (Figure 2b).

The effect of the time of application of quercetin relative to TPA treatment on skin tumor promotion was also determined. In these experiments, quercetin was applied at the indicated times after each twice weekly application of TPA to skin (Figure 3). There was a comparable reduction in both the incidence and tumor multiplicity, throughout the length of the experiment, when quercetin was applied 1 hr before (Figure 3a) and 1 hr after (Figure 3b) TPA. At 20 weeks of promotion, the tumor multiplicity was inhibited by 72% and 62% when quercetin was applied 1 hr before (Figure 3a) and 1 hr after (Figure 3b) TPA, respectively. In contrast, the application of quercetin 24 hrs after TPA treatment had no effect on either incidence or papillomas/mouse (Figure 3b).

In the same tumor induction experiment (Figure 3), the incidence of carcinomas was determined by extending tumor promotion treatment to 35 weeks. As shown in Table I, quercetin also inhibited the incidence of carcinomas when applied either 1 hr before or 1 hr after TPA. However, quercetin, when administered 24 hrs after TPA, failed to inhibit carcinoma incidence (Table I).

An inhibitory effect of quercetin on mouse skin tumor promotion has previously been reported by Kato et al. (37) and Horiuchi et al. (38). However, in their studies (37,38), the effect of the time of application of a dose of quercetin on mouse skin carcinoma incidence was not determined.

Effect of quercetin on mouse skin cancer elicited by DMBA by a complete carcinogenesis protocol. We have shown that the biology of mouse skin tumor formation by the initiation-promotion protocol differs from that of tumor formation by the complete carcinogenesis by DMBA. In this context, it is noteworthy that the agents such as retinoic acid, which are potent inhibitors of tumor promotion by TPA, fail to inhibit complete carcinogenesis by DMBA (32).

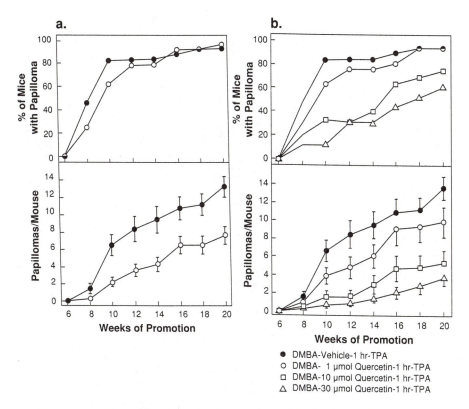

Figure 2.   Effect of application of quercetin to skin on the initiation of mouse skin tumors with DMBA and promotion with TPA. Female CD-1 mice were initiated by application of 0.2 µmol DMBA to their shaved backs; 2 weeks later, mice were treated twice a week with 10 nmol of TPA for the entire duration of the experiment.

a)   Quercetin (30 µmol) was applied in conjunction with DMBA-initiation. (●), initiation with DMBA and promotion with TPA; (○), quercetin was applied 1 hr before and 1 hr after initiation with DMBA followed by promotion with TPA.

b)   Either vehicle [acetone:ethanol, 4:1 (v/v)] (●) or quercetin at doses of 1 µmol (○), 10 µmol (□) or 30 µmol (△) were applied 1 hr prior to the twice weekly application of 10 nmol TPA to skin for the duration of the experiment. The individual values for the average number of papillomas/ mouse represents the mean ± S.E. of 24 mice.

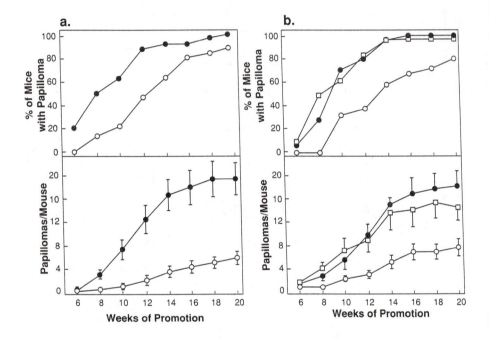

Figure 3.   Effect of time of application of quercetin relative to TPA treatment on mouse skin tumor promotion. Female CD-1 mice were
initiated by application of 0.2 μmol DMBA to their shaved
backs; two weeks later, the mice were treated with:
a) Either vehicle (●) or 30 μmol quercetin (O) 1 hr prior to
each TPA treatment for the duration of the experiment; or
b) Either vehicle (●) or 30 μmol quercetin was applied 1 hr
(O) or 24 hr (□) after each twice weekly applications of 10
nmol TPA to skin for the entire duration of the experiment.
The individual values for the average number of
papillomas/mouse represents the mean ± S.E. of 24 mice.

These findings prompted us to examine the effect of quercetin on skin tumor formation by DMBA. In this experiment, quercetin treatments, in conjunction with DMBA, were started after the initiation of mouse skin tumors with DMBA. As shown in Figure 4, quercetin elicited slight inhibitory effects, on the induction of both skin papillomas (Figure 4a) and carcinomas (Figure 4b).

Table I.   Effect of topical application of quercetin to mouse skin carcinoma formation

Treatments[a]		% Carcinoma at the following weeks of promotion			
Initiation	Promotion	24	28	32	35
DMBA	Vehicle - 1 hr - TPA	4	12	25	38
DMBA	Quercetin - 1 hr - TPA	0	0	4	8
DMBA	TPA - 1 hr - Vehicle	4	4	17	21
DMBA	TPA - 1 hr - Quercetin	0	0	4	12
DMBA	TPA - 24 hr - Quercetin	4	21	29	42

[a]As described in Figure 3, mice were initiated with 0.2 μmol DMBA, promoted twice weekly with 10 nmol TPA and treated with 30 μmol of quercetin at the times indicated. Survival at 35 weeks varied between 83 and 96%. Carcinoma data are expressed as percentage of the effectual total. The effectual total is defined as the number of mice in each group at the time of appearance of the first carcinoma in any group.

Effect of dietary quercetin on mouse skin tumor initiation and promotion. As shown in Figure 6, administration of quercetin in the diet either in conjunction with initiation with DMBA or promotion with TPA failed to affect the induction of skin papillomas. Five-percent dietary quercetin had no effect on initiation when given in association with DMBA (Figure 5). Similarly, quercetin at doses of 2% and 5% fed to mice during promotion with TPA did not significantly ($p > 0.1$) effect the incidence or tumor multiplicity compared to the control (Figure 6). Furthermore, dietary quercetin (2%) did not elicit any toxic side effects or carcinogenicity in mice (data not shown). Similarly, Horiuchi *et al.* (*38*) have shown that quercetin administered in the diet either at 1 or 4% did not significantly inhibit mouse skin tumor promotion by teleocidine.

Inhibition of TPA-induced phospholipid biosynthesis by Quercetin. The effect of quercetin on TPA-induced phospholipid biosynthesis was analyzed when quercetin was applied 1 hr before TPA application. The quantity of

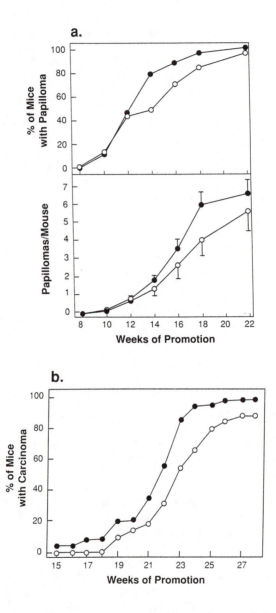

Figure 4.   Effect of quercetin on mouse skin carcinogenesis by DMBA.
Groups of mice were initiated by application of 0.2 μmol of
DMBA to skin. Two weeks after initiation, vehicle (●) or
30 μmol quercetin (○) was applied 1 hr before each twice
weekly applications of 50 nmol of DMBA.
  a) Effect of quercetin on the induction of skin papillomas;
  b) Effect of quercetin on the incidence of skin carcinomas.
The individual values for the average number of papillomas/
mouse represents the mean ± S.E. of 24 mice.

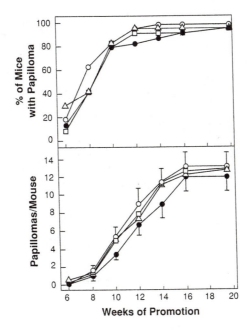

Figure 5.   Effect of dietary quercetin on skin tumor initiation and promotion. All mice were initiated with 0.2 μmol DMBA. Two weeks after initiation 10 nmol TPA was applied twice weekly for the duration of the experiment. Quercetin was fed to mice in association with initiation or promotion. The mice were started on either control (●, ○, □) or 5% quercetin diet (▲) 1 week prior to initiation. One week after initiation a group of mice were continued on control diet (●), while the other groups were switched to 2% quercetin (○), 5% quercetin (□) or control diet (▲). The individual values for the average number of papillomas/mouse represents the mean ± S.E. of 24 mice.

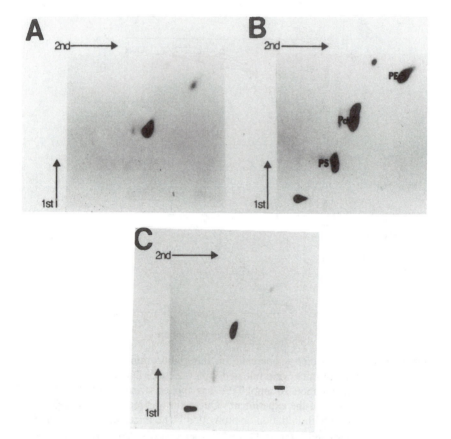

Figure 6.   Effect of quercetin on TPA induced ^{32}P-incorporation into phospholipids. Shown are autoradiographs of two-dimensional, thin-
layer chromatography of epidermal phospholipids. Samples were
processed as described in the "Materials and Methods."
Quercetin (30 µmol) or the vehicle was applied 1 hr before
application of 10 nmol of TPA to skin. A) Vehicle
[acetone:ethanol (4:1)] - 1 hr - acetone; B) Vehicle - 1 hr - TPA;
and C) Quercetin - 1 hr - TPA.

^{32}P-incorporated into each phospholipid class was determined. Application of TPA after the vehicle acetone:ethanol (4:1) significantly (p<0.05) increased the incorporation of ^{32}P into total phospholipids as compared to the control. Quercetin administered 1 hr before TPA significantly (p<0.05) inhibited the ^{32}P-incorporation into phospholipids by 66% (data not shown). The effect of quercetin applied 1 hr before TPA on the classes of individual phospholipids is represented in Figure 6. TPA appears to mainly induce the synthesis of PC and PE. Quercetin suppressed the TPA-induced induction of both PC and PE when applied 1 hr before TPA (Figure 6).

Discussion

The results presented herein confirm previous findings (*27,28,30,37,38*) and also present new experimental results supporting the conclusion that quercetin is anti-carcinogenic in the two-stage mouse skin carcinogenesis model. Both initiation with DMBA and promotion with TPA were significantly inhibited by quercetin application. The effect on initiation may be due to alterations in epidermal metabolism of DMBA and/or inhibition of the interaction of the metabolite(s) of DMBA with DNA. Das *et al.* (*23,24*) address both questions and indicate significant inhibition of epidermal aromatic hydrocarbon hydroxylase by quercetin as well as reductions in the 7-ethoxycoumarin O-deethylase and 7-ethoxyresonitin *in vitro*. Similar results were observed *in vivo* by topical application of quercetin to the backs of Senicar mice (*23*). In addition, Das *et al.* (*24*) also demonstrate that quercetin inhibits DNA-adduct formation by *in vivo* treatment with DMBA. The extent of polycyclic aromatic hydrocarbon binding to protein was also reduced by quercetin (*24*). These observations support the results of the effects of quercetin on mouse skin tumor initiation with DMBA.

In contrast to the apparent clarity by which quercetin alters initiation, the process by which quercetin affects promotion is unclear. Quercetin may affect either the binding of TPA to its receptor (PKC) or inhibit PKC activity. However, since quercetin applied 1 hr after TPA still inhibited tumor promotion (Figure 3b), it appears unlikely that quercetin affects TPA interaction with PKC. These results indicate that quercetin affects a stage of promotion post-TPA binding to the phorbol ester receptor. The data is supported by Nishino *et al.* (*28*) and Horiuchi *et al.* (*33*) who demonstrated that quercetin did not affect the specific ^{3}H-TPA binding to the receptor. Quercetin did, however, inhibit the activity of protein kinase C isolated from mouse brain cytosol (*39*) and mouse epidermis (*38*). Thus the anti-carcinogenic properties of quercetin appear to affect a step distant to protein kinase C activation, such as effect on gene expression.

The inhibition of TPA-induced ^{32}P-incorporation into total phospholipids by quercetin has been documented (*28,29*). However, the effect of quercetin on individual classes of phospholipids has not been evaluated. The primary phospho-lipid induced by TPA application was PC. These results are in accord with previous findings in mouse skin (*40*) and HeLa cells (*41*). Guy and Murray (*41*) state that perturbation of PC turnover by promoters occurs rapidly and is a prime candidate for the initial event triggered by promoters. Our results (Figure 6) indicate that quercetin inhibits the biosynthesis of PC and PE when given 1-hr before TPA application.

Quercetin appears to antagonize numerous other tumor promoter-induced events in addition to ^{32}P-incorporation into phospholipids in HeLa cells and human embryo fibroblasts (28,29), ornithine decarboxylase activity in mouse epidermis (30,37) and activation of protein kinase C (32,33). Quercetin also inhibits lipoxygenase activity in mouse epidermal cells (42,43) and TPA-induced release of superoxide anion radicals from human peripheral leukocytes (44). All of these properties have been associated with tumor promotion, however which, if any, may be the primary mechanism by which quercetin inhibits skin tumor formation has yet to be delineated.

Quercetin failed to elicit dramatic inhibitory effects on complete carcinogenesis by DMBA (Figure 4). We have shown that retinoic acid applied in conjunction with TPA following initiation inhibits the formation of skin papillomas. In contrast, retinoic acid applied in conjunction with each weekly application of DMBA did not inhibit DMBA-induced tumors (32). The results (Fig. 4) further strengthen the concept that the promotion of skin carcinogenesis by the complete carcinogen DMBA may be mechanistically different from that of the tumor promoter TPA (32).

In summary, the results presented indicate that quercetin, a ubiquitous component of edible plants, despite its mutagenicity in bacterial assay systems, is not carcinogenic. Quercetin is neither an initiator nor a promoter of mouse skin tumor formation. Topical application of quercetin to mouse skin inhibited both the initiation and the promotion steps of mouse skin tumor formation. We also found that dietary quercetin (Figure 4), which failed to inhibit skin tumor promotion by TPA, inhibited the induction of rat mammary cancer both by DMBA and NMU (45). Deschner et al. (46) found that mice fed quercetin at 2% concentration had significantly fewer azoxymethane-induced colonic tumors than mice fed the control diet (46). The lack of inhibition of tumor promotion by dietary quercetin in mouse skin may be related to its pharmacokinetics.

Acknowledgments

I am thankful to Mohamad Ali and Jeffrey A. Johnson for assistance in the tumor-induction experiments. The investigation was supported by the American Institute of Cancer Research Grant 86-A45-REV.

Literature Cited

1. Brown, J. P. Mutat. Res. 1980, 75, 243-277.
2. MacGregor, J. T. Prog. Clin. Biol. Res. 1986, 213, 411-424.
3. Bjeldanes, L. F.; Chang, G. W. Science 1977, 197, 577-578.
4. Hatcher, J. F.; Bryan, G. T. Mutat. Res. 1985, 148, 13-23.
5. MacGregor, J. T. In Nutritional and Toxicological Aspects of Food Safety; Friedman, M., Ed.; Plenum: New York, New York, 1986, Vol. 177, pp. 497-526.

6. Carver, J. H.; Carrano, A. V.; MacGregor, J. F. *Mutat. Res.* **1983**, *113*, 45-60.
7. Pamukcu, A. M.; Wong, C. Y.; Hatcher, J.; Bryan, G. T. *J. Natl. Cancer Inst.* **1980**, *65*, 131-135.
8. Ertürk, E.; Hatcher, J. F.; Nunoya, T.; Pamukcu, A. M.; Bryan, G. T. *Proc. Am. Assoc. Cancer Res.* **1984**, *25*, 95.
9. Morino, K.; Matsukura, N.; Kawachi, T.; Ohgaki, H.; Sugimura, T.; Hirono, I. *Carcinogenesis* **1982**, *3*, 93-97.
10. Hosaka, S.; Hirono, I. *Gann* **1981**, *72*, 327-328.
11. Saito, D.; Shirai, A.; Matsushima, T.; Sugimura, T.; Hirono, I. *Teratogenesis, Carcinogenesis and Mutagenesis* **1980**, *1*, 213-221.
12. Takanashi, H.; Aiso, S.; Hirono, I.; Matsushima, T.; Sugimura, T. *J. Food Sci.* **1983**, *5*, 55-60.
13. Stoewsand, G. S.; Anderson, J. L.; Boyd, J. N.; Hrazdina, G.; Babish, J. G.; Walsh, K. M.; Losco, P. *J. Toxicol. Environ. Health* **1984**, *14*, 105-114.
14. Hirose, M.; Fukushima, S.; Sakata, T.; Inui, M.; Ito, N. *Cancer Lett.* **1983**, *21*, 23-27.
15. Kato, K.; Mori, H.; Tanaka, T.; Fujii, M.; Kawai, T.; Nishikawa, A.; Takahashi, M.; Hirono, I. *Ecotoxicol. Environ. Saf.* **1985**, *10*, 63-69.
16. Kato, K.; Mori, H.; Fujii, M.; Bunai, Y.; Nishikawa, A.; Shima, H.; Takahashi, M.; Kawai, T.; Hirono, I. *J. Toxicological Sci.* **1984**, *9*, 319-325.
17. Harada, T.; Maita, K.; Odanaka, Y.; Shirasu, Y. *Jpn. J. Vet. Sci.* **1984**, *46*, 527-532.
18. Nishino, H.; Naito, E.; Iwashima, A.; Tanaka, K.; Matsuura, T.; Fujiki, H.; Sugimura, T. *Gann* **1984**, *74*, 311-316.
19. Ueno, I.; Kohno, M.; Haraikawa, K.; Hirono, I. *J. Pharm. Dyn.* **1984**, *7*, 798-803.
20. Buening, M. K.; Chang, R. L.; Huang, M.; Fortner, J. G.; Wood, A. W.; Conney, A. H. *Cancer Res.* **1981**, *41*, 67-72.
21. Lasker, J. M.; Huang, M.; Conney, A. H. *J. Pharmacol. Exp. Ther.* **1984**, *229*, 162-170.
22. Huang, M.; Johnson, E. F.; Muller-Eberhard, U.; Koop, D. R.; Coon, M. J.; Conney, A. H. *J. Biol. Chem.* **1981**, *256*, 10897-10901.
23. Das, M.; Mukhtar, H.; Bik, D. P.; Bickers, D. R. *Cancer Res.* **1987**, *47*, 760-766.
24. Das, M.; Khan, W. A.; Asokan, P.; Bickers, D. R.; Mukhtar, H. *Cancer Res.* **1987**, *47*, 767-773.
25. Fujiki, H. *Jpn. J. Cancer Chemother.* **1986**, 3384-3391.
26. Chang, R. L.; Huang, M.; Wood, A. W.; Wong, C.; Newmark, H. L.; Yagi, H.; Sayer, J. M.; Jerina, D. M.; Conney, A. H. *Carcinogenesis* **1985**, *6*, 1127-1133.
27. Nishino, H.; Iwashima, A.; Fujiki, H.; Sugimura, T. *Gann* **1984**, *75*, 113-116.

28.   Nishino, H.; Nagao, M.; Fujiki, H.; Sugimura, T. *Cancer Lett.* **1983**, *21*, 1-8.
29.   Nishino, H.; Nishino, A.; Iwashima, A.; Tanaka, K.; Matsuura, T. *Oncology* **1984**, *41*, 120-123.
30.   Nakadate, T.; Yamamoto, S.; Aizu, E.; Kato, R. *Gann* **1984**, *75*, 214-222.
31.   Verma, A. K.; Boutwell, R. K. *Carcinogenesis* **1980**, *1*, 271-276.
32.   Verma, A. K.; Conrad, E. A.; Boutwell, R. K. *Cancer Res.* **1982**, *42*, 3519-3525.
33.   Markwell, M. K.; Haas, S. M.; Bieber, L. L.; Tolbert, N. E. *Anal. Biochem.* **1978**, *87*, 206-210.
34.   Nishino, H.; Fujiki, H.; Terada, M.; Sato, S. *Carcinogenesis* **1983**, *4*, 107-110.
35.   Dittmer, J. C.; Lester, R. L. *J. Lipid Res.* **1964**, *5*, 126-127.
36.   Sato, H.; Takahashi, M.; Furukawa, F.; Miyakawa, Y.; Hasegawa, R.; Toyoda, K.; Hayashi, Y. *Cancer Lett.* **1987**, *38*, 49-56.
37.   Kato, R.; Nakadate, T.; Yamamoto, S.; Sugimura, T. *Carcinogenesis* **1983**, *4*, 1301-1305.
38.   Horiuchi, T.; Fujiki, H.; Hakii, H.; Saganuma, M.; Yanashita, K.; Sugimura, T. *Jpn. J. Cancer Res. (Gann)* **1986**, *77*, 526-531.
39.   Gschwendt, M.; Horn, F.; Kittstein, W.; Marks, F. *Biochem. Biophys. Res. Comm.* **1983**, *117*, 444-447.
40.   Rohschneider, L. R.; O'Brien, O. H.; Boutwell, R. K. *Biochim. Biophys. Acta* **1972**, *280*, 57-70.
41.   Guy, G. R.; Murray, A. W. *Cancer Res.* **1983**, *43*, 5564-5569.
42.   Wheeler, E. L.; Berry, D. L. *Carcinogenesis* **1986**, *7*, 33-36.
43.   Nakadate, T.; Yamamoto, S.; Aizu, E.; Kato, R. *Jpn. J. Pharmacol.* **1985**, *38*, 161-168.
44.   Shwarz, M.; Peres, G.; Kunz, W.; Fürstenberger, G.; Kittstein, W.; Marks, F. *Carcinogenesis* **1984**, *5*, 1663-1670.
45.   Verma, A. K.; Johnsnon, J. A.; Gould, M. N.; Tanner, M.A. *Cancer Res.* **1988**, *48*, 5754-5758.
46.   Deschner, E. E.; Ruperto, J.; Wong, G.; Newmark, H. L. *Carcinogenesis* **1991**, *12*, 1193-1196.

RECEIVED December 17, 1991

Chapter 19

Dietary Quercetin and Rutin

Inhibitors of Experimental Colonic Neoplasia

Eleanor E. Deschner

Memorial Sloan–Kettering Cancer Center, 1275 York Avenue, New York, NY 10021

The phenolic flavonoids quercetin (QU) and rutin (RU) found in the outer leaves of vegetables and fruits were examined as modifiers of an experimental model of colon cancer. CF1 mice were fed an AIN-76A diet to which was added various levels of QU or RU. Acute studies revealed that, in control mice, QU and RU had no effect on colonic epithelial cell proliferation but 2% QU and 4% RU significantly reduced azoxymethanol (AOM) induced hyperproliferation of these cells. The same doses also significantly inhibited the development of focal areas of dysplasia (FADs). Compared with controls, 2% QU for 50 weeks significantly depressed colon tumor incidence while both 2% QU and 4% RU suppressed tumor multiplicity. Thus, under conditions of low dietary fat intake, these flavonoids have considerable activity to suppress the hyperproliferation of colonic epithelial cells, thereby reducing FADs and ultimately colon tumor incidence.

Flavonoids such as QU and RU occur widely in edible fruits and vegetables. To date, over two thousand flavonoids have been chemically characterized and most occur in nature in a glycoside form. Dietary intake of QU is thought to range between 50 and 500 mg per day (*1*).

Some in vitro assays employing QU have erroneously reported the mutagenicity of this compound due to the production of hydrogen peroxide by its interaction with trace metals (i.e. copper and iron) in the media. The latter act on the catechol group of the flavonoids exposed to atmospheric oxygen in a classical reaction producing the toxic hydrogen peroxide (*2*). This reaction does not occur in biological in vivo systems.

0097–6156/92/0507–0265$06.00/0

Dietary Quercitin. Ingestion of a known quantity of QU revealed that less than 1% was absorbed in the gut. More than 50% of the dose given was degraded by microbes in the colon while the remainder was lost in the feces (*3*). Urine samples over 24 hours in volunteers given 4g of oral QU contained no QU or QU conjugates.

The presence of unabsorbed dietary QU in the colon or alternatively the release of QU from the glycoside RU when it is cleaved by microflora in the colon (*4*) suggest that, in either case, QU in close proximity with the colonic mucosa could act as an agent which might protect, prevent or inhibit carcinogenesis there. Furthermore, studies of experimentally induced mammary cancer in rats fed dietary QU revealed a diminution in tumor incidence (*5*) although the amount of QU delivered to the mammary gland by this route in this instance is questionable.

Model Testing Efficacy. To evaluate the possible protective effect of dietary QU (0.1, 0.5 and 2% of diet) and RU (1.0 and 4.0% of diet), a model for experimental colon cancer in CF1 mice was employed using the carcinogen AOM (*6*). 2% QU is approximately equimolecular to 4% RU.

Three tests carried out helped determine the efficacy of QU and RU against AOM induced carcinogenesis. Two were short term studies and one a long term investigation. The two acute studies evaluated the incidence of FADs and the modification of carcinogen induced elevated levels of colonic epithelial cell proliferation. The chronic study with QU and RU assessed the tumor incidence and tumor multiplicity levels when compared to mice not fed the flavonoids.

Acute Studies. AOM is an active metabolite of the well studied carcinogen 1,2-dimethylhydrazine (DMH) (*7*). Six injections of DMH, once weekly at a dose of 20 mg/kg body weight, induce FADs in the colon. They appear as early as 27 days after delivery of the first few injections. These FADs arise in the upper two-thirds of the colonic crypts of the otherwise normal appearing mucosa. FADs are composed of epithelial cells exhibiting a loss of mucin, loss of polarity of nuclei as well as nuclei of various size and shape (*8,9*). The number of FADs present in 500 microns of serially sectioned distal colon was determined. In mice after 6 weekly injections of AOM at a dose of 10 mg/kg body weight and concomitant dietary intake of 2% QU and 4% RU, a significantly reduced incidence of FADs was observed compared with the number present in mice fed the control diet (*10*).

Prior to and simultaneously with the development of AOM induced FADs, alterations of colonic epithelial cell proliferation are initiated by the first and repeated injections of the carcinogen (*9*). To analyze these induced abnormalities in cell proliferation, tritiated thymidine is injected one hour before sacrifice of the mice to allow incorporation into newly synthesized DNA. Microscopic observation of slides treated by autoradiography allow the number and position of tritiated thymidine labeled cells to be noted as well as the total number of cells per crypt column or one half of a crypt or gland.

Dietary QU and RU had no effect on colonic epithelial cell proliferation in mice not injected with the carcinogen. However, AOM treated mice fed the 2% QU and 4% RU had significantly fewer cells/crypt column (i.e. reduction of hyperplasia) than carcinogen treated mice receiving the control diet. Moreover, the population of DNA synthesizing cells was significantly reduced (i.e. reduction of hyperproliferation) in the colonic mucosa of mice fed 2% QU and 4% RU compared with those treated by AOM alone (*10*).

Chronic Dietary Study. The long term tumor incidence investigation involved feeding QU and RU diets for a total of fifty weeks (*6*). Diets were provided two weeks before carcinogen treatment was initiated and protocol called for three weekly subcutaneous injections of AOM at a level of 10 mg/kg body weight. The highest level of dietary QU (2%) significantly reduced colonic tumor incidence, a four fold inhibition, when compared with those mice receiving the control diet. When the 3 QU dietary groups were combined and 2 RU groups combined and then compared with the AOM treated control group, it was determined that both QU and RU significantly reduced the number or multiplicity of tumors/tumor bearing mouse (*10*).

General Conclusions. Previous studies examining the ability of QU to inhibit the formation of nuclear aberrations 24 hours following DMH injection demonstrated that QU was inactive at this stage of carcinogenesis (*11*). Rather than the initiation phase, the acute studies with QU and RU described here, showing reduced FAD incidence and a depression in AOM induced hyperproliferation, suggest instead that it is the promotional stage which is more intensely affected by dietary consumption of these phenolics.

These experiments have not elucidated a mechanism by which QU inhibits or modifies the development of neoplasia in the colon. Nevertheless it has become apparent that this flavonoid has significant antitumoral activity there and that further experimental studies with it should be undertaken.

Literature Cited
1. Jones, E.; Hughes, R.E. *Expt. Gerontol.* **1982**, 117, 213-217.
2. Ueno, I.; Haraikama, K.; Kohno, M.; Hinomoto, T.; Ohya-Nishiguchi, H.; Tomatsuri T.; Yoshihira K. In: *Plant Flavonoids in Biology and Medicine*; Cody, V., Middleton E., Harborne J.B., Eds; Alan R. Liss: New York, NY, **1986**, pp 425-428.
3. Gugler, R.; Leschick. M., Dengler. *Europe. J. Clin. Pharmacol.* **1975**, 9, 229-234.
4. Goldin, B.R.; Lichtenstein, A.H.; Gorbach, S.L. In: *Modern Nutrition in Health and Disease*; Shils, M.E., Young, V.R. 7th edition. Lea and Febiger: Philadelphia, PA, **1988**; pp 503.

5.   Verma, A.K.; Johnson, J.J.; Gould, M.N.; Tanner, M.A. *Cancer Res.*
     1988, 48, 5754-5758.
6.   Deschner, E.E.; Lytle, J.S.; Wong, G.; Ruperto J.F.; Newmark, H.L.
     Cancer **1990**, 66, 2350-2356.
7.   Fiala, E.S. *Cancer* **1975**, 36, 2407-2412.
8.   Deschner, E.E. *Cancer* **1974**, 34, 824-828.
9.   Deschner, E.E. *Z. Krebsforsch.* **1978**, 91, 205-216.
10. Deschner, E.E.; Ruperto, J.; Wong, G.; Newmark, H.L.
     Carcinogenesis. **1991**, 12, 1193-1196.
11. Wargovich, M.J.; Newmark, H.L. *Mutat. Res.* **1983**, 121, 77-80.

RECEIVED December 2, 1991

Chapter 20

Carcinogenicity and Modification of Carcinogenic Response by Plant Phenols

N. Ito, M. Hirose, and T. Shirai

First Department of Pathology, Nagoya City University Medical School, 1 Kawasumi, Mizuho-cho, Mizuho-ku, Nagoya 467, Japan

Since the carcinogenicity of BHA in rodent forestomach was first shown, several plant phenols have been demonstrated to exert carcinogenic potential in rats, e.g., caffeic acid (forestomach and kidney), sesamol (forestomach), catechol (glandular stomach) and hydroquinone (kidney). In the second stage of 2-stage carcinogenesis, these compounds demonstrate different modifying activities depending on the organ. For example, catechol enhances forestomach, glandular stomach, tongue and esophageal carcinogenesis while inhibiting colon, lung and hepato-carcinogenesis. Hydroquinone and caffeic acid enhance kidney and forestomach carcinogenesis, respectively. Several other phenolic antioxidants have been examined for their modifying effects using multi-organ carcinogenesis models. The results indicate that catechins clearly inhibit rat small intestinal carcinogenesis with the effects being most obvious when applied during the initiation stage. On the other hand, combined treatment with sodium nitrite and phenolic compounds such as catechol and 3-methoxycatechol strongly induces forestomach proliferative lesions which are not observed with individual treatments. The results suggest forestomach carcinogenicity of these compounds in the presence of sodium nitrite.

The synthetic antioxidant butylated hydroxyanisole (BHA) has been shown to induce forestomach carcinomas in rats and hamsters (1-3). It strongly enhances forestomach carcinogenesis in rats pretreated with N-methyl-N'-nitro-N-nitrosoguanidine (MNNG), N-methylnitrosourea (MNU) or N,N-dibutylnitrosamine (DBN), and enhances urinary bladder carcinogenesis induced by N-butyl-N-(4-hydroxybutyl)nitrosamine or DBN. On the other hand, it inhibits liver, lung, and mammary carcinogenesis induced by carcinogens (4-6). BHA rapidly induces cell proliferation in the forestomach epithelium of both rats (3) and hamsters (7); this increase in DNA synthetic activity presumably plays an important role in the observed carcinogenic effect.

0097–6156/92/0507–0269$06.00/0

Recently, some plant phenols, i.e. caffeic acid, sesamol and catechol, have been shown to also induce strong cell proliferation in either the forestomach or glandular stomach epithelium of rats (8) and hamsters (9). These antioxidants similarly appear to exert carcinogenic potential in their target organs, enhance stomach carcinogenesis, or modify carcinogenesis in other organs.

Recently we developed multi-organ carcinogenesis models to examine modifying activity of chemicals (10-13). In these models complex (enhancing or inhibitory) modifying effects of chemicals in over ten organs could be detected in a single experiment. Thus these models could be applied to distinguish chemopreventors which do not exert enhancing effects in any organ.

Catechins, which are major components of green tea tannins, have been suggested as potential chemopreventors since they inhibit teleocidine-induced promotion of mouse skin carcinogenesis initiated by DMBA (14), N-ethyl-N'-nitro-N-nitrosoguanidine-induced mouse duodenal carcinogenesis (15) and spontaneously-induced mammary carcinomas in C3H/HeN mice (16).

It is well known that the reaction of secondary amines and sodium nitrite in acidic conditions results in formation of carcinogenic nitroso compounds. Several phenolic compounds and antioxidants such as ascorbic acid and α–tocopherol have been demonstrated to block formation of carcinogenic nitroso compounds by reducing nitrous acid to nitric oxide (17). On the other hand, interaction of phenolic compounds such as phenol (18), butylated hydroxyanisole (19), catechol and 3-methoxycatechol (20), and $NaNO_2$ under acidic conditions is itself associated with generation of genotoxic substances. These genotoxic compounds may produce tumors in the rat stomach, where they directly contact the epithelium.

In the present paper carcinogenicities of caffeic acid, sesamol, catechol and hydroquinone (an isomer of catechol), modification of carcinogenesis by these phenolic compounds, modification of carcinogenesis by green tea catechins using rat multi-organ carcinogenesis model, and combined effects of phenolic compounds and $NaNO_2$ on rat gastric epithelium are reviewed.

Carcinogenicities of Phenolic Compounds in Rats

Groups of 30, 6-week-old F344 rats of both sexes (Charles River Japan Inc., Kanagawa) were treated with 2% caffeic acid (Tokyo Kasei Kogyo Co., Tokyo, purity > 98%), 2% sesamol (Fluka Chemie AG, Switzerland, purity >98%) or 0.8% catechol (Wako Pure Chemical Industies, Osaka, purity > 99%) in Oriental MF powdered basal diet (Oriental Yeast Co., Tokyo) or basal diet alone for 104 weeks. Food and water were given *ad libitum*. Chemicals were incorporated into powdered diet using a mixer and the diets stored at 4° C until use. Animals which died during the experiment were necropsied and all surviving animals were killed under anesthesia and subjected to complete autopsy at the end of week 104. Livers and kidneys were weighed before fixation in 10% buffered formalin solution. Formalin was injected into the stomachs, which were then opened via an incision along the greater curvative. Three sections each were cut from the anterior and posterior walls of the forestomach and six sections from each glandular stomach. Tissues were processed in the usual way for histopathological examination. Animals which survived more than 77 weeks (caffeic acid, rat), 82 weeks (sesamol, rat), 26 weeks (catechol, male rats), 52 weeks (catechol, female rats), when the first tumor appeared, were included in the effective numbers. Student's *t* test and Fisher's exact probability test were used for statistical evaluation of the data.

At the end of the experiment, body weights of animal treated with phenolic compounds were generally lower than in the controls, particularly for animals treated with catechol (17.1%-41.1% reduction compared to controls). The relative liver and kidney weights, however, were all higher in animals receiving the phenolic compounds.

Histopathologically, changes in the forestomach were classified into papilloma and squamous cell carcinoma categories as previously reported. The results are summarized in Table I. Caffeic acid induced significant increases in the incidences of papillomas and squamous cell carcinomas in male and female rats. Sesamol also induced, 34 and 47%, incidences of papillomas in male and female rats, respectively. Significant increase in the development of squamous cell carcinoma was, however, observed only in male rats (p<0.001). Squamous cell carcinomas induced in rats by these compounds were mostly of well to moderately differentiated type. No significant increase in forestomach tumor incidence was found in rats given catechol. Lesions of the glandular stomach were classified into adenoma and adenocarcinoma categories as previously reported. The results are summarized in Table I. Catechol induced adenomas in all male and female rats. Well differentiated adenocarcinomas were induced in 54 and 43% of male and female rats, respectively. No neoplastic lesions were observed in the fundic region of the glandular stomach of rats treated with these phenolic compounds (*21*).

In addition, neoplastic and preneoplastic lesions were found in the kidneys of some animals treated with caffeic acid and hydroquinone, as summarized in Table I. Seventy three and 20% of males and females, respectively, treated with caffeic acid had hyperplasia, and 13% of males had adenoma. In the case of hydroquinone, 100% of the males and 7% of the females had tubular hyperplasia. In addition, 47% (p<0.001) of male rats treated with hydroquinone had renal adenomas while no such tumorous lesions were in females.

Modifying Effects of Phenolic Compounds on Second Stage Carcinogenesis

Separate groups of 6 - 7 week old F344 male or SD female rats were treated with carcinogens as initiators as following: *N*-methyl-*N*'-nitro-*N*-nitrosoguanidine (MNNG, 150 mg/kg bw ig, single), *N*-methyl-*N*-amylnitrosamine (MNAN, 25 mg/kg bw sc, 1/week x 3), 1,2-dimethylhydrazine (DMH, 20 mg/kg bw sc, 1/week x 4), *N*-ethyl-*N*-hydroxyethylnitrosamine (EHEN, 0.1% in drinking water, 2 weeks), 2,2-dihydroxy-di-*n*-propylnitrosamine (DHPN, 0.1% in drinking water, 2 weeks), *N*-butyl-*N*-(4-hydroxybutyl)nitrosamine (BBN, 0.05% in drinking water, 2 weeks) or 7,12-dimethylbenz[*a*]anthracene (DMBA, 50 mg/kg bw ig, single). Then they were administered diet containing 1% caffeic acid, 0.8% catechol or 0.8% hydroquinone for 30 - 51 weeks. Further groups of 10 - 15 animals were treated with carcinogen or test chemical, or basal diet alone. The animals were sacrificed under ether anesthesia at weeks 32 - 52 and the target organs for each carcinogen were examined histopathologically.

Incidences of squamous cell carcinomas and adenocarcinomas of forestomach and glandular stomach, respectively, in rats pretreated with MNNG followed by caffeic acid, catechol or hydroquinone are summarized in Figure 1. All animals treated with 1% caffeic acid or 0.8% catechol after MNNG exposure had forestomach squamous cell carcinomas (SCC) whereas 0 to 26.3% of animals treated with MNNG alone had SCC. The incidence of SCC in animals treated with hydroquinone (25%)

Table I. Incidences of Lesions in the Stomach and Kidney Treated with Chemicals for 104 Weeks in F344 Rats

Chemicals	Dose (%)	Sex	No. of rats	Forestomach		Gl. stomach		Kidney	
				Papilloma	Squamous cell carcinoma	Adenoma	Adeno-carcinoma	Tubular hyperplasia	Adenoma
Caffeic acid	2	M	30	23 (77)***	17 (56.7)***	1 (3)	0	22 (73)***	4 (13.3)
		F	30	24 (80)***	15 (50)***	0	0	6 (20)*	0
Sesamol	2	M	29	10 (34)***	9 (31.0)***	1 (3)	0	0	0
		F	30	14 (47)***	3 (10)	0	0	0	0
Catechol	0.8	M	28	0	0	28 (100)***	15 (53.6)***	0	0
		F	28	0	0	28 (100)***	12 (42.9)***	0	0
Hydroquinone	0.8	M	30	0	0	0	0	30 (100)***	14 (46.6)***
		F	30	0	0	0	0	2 (7)	0
Control	-	M	30	0	0	0	0	1 (3)	0
		F	30	0	0	0	0	0	0

*P<0.02, ***P<0.001 vs respective control group value

was not different from the control value (20%). The incidence of glandular stomach adenocarcinoma was significantly enhanced only in rats treated with MNNG followed by catechol (94.7% vs 0% in control) (*7, 21, 22*). In organs other than those targeted for carcinogenicity, different modifying effects (promotion and inhibition) were observed depending on the individual organ as shown in Table II. For example, catechol promoted esophageal carcinogenesis but inhibited colon, lung, and mammary gland carcinogenesis; hydroquinone enhanced kidney carcinogenesis and weakly promoted esophageal carcinogenesis (*22, 23*).

Effects of Green Tea Catechins in a Rat Multi-organ Carcinogenesis Model

Groups of 15 male F344 rats, 6 weeks old at the commencement of the study (Charles River Japan, Inc., Atsugi, Kanagawa), were treated with combined single ip administration of 100 mg/kg bw N-diethylnitrosamine (DEN), 4 ip administrations of 20 mg/kg bw N-methylnitrosourea (MNU), 4 sc doses of 40 mg/kg bw N,N-dimethylhydrazine (DMH), 0.05% BBN in drinking water for 2 weeks or 0.1% DHPN for 2 weeks during the initial 4 week period for wide spectrum initiation. Experimental groups were formed as follows: Group 1, 1% catechins in Oriental MF powdered basal diet administered from one day before and during carcinogen exposure. Group 2, 1% catechins in diet administered after carcinogen exposure. Group 3, carcinogens alone. Group 4, 1% catechins in diet alone and group 5, basal diet alone during the experiment. All animals were killed under ether anesthesia at the end of week 36, and complete autopsies were performed. Tissues were processed routinely for histopathological examination.

Final average body weights of rats treated with catechins were slightly higher than those in the initiation alone group. Incidences and average number of tumors in the small intestine are summarized in Table III. Incidences of adenomas in groups 1 and 2 were significantly lower (7 and 8%, respectively, $p<0.001$) than in group 3 (50%), and also average numbers of adenomas per rat were significantly lower in these groups (0.1 ± 0.3, $p<0.001$) than in group 3 (0.7 ± 0.8). Values for incidence and average number per rat of adenocarcinomas in group 1 (7%, 0.1 ± 0.3, $p<0.005$) were also significantly lower than in group 3 (42%, 0.6 ± 0.8), and those in group 2 tended to be lower than in group 3 but not significantly. Incidence and average number of total tumors (including adenomas and carcinomas) were also clearly lower in group 1 (13%, 0.1 ± 0.4, $p<0.01$) than in group 3 (67%, 1.3 ± 1.2), but in group 2, significant inhibition was observed only for the number of tumors per rat ($p<0.05$).

Incidences of tumors in other organs are presented in Table IV. No significant enhancement or inhibition of tumor development was evident in the esophagus, forestomach, colon, liver, lung, kidney, urinary bladder, hematopoietic system and thyroid glands.

Effect of Combined Treatment with Sodium Nitrite and Phenolic Compounds

Groups of 10 F344 male rats at 10 weeks of age (Charles River Japan Inc., Kanagawa) were treated with 0.3% $NaNO_2$ in the drinking water and simultaneously administered catechol or 3-methoxycatechol in Oriental MF powdered basal diet (Oriental Yeast Co., Tokyo) for 24 weeks. Additional groups received the single

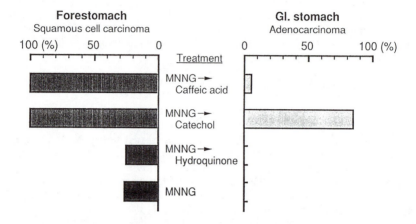

Figure 1. Incidences of forestomach and glandular stomach tumors in rats pretreated with MNNG followed by phenolic compounds.

Table II. Modifying Effects of Phenolic Compounds Carcinogenesis in Rats Pretreated with Different Carcinogens

Target organ (carcinogen)	Catechol	Hydroquinone
Esophagus (MNAN)	↑	↑
Colon (DMH)	↓	NE
Liver (EHEN)	→	→
Lung (DHPN)	↓	→
Kidney (EHEN)	→	↑
Urinary bladder (BBN)	→	→
Mammary gland (DMBA)	↓	NE
Thyroid gland (DHPN)	→	NE

NE, not examined

↑, enhancement; →, no effect, ↓, inhibition

Table III. Incidences and Numbers of Tumors in The Small Intestine

Group	Treatment	No. of rats	Adenoma Incidence[a]	Adenoma No./rat[b]	Carcinoma Incidence	Carcinoma No./rat	Total Incidence	Total No./rat
1	Initiation + Catechins	15	1 (7)*	0.1±0.3***	1 (7)*	0.1±0.3*	2 (13)**	0.1±0.4**
2	Initiation → Catechins	13	1 (8)*	0.1±0.3***	3 (23)	0.2±0.4	4 (31)	0.3±0.5*
3	Initiation alone	12	6 (50)	0.7±0.8	5 (42)	0.6±0.8	8 (67)	1.3±1.2

[a](%), [b]Mean±SD
***p<0.001, **p<0.01, *<0.05 vs respective initiation alone group value

Table IV. Incidences of Tumors in Other Organs

Sites and Lesions	Group 1 Initiation+Catechins n = 15	Group 2 Initiation → Catechins n=13	Group 3 Initiation alone n=12
Esophagus Papilloma	1 (7)	0	0
Forestomach Papilloma	1 (7)	0	0
Liver Hyperplastic nodule	0	0	1 (8)
Colon Adenoma	9 (60)	7 (53)	7 (58)
Carcinoma	3 (20)	6 (46)	3 (25)
Lung Adenoma	9 (60)	8 (62)	5 (41)
Carcinoma	2 (13)	1 (8)	0
Kidney Adenoma	5 (33)	4 (31)	4 (33)
Nephrobrastoma	6 (40)	7 (54)	8 (67)
Urinary bladder Papilloma	0	3 (23)	2 (17)
Hematopoietic Lymphoma/Leukemia	4 (27)	1 (8)	0
Thyroid glands Adenoma	5 (33)	3 (23)	2 (17)
Carcinoma	2 (13)	0	1 (8)

Numbers in parentheses: %

0.3% $NaNO_2$, 0.8% catechol, 2% 3-methoxycatechol treatments or basal diet alone for the same period. All chemicals were obtained from Wako Pure Chemical Industries, Osaka. Animals were kept 5 to a plastic cage in an air-conditioned room at $24 \pm 2°$ C, and food and water were given *ad libitum*. Surviving animals were killed under ether anesthesia at the end of the experimental period of 24 weeks. Stomachs were removed and injected with 10% buffered formalin solution. After opening along the greater curvature, six strips each were cut from the anterior and posterior walls of the forestomach and 6 strips were cut from the pyloric region of the glandular stomach. Tissues were processed routinely and stained with hematoxylin and eosin. The Fisher exact test and Student's *t* test were used for the statistical analysis of the data.

Final body weights of animals treated with phenolic compounds or $NaNO_2$ were significantly lower than those of rats receiving basal diet alone. Combined treatment with either of the phenolic compounds and $NaNO_2$ futher reduced the body weights as compared to the phenolic compound alone. Food consumption of animals treated with phenolic compound and $NaNO_2$ was also slightly lower than that of animals on phenolic compounds alone. Grossly, the stomach contents and forestomach epithelia of rats treated with $NaNO_2$ plus phenolic compound demonstrated blackening and marked papillary projections with dense keratin-like material being evident scattered throughout the epithelium. No such abnormalities were observed in the forestomach of rats treated with catechol or 3-methoxycatechol alone. The pyloric region of the glandular stomach of rats treated with catechol or 3-methoxycatechol was thickened nodularly, but the changes were much less in animals treated with $NaNO_2$ plus catechol or $NaNO_2$ plus 3-methoxycatechol. Histological changes in the forestomach were classified into hyperplasias and papillomas. As shown in Figure 2, one rat and 5 rats, respectively, treated with $NaNO_2$ plus catechol and $NaNO_2$ plus 3-methoxycatechol, had papilomas. Severe hyperplasia was not found in rats receiving catechol or 3-methoxycatechol alone. Simultaneous treatment with $NaNO_2$, however, was associated with its induction, i.e., 80% of rats treated with catechol plus $NaNO_2$ had severe hyperplasia and all rats treated with 3-methoxycatechol plus $NaNO_2$ had severe hyperplasia. In the glandular stomach, lesions were classified into submucosal hyperplasia and adenoma categories. All rats treated with catechol had submucosal hyperplasia and adenomas, and 9 and 7 rats treated with 3-methoxycatechol had submucosal hyperplasia and adenomas, respectively. The incidences of these lesions were remarkably reduced by the combined treatment with $NaNO_2$ (25).

Discussion

Previously, phenolic antioxidants such as BHA and BHT have been considered to be good chemopreventors since they generally inhibit carcinogenesis when applied to animals simultaneously with or shortly before carcinogen treatment, without showing significant toxicity. BHA, however, was demonstrated to induce squamous cell carcinomas in rat and hamster forestomach epithelium (1-3), while enhancing forestomach and urinary bladder carcinogenesis when administered both during and after carcinogen exposure (26). On the other hand, BHA inhibited carcinogenesis in the rat colon, lung and mammary gland (4-6, 23). In the present series of experiments, some plant phenols could be demonstrated to have a wide variety of actions like BHA with regard to chemical carcinogenesis. Caffeic acid and sesamol were shown to be carcinogenic to rat forestomach epithelium, catechol to rat glandular stomach epithelium and caffeic acid and hydroquinone were tumorigenic for the rat kidney. Chemical structures of carcinogenic antioxidants are presented in Figure 3. These chemicals are distributed widely in our environment and exposure to man is likely through food ingestion, cigarette smoke and cosmetic application. Caffeic acid is present in potatoes, lettuce, apples, coffee beans and soybeans, sesamol is a minor component of sesame seed oil, catechol has been demonstrated in cigarette smoke, wood smoke condensates, hair dye and coffee, and hydroquinone has been found in many kinds of plants as a component of the glucoside arbutin as well as in wood smoke condensates, hair dyes and skin lighteners (27, 28). Of these chemicals, catechol seems to be potentially the most important since the human stomach resembles the rat glandular stomach which is its target. Actually, catechol or its conjugates are excreted in human urine at levels up to 30 mg/day (29).

In the 2-stage carcinogenesis models, phenolic compounds showed inhibitory or enhancing effects on different organs. Generally, they enhanced carcinogenesis in the target organs of the initiating carcinogenic compound, caffeic acid and catechol treatments being associated with increased development of forestomach and glandular stomach lesions, respectively, when given after MNNG initiation. Similarly, hydroquinone is a renal carcinogen and enhanced EHEN-induced renal carcinogenesis. These enhancing effects on the second stage of carcinogenesis might be closely related to cell proliferation or an increase in DNA synthesis. Thus catechol strongly enhanced forestomach carcinogenesis at a dose 0.8%, at which level only hyperplasias and a small incidence of papillomas were induced. Catechol which promoted rat esophageal carcinogenesis also induced increase in the labeling index in rat esophageal epithelium (30). Similar relationships between DNA synthesis and promotion of carcinogenesis have been demonstrated for other phenolic compounds (24). On other hand, the mechanisms whereby phenolic compounds inhibit the second stage of 2-stage carcinogenesis are not known.

For the primary prevention of human cancer, it is clearly of importance that carcinogens or enhancers be eliminated, and ingestion of chemopreventors be increased. Recently several multi-organ carcinogenesis models have been developed in which promotion as well as inhibition by chemicals over multiple organs could be studied in a single experiment (10-12). Since chemopreventors should ideally not have any enhancing effects, these animal models should be useful for their detection. In the present multi-organ carcinogenesis model, catechins inhibited rat small intestinal carcinogenesis when they were administered during as well as after carcinogen exposure. Catechins did not, however, inhibit carcinogenesis in the colon which is a target of DMH. These results might indicate that catechins do not modify metabolism of carcinogens but rather directly inhibit cell proliferation or DNA synthesis in the small intestinal epithelium. Similar inhibitory effects of (-)-epigallocatechin gallate on N-ethyl-N'-nitro-N-nitrosoguanidine-induced mouse duodenal carcinogenesis have been previously reported (15). This chemical did not enhance carcinogenesis in any other organ as evaluated in terms of tumor incidences.

Phenolic compounds have been shown to inhibit nitrosamine formation by the interaction of amines and $NaNO_2$ (14-17). The present experiment, however, clearly demonstrated that the target organ for induction of cell proliferation by phenolic compounds can be altered by combined treatment with $NaNO_2$. Thus, in the present study catechol and 3-methoxycatechol caused strong cell proliferation in the glandular stomach epithelium as evidenced by the induction of submucosal hyperplasia and adenomas, but combined administration of $NaNO_2$ strong reduced their activity. On the other hand $NaNO_2$ remarkably enhanced the development of proliferation-

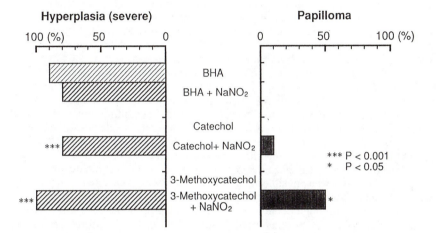

Figure 2. Incidences of forestomach lesions in rats treated with phenolic compounds with or without the presence of NaNO$_2$— significantly different at ***P<0.001 and *P<0.05 as compared with phenolic compound alone group.

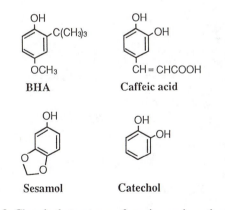

Figure 3. Chemical structures of carcinogenic antioxidants.

associated lesions in the forestomach of rats treated with catechol or 3-methoxycatechol. Considering the positive correlation between strong cell proliferation and carcinogenicity as evidenced by the results with BHA, caffeic acid, sasamol and catechol, these phenolic compounds may cause forestomach carcinogenicity in the presence of $NaNO_2$. Although the decreased incidences of adenomas and submucosal hyperplasia in rats treated with phenolic compounds and $NaNO_2$ could be partly due to reduced food consumption due to $NaNO_2$ and therefore decreased intake of phenolic compounds, the observed shift in target site favors the interpretation that alternative reactive metabolites exist. Thus, in place of those which induce glandular stomach cell proliferation, other metabolites, targeting the forestomach epithelium and associated with blackening, appeared to be formed by the reaction with $NaNO_2$. It has been shown that phenol reactions with nitrite to produce nitrosophenol, p-diazoquinone and o-diazoquinone (18, 31). Interestingly, p-diazoquinone proved to be mutagenic in S. typhimurium TA 98 and TA 100 without metabolic activation (18). Similarly, Ohshima et al. reported that 3-methoxycatechol and catechol could produce direct-acting mutagenic diazonium compounds after nitrosation under acidic conditions in vitro (20). Recently we demonstrated DNA adducts in the forestomach epithelium of rats treated with catechol and $NaNO_2$ (unpublished results). On the other hand, reaction of BHA and $NaNO_2$ under acidic conditions did not produce diazonium compounds (19). Thus genotoxic diazonium compounds might be responsible for the development of forestomach lesions by phenolic compounds and $NaNO_2$.

In conclusion, plant phenols exert various beneficial and hazardous effects either by themselves or in combination with other chemicals. More care in the assessment of the actions of these plant phenols in man is clearly required.

Literature Cited

1. Ito, N.; Fukushima, S.; Hagiwara, A.; Shibata, M.; Ogiso, T. J. Natl. Cancer Inst. 1983, 70, 343.
2. Ito, N.; Fukushima, S.; Tamano, S.; Hirose, M.; Hagiwara, A. J. Natl. Cancer Inst. 1986, 77, 1261.
3. Masui, T.; Hirose. M.; Imaida, K.; Fukushima, S.; Tamano, S.; Ito, N. Jpn. J. Cancer Res. 1986, 77, 1083.
4. Ito, N.; Hirose, M. Jpn. J. Cancer Res. 1987, 78, 1011.
5. Ito, N.; Hirose, M.; Fukushima, S.; Tsuda, H; Shirai, T.; Tatematsu, M. Fd. Chem. Toxicol. 1986, 24, 1071.
6. Ito, N.; Hirose, M. Adv. Cancer Res. 1989, 53, 247.
7. Hirose, M.; Masuda, A.; Kurata, Y.; Ikawa, E.; Mera, Y.; Ito, N. J. Natl. Cancer Inst., 1986, 76, 143.
8. Hirose, M.; Masuda, A.; Imaida, K.; Kagawa, M.; Tsuda, H.; Ito, N. Jpn. J. Cancer Res. 1987, 78, 317.
9. Hirose, M.; Inoue, T.; Asamoto, M.; Tagawa, Y.; Ito, N. Carcinogenesis 1986, 7, 1285.
10. Ito, N.; Imaida, K.; Tsuda, H.; Shibata, M.; Aoki, T.; de Camargo, J.L.V.; Fukushima, S. Jpn. J. Cancer Res. 1988, 79, 413.
11. Thamavit, W.; Fukushima, S.; Kurata, Y.; Asamoto, M.; Ito, N. Cancer Lett. 1989, 45, 93.
12. Shibata, M.; Fukushima, S.; Takahashi, S.; Hasegawa, R.; Ito, N.; Carcinogenesis 1990, 11, 1027.

13. Fukushima, S.; Hagiwara, A.; Hirose, M.; Yamaguchi, S.; Tiwawech, D.; Ito, N. *Jpn. J. Cancer Res.* **1991**, *82*, 642.
14. Yoshizawa, S.; Horiuchi, T.; Fujiki, H.; Yoshida, T.; Okuda, T.; Sugimura, T. *Phytother. Res.* **1987**, *1*, 44.
15. Fujita, Y.; Yamane, T.; Tanaka, M.; Kuwata, K.; Okuzumi, J.; Takahashi, T.; Fujiki, H.; Okuda, T.; *Jpn. J. Cancer Res.* **1989**, *80*, 503.
16. Asai, H.; Hara, Y.; Nakamura, K.; Hosaka, H.; Kokue, A. *Proc. Jpn. Cancer Assoc.* **1988**, *47*, 139.
17. Mirvish, S.S. *Cancer* **1986**, *58*, 1842.
18. Kikugawa, K.; Kato, T. *Fd. Chem. Toxicol.* **1988**, *26*, 209.
19. Mizuno, M.; Ohara, A.; Danno, G.; Kanazawa, K; Natake, M. *Mutation Res.* **1987**, *176*, 179.
20. Ohshima, H.; Friesen, M.; Malaveille, C.; Brouet, I.; Hautefeuille, A.; Bartsch, H. *Fd. Chem. Tocicol.* **1989**, *27*, 193.
21. Hirose, M.; Fukushima, S.; Shirai, T.; Hasegawa, R.; Kato, T.; Tanaka, H.; Asakawa, E.; Ito, N. *Jpn. J. Cancer Res.* **1990**, *81*, 207.
22. Yamaguchi, S.; Hirose, M.; Fukushima, S.; Hasegawa, R.; Ito, N. *Cancer Res.* **1989**, *49*, 6015.
23. Hasegawa, R.; Furukawa, F.; Toyoda, K.; Takahashi, M.; Hayashi, Y.; Hirose, M.; Ito, N. *Jpn. J. Cancer Res.* **1990**, *81*, 871.
24. Hirose, M.; Yamaguchi, S.; Fukushima, S.; Hasegawa, R.; Takahashi, S.; Ito, N. *Cancer Res.* **1989**, *49*, 5143.
25. Hirose, M.; Fukushima, S.; Hasegawa, R.; Kato, T.; Tanaka, H.; Ito, N. *Jpn. J. Cancer Res.* **1990**, *81*, 857.
26. Ito, N.; Hirose, M.; Shibata, M.; Tanaka, H.; Shirai, T. *Carcinogenesis* **1989**, *10*, 2255.
27. Stich, H.F.; Rosin, M.P. *Adv. Exp. Biol.* **1984**, *177*, 1.
28. IARC Monographs on the Evaluation of Carcinogenic Risk of Chemicals to Humans "Some Fumigants, the Herbicides 2,4-D and 2,4,5-T, Chlorinated Dibenzodioxins and Miscellaneous Industrial Chemicals"; IARC: Lyons, **1977**, Vol. 15; pp. 155-175.
29. Carmella, S.G.; La Voie, E.J.; Hecht, S.S.; *Fd. Chem. Toxicol.* **1988**, *20*, 587.
30. Mirvish, S.S.; Salmasi, S.; Lawson, T.A.; Pour, P.; Sutherland, D. *J. Natl. Cancer Inst.* **1985**, *74*, 1283.
31. Challis, B.C. *Nature* **1973**, *244*, 466.

RECEIVED February 11, 1992

GREEN TEA AND CANCER PREVENTION

Chapter 21

Inhibitory Effect of Green Tea on Tumorigenesis and Tumor Growth in Mouse Skin

Allan H. Conney[1], Zhi Yuan Wang[1], Chi-Tang Ho[2], Chung S. Yang[1], and Mou-Tuan Huang[1]

[1]Laboratory for Cancer Research, College of Pharmacy, Rutgers, The State University of New Jersey, P.O. Box 789, Piscataway, NJ 08855–0789
[2]Department of Food Science, Cook College, Rutgers, The State University of New Jersey, New Brunswick, NJ 08903

Topical application of a green tea polyphenol fraction inhibited 12-O-tetradecanoylphorbol-13-acetate-induced tumor promotion in mouse skin. Oral administration of green tea inhibited ultraviolet B light-induced tumorigenesis in mouse skin, and tumors from tea-treated animals were smaller than tumors from water-treated control animals. Oral administration of green tea to papilloma-bearing mice inhibited tumor growth and in some experiments caused a decrease in tumor size.

Green tea is an important beverage in Asian countries such as China and Japan, whereas black tea is more popular in North America and Europe. Recent investigations have shown that constituents of green tea have an inhibitory effect on teleocidin-induced and 7,12-dimethylbenz[a]anthracene (DMBA)-induced tumorigenesis in mouse skin and on N-ethyl-N'-nitro-N-nitrosoguanidine-induced tumorigenesis in mouse duodenum (1-3). Additional studies have shown an inhibitory effect of green and black tea on N-nitrosomethylbenzylamine-induced esophageal tumors in rats (4) and an inhibitory effect of green tea on ultraviolet B light (UVB)-induced and 12-O-tetradecanoylphorbol-13-acetate (TPA)-induced skin tumors in mice (5-7). In the present report, we summarize our studies showing an inhibitory effect of a topically applied green tea polyphenol fraction on tumor promotion by TPA and an inhibitory effect of orally administered green tea on UVB-induced tumorigenesis in mouse skin. We also describe an inhibitory effect of orally administered green tea on the growth of experimentally-induced papillomas in mouse skin.

Materials and Methods

Animals. Female CD-1 or SKH-1 (hairless) mice (6 to 8 weeks old) were purchased from Charles River Laboratories (Kingston, NY). The animals were kept in our animal facility for at least 1 week before use. Mice were given water and food ad libitum and kept on a 12 hour light, 12 hour dark cycle. Purina Laboratory Chow 5001 diet (Ralston-Purina Co., St. Louis, MO) was used for all studies.

0097–6156/92/0507–0284$06.00/0

Preparation and Composition of Green Tea Polyphenol Fraction. One hundred grams of green tea leaves were extracted three times with 300 ml of methanol at 50°C for 3 hours, and the samples were filtered after each extraction. Solvent was removed from the combined extract with a vacuum rotary evaporator. The residue was dissolved in 500 ml of water (50°C) and extracted three times with 200 ml hexane and three times with 200 ml chloroform. The aqueous phase was extracted three times with 180 ml ethyl acetate, and the ethyl acetate was evaporated under reduced pressure. The residue was redissolved in 300 ml water and lyophilized to obtain 8 to 9 g of green tea polyphenol fraction. HPLC analysis of the green tea polyphenol fraction revealed the following composition: (-)-epigallocatechin gallate (49.5%), (-)-epicatechin gallate (11%), (-)-epicatechin (6.1%), (-)-epigallocatechin (11.5%) and caffeine (7.6%).

Preparation and Composition of Green Tea Aqueous Extract. Green tea leaves (12.5 g) were added to 500 ml of boiling water and were steeped for 15 minutes. The infusion was cooled to room temperature in an ice bath and then filtered. The tea leaves were extracted a second time, and the two filtrates were combined to obtain a 1.25% green tea water extract (1.25 g tea leaves/100 ml water). The resulting clear solution is similar to tea brews consumed by humans. In some experiments, the 1.25% green tea water extract was diluted 1:1 with water to make a 0.63% green tea extract. During specified experimental periods, the tea extract was given *ad libitum* as the sole source of drinking water.

The amount of solids present in the 1.25% green tea extract was determined to be 4.69 mg per ml by drying samples in an air convection oven (18 hours at 65°C) and weighing the dry residue. Analysis of the green tea extract by HPLC indicated that the total green tea solids in the extract contained (-)-epigallocatechin gallate (15.1%), (-)-epigallocatechin (6.9%), (-)-epicatechin gallate (3.0%), (-)-epicatechin (1.8%), (+)-catechin (0.5%), caffeine (8.1%) and theobromine (0.4%). The compounds that were measured accounted for 36% of the total green tea solids present in the water extract. Storage of the green tea water extract for 72 hours at room temperature resulted in only a small decrease (8 to 12%) in the concentration of tea catechins.

Ultraviolet Light. Ultraviolet lamps (FS72T12-UVB-HO) that emit ultraviolet B light (UVB) (280 to 320 nm; 75 to 80% of total energy) and ultraviolet A light (UVA) (320 to 375 nm; 20 to 25% of total energy) were obtained from the Voltare Co. (Fairfield, CT). The dose of UVB was quantified with an UVB Spectra 305 Dosimeter that was obtained from the Daavlin Co. (Bryan, OH). The radiation was further calibrated with an IL1700 Research Radiometer/Photometer from International Light Inc. (Neburgport, MA).

For exposure to UV (UVA+UVB), 10 mice were housed in a 25.4 cm (width) x 45.7 cm (length) plastic box. Six boxes (without tops) were placed under eight UV lamps; 50.8 cm (width) x 182.9 cm (length), and the boxes were systematically rotated during the course of the study to compensate for possible small differences in flux at various positions under the lamps. The distance between the UV lamps and the backs of the mice or the UVB detector was 43.2 cm. The amount of exposure to UVB was controlled by a Spectra 305 Dosimeter. The exposure time for a 180 mJ/cm^2 dose of UVB was 130 to 160 sec. Although all data are expressed as exposure to UVB, some additional exposure to UVA also occurred as indicated above.

Results and Discussion

Inhibitory Effect of Topically Applied Green Tea Polyphenol Fraction on TPA-induced Tumor Promotion in Mouse Skin. Mice that were initiated with 200 nmol of DMBA and promoted with 5 nmol of TPA twice weekly for 20 weeks developed 20.7 skin tumors per mouse, and 93% of the mice had skin tumors (Table I). Topical application of 3.6 mg of green tea polyphenol fraction together with each treatment of TPA markedly inhibited the percentage of mice with tumors and the number of tumors per mouse (Table I).

Table I. Effect of Topical Application of a Green Tea Polyphenol Fraction on TPA-induced Tumor Promotion in Mouse Skin

	12 Weeks		20 Weeks	
Treatment	Tumors per mouse	Percent mice with tumors	Tumors per mouse	Percent mice with tumors
DMBA + Acetone	0[a]	0	0[a]	0
DMBA + TPA	15.3 ± 2.4	90	20.7 ± 2.9	93
DMBA + TPA + GTP (3.6 mg)	0.6 ± 0.4[a]	17	1.2 ± 0.5[a]	27

Female CD-1 mice were treated topically with 200 nmol of DMBA in 200 μl acetone followed a week later by acetone, TPA (5 nmol) or TPA (5 nmol) together with a green tea polyphenol fraction (GTP, 3.6 mg) in acetone twice weekly for 20 weeks. Skin tumors with a diameter greater than 1 mm were recorded. The data were taken from ref. 7 and represent the mean ± S.E. from 30 mice per group.
[a]Statistically different from DMBA + TPA group ($p<0.001$)

Inhibitory Effect of Orally Administered Green Tea on UVB-induced Tumorigenesis in Mouse Skin. Treatment of the backs of SKH-1 mice with UVB (180 mJ/cm^2) once daily for 10 days followed by topical application of TPA twice weekly for 25 weeks resulted in the development of skin tumors. The time of appearance of tumors was delayed, and the number of tumors per mouse was decreased when green tea extract was given in the drinking water for 2 weeks before and until 1 week after UVB treatment (Table II). In experiment one, the animals were given 1.25% green tea extract starting 2 weeks before UVB treatment, and this resulted in a 10 to 18% decrease in body weight compared with water-treated control mice during the first 3 to 8 weeks of the study. In a second experiment, the mice were administered 25, 50, 75, and 100% of full strength 1.25% green tea extract starting on days 1, 3, 5, and 7 (2 days at each dose), and full strength green tea was then administered for a week before and until 1 week after UVB treatment. Green tea was discontinued 1 week after the last dose of UVB, and the animals were treated with TPA twice weekly for 25 weeks. No significant effect of green tea on body weight was observed in this second study. Administration of 1.25%

Table II. Inhibitory Effect of Oral Administration of Green Tea on the Initiation of Skin Tumors by UVB

Treatment	5 weeks of TPA promotion		15 weeks of TPA promotion		25 weeks of TPA promotion	
	Tumors per mouse	Percent of mice with tumors	Tumors per mouse	Percent of mice with tumors	Tumors per mouse	Percent of mice with tumors
Experiment 1						
Water	0.1 ± 0.1	13	0.2 ± 0.1	20	1.1 ± 0.2	57
Tea (1.25%)	0^a	0	0^a	0	0.2 ± 0.1^a	13
Experiment 2						
Water	0.3 ± 0.1	22	0.9 ± 0.2	41	2.5 ± 0.7	81
Tea (1.25%)	0^a	0	0.3 ± 0.1^a	26	0.9 ± 0.2^a	48

In experiment 1, female SKH-1 mice (30 per group) were given 1.25% green tea extract as the sole source of drinking water for 2 weeks prior to and during 10 days of UVB treatment (180 mJ/cm^2/day) and for 1 week after UVB treatment. The mice were then given only water and treated topically with 10 nmol of TPA twice weekly for 15 weeks and with 16 nmol TPA twice weekly for another 10 weeks. In experiment 2, female SKH-1 mice (30 per group) were given water or green tea extract in the drinking water with a step-wise increase in dose from 25, 50, 75 and 100% of the final dose of 1.25% green tea extract on days 0, 2, 4, and 6, respectively, starting 2 weeks before the administration of UVB. The mice continued on their final dose of 1.25% green tea extract for an additional week prior to UVB treatment, during 10 days of UVB treatment (180 mJ/cm^2/day) and for 1 week after UVB treatment. The mice were then given only water and treated topically with 16 nmol of TPA twice weekly for 25 weeks. The data were taken from ref. 6 and represent the mean ± S.E. from 26-30 mice.

[a]Statistically different from the corresponding water group (p< 0.05).

green tea extract decreased the number of tumors per mouse by 82% in experiment 1 and by 64% in experiment 2 (Table II). The percentage of tumor-bearing mice decreased by 77% in experiment 1 and 41% in experiment 2 (Table II).

Although the topical application of 200 nmol of DMBA to SKH-1 mice did not result in the formation of skin tumors (data not presented), subsequent treatment of these mice with UVB (30-180 mJ/cm^2) twice weekly for 30 weeks produced large numbers of tumors. The formation of these tumors was inhibited by administration of green tea just prior to and during the UVB treatment period (Table III). In three separate experiments, we treated SKH-1 mice with 200 nmol of DMBA. One week later, the mice were given green tea in the drinking water, and 2 weeks after starting green tea administration the mice were treated with 30, 60 or 180 mJ of UVB per cm^2 twice weekly for 30 weeks. In these experiments, the dose of green tea was gradually increased during the first week of tea treatment, and full strength green tea was administered for 1 week prior to UVB administration and until the end of the study. No significant effect of green tea on body weight was observed using this dosing regimen. Under these conditions, 1.25% green tea extract decreased by 41%, 67% and 59%, respectively, the number of tumors per mouse that were induced by 180, 60 or 30 mJ of UVB per cm^2 twice weekly for 30 weeks (Table III). In these studies, oral administration of 0.63% green tea decreased by 53% and 38%, respectively, the number of tumors per mouse that resulted from administration of 60 or 30 mJ of UVB per cm^2 twice weekly in DMBA-initiated mice (Table III). These results indicate that oral administration of 0.63% or 1.25% green tea extract as the sole source of drinking water inhibits UVB-induced tumorigenesis in DMBA-initiated mice. Our results with green tea in the drinking water are in accord with those obtained in a recent study indicating an inhibitory effect of an orally administered green tea polyphenol fraction on ultraviolet light-induced tumorigenesis (5).

Administration of green tea not only decreased the number of UVB-induced tumors in DMBA-initiated mice, but tumor size was also markedly decreased (Table III). Mice that were initiated with DMBA and treated twice weekly with 30, 60 or 180 mJ/cm^2 of UVB twice a week for 30 weeks had an average tumor size of 31 mm^3, 62 mm^3, and 84 mm^{3,} respectively (Table III). Comparable mice treated with 1.25% green tea in the drinking water had an average tumor size of 5 mm^3, 14 mm^3, and 21 mm^3, respectively, and a substantial decrease in tumor size was also observed in mice treated orally with 0.63% green tea (Table III).

Inhibitory Effect of Orally Administered Green Tea on the Growth of Experimentally Induced Skin Papillomas. During the course of our chemoprevention studies on the effects of green tea on UVB-induced tumorigenesis, we observed that oral administration of green tea not only inhibited the formation of tumors, but we also found that tumors in tea-treated animals were smaller than tumors in control animals (Table III). Because of these observations, we have examined the effect of green tea administration on the growth of tumors in tumor-bearing mice. In these studies, papilloma-bearing animals were generated by treatment of mice with DMBA or UVB followed by topical application of TPA twice a week for 10-25 weeks in DMBA-initiated mice or 30 weeks in UVB-initiated mice. TPA administration was stopped and papilloma-bearing mice were treated with 0.63% or 1.25% green tea as the sole source of drinking water for 6 weeks (Table IV). Control animals received water. In experiment 1, tumors in control animals increased in size by 420%, whereas, tumors in animals treated with 0.63%

Table III. Effect of Oral Administration of Green Tea on the Number and Size of Skin Tumors Induced by UVB in DMBA-initiated SKH-1 Mice

Treatment	Number of mice	Percent of mice with tumors	Tumors per mouse	Tumor volume per mouse (mm^3)	Average volume per tumor (mm^3)
30 mJ/cm^2 twice weekly					
Water	30	83	3.9 ± 0.4	121 ± 56	31 ± 13
Tea (0.63%)	28	70	2.4 ± 0.6[a]	34 ± 19	14 ± 7
Tea (1.25%)	29	66	1.6 ± 0.3[c]	8 ± 4[a]	5 ± 2[a]
60 mJ/cm^2 twice weekly					
Water	30	100	12.1 ± 1.1	750 ± 313	62 ± 24
Tea (0.63%)	29	86	5.7 ± 0.8[b]	51 ± 15[a]	9 ± 2[a]
Tea (1.25%)	30	77	4.0 ± 0.9[b]	56 ± 18[a]	14 ± 4[a]
180 mJ/cm^2 twice weekly					
Water	25	100	11.1 ± 1.0	932 ± 324	84 ± 24
Tea (1.25%)	26	96	6.5 ± 1.1[b]	136 ± 29[a]	21 ± 5[a]

Female SKH-1 mice (6-8 weeks old; 30 mice per group) were treated topically with DMBA (200 nmol). One week later, the mice were given 1/4 of the final concentration of green tea (GT) for 2 days, 1/2 of the final concentration of GT for 2 days, 3/4 of the final concentration of GT for 2 days and full strength GT as the sole source of drinking water for an additional 8 days prior to and during treatment with UVB twice weekly for 30 weeks. The data were taken from ref. 6 and are expressed as the mean ± S.E.

[a]Statistically different from water group (p<0.05)
[b]Statistically different from water group (p<0.01)
[c]Statistically different from water group (p<0.001)

Table IV. Effect of Oral Administration of Green Tea on the Growth of Skin Papillomas

Treatment[a]	Duration (wk)	Number of mice	Body weight (g)			Tumor volume per mouse (mm^3)		
			Initial	Final	% Change	Initial	Final	% Change
Experiment 1. DMBA/TPA								
Water	6	30	34.1 ± 0.7	35.9 ± 0.7	5.3	54 ± 9	281 ± 101	420
Tea (0.63%)	6	31	34.9 ± 0.7	35.9 ± 0.6	2.9	44 ± 9	150 ± 34	241
Tea (1.25%)	6	29	35.9 ± 0.7	33.1 ± 0.4	-7.8	41 ± 8	129 ± 33	215
Experiment 2. DMBA/TPA								
Water	6	21	42.0 ± 1.4	46.1 ± 0.2	9.8	340 ± 95	1804 ± 528	431
Tea (1.25%)	6	21	41.1 ± 1.1	42.2 ± 0.8	2.7	274 ± 94	203 ± 35	-26
Experiment 3. UVB/TPA								
Water	6	8	39.0 ± 2.3	36.8 ± 1.8	-5.6	119 ± 63	270 ± 101	127
Tea (1.25%)	6	8	38.6 ± 1.6	37.3 ± 1.3	-3.4	117 ± 37	126 ± 54	8

[a]In experiment 1, female CD-1 mice were initiated with 200 nmol DMBA and one week later they were promoted with topical applications of 5 nmol of TPA twice a week for 10 weeks. One week after the last dose of TPA, tumor bearing mice were treated with gradually increasing doses of green tea for one week (1/4 full dose for 2 days, 1/2 full dose for 2 days, and 3/4 full dose for 2 days) followed by the full dose level for an additional 5 weeks. In experiment 2, female CD-1 mice were initiated with 200 nmol DMBA and one week later they were promoted with 5 nmol of TPA twice a week for 25 weeks. One week after discontinuing TPA administration, the mice were treated with 1.25% green tea for 6 weeks. In experiment 3, female CD-1 mice were treated with 180 mJ/cm^2 of UVB once daily for 10 days followed one week later by topical application of 16 nmol TPA twice a week for 30 weeks. One week later, the mice were treated with 1.25% green tea for 6 weeks. The body weights and tumor volumes were determined before and after administration of green tea or water for 6 weeks. Each value represents the mean ± S.E.

or 1.25% green tea increased in size by 241% or 215%, respectively. In experiment 2, tumors in control mice increased in size by 431%, whereas, tumors in animals treated with 1.25% green tea *decreased* in size by 26%. In experiment 3, tumors from control animals increased in size by 127%, whereas, tumors from mice treated with 1.25% green tea in the drinking water increased in size by only 8%. An inhibitory effect of green tea on body weight gain was observed in some but not all experiments. Additional studies in other laboratories have indicated an inhibitory effect of orally or intraperitoneally administered green tea or green tea polyphenol fraction on the growth of transplanted Ehrlich ascites tumor or sarcoma 180 (*8-10*). The results of our studies with experimentally induced tumors and the studies of others with transplanted tumors indicate that green tea has an inhibitory effect on tumor growth.

Acknowledgments

The research described here was supported in part by Grant No. CA 49756 from the National Institutes of Health. We thank Ms. Deborah Bachorik and Ms. Diana Lim for their excellent help in the preparation of this manuscript.

References

1. Yoshizawa, S.; Horiuchi, T.; Fujiki, H., Yoshida, T.; Okuda, T.; Sugimura, T. *Phytotherapy Res.* 1987, *1*,44-47.
2. Fujita, Y.; Yamane, T.; Tanaka, M.; Kuwata, K.; Okuzumi, J.; Takahashi, T; Fujiki, H.; Okuda, T. *Jpn. J. Cancer Res.* 1989, *80,* 503-505.
3. Wang, Z. Y.; Khan, W.A.; Bickers, D.R.; Mukhtar, H. *Carcinogenesis (Lond.)* 1989, *10*, 411-415.
4. Chi, H.; Yong, X. *Biomed. Environ. Sci.*1990, *3*, 35.
5. Wang, Z.Y.; Agarwal, R.; Bickers, D.R.; Mukhtar, H. *Carcinogenesis (Lond.)* 1991, *12*, 1527-1530.
6. Wang, Z.Y.; Huang, M-T.; Ferraro, T.; Wong, C-Q.; Lou, Y-R.; Reuhl, K.; Iatropoulos, M.; Yang, C.S.; Conney A.H. *Cancer Res.* 1992, *52*, 1162-1170.
7. Huang, M-T., Ho, C-T., Wang Z.Y., Ferraro, T., Finnegan-Olive, T., Lou, Y-R., Mitchell, J. M., Laskin, J. D., Newmark, H., Yang, C. S. and Conney, A. H.: *Carcinogenesis,* in press.
8. Yi, Z.X.; Yi, C.S.; Li, Z.Y.; Wu, P.Y.; Fang, K.; Fu,Q.; Dai, Z.L.; Yang, X.Y.; Lu, C.H.; Shuang, W.H.; Chao, W.Z. *Tumor* 1984, *4,*128.
9. Oguni, I.; Nasu, K.; Yamamoto, S.; Nomura, T. *Agric. Biol. Chem.* 1988, *52*, 1879-1880.
10. Hara, Y.; Matsuzaki, S.; Nakamura, K. *J. Jpn. Soc. Nutr. Food Sci.* 1989, *42*:39-45.

RECEIVED April 20, 1992

Chapter 22

Inhibition of Nitrosamine-Induced Tumorigenesis by Green Tea and Black Tea

Zhi Yuan Wang, Jun-Yan Hong, Mou-Tuan Huang, Allan H. Conney, and Chung S. Yang[1]

Laboratory for Cancer Research, Department of Chemical Biology and Pharmacognosy, College of Pharmacy, Rutgers, The State University of New Jersey, Piscataway, NJ 08855-0789

Oral administration of green tea infusion (12.5 g tea leaves in 1 liter) as the sole source of drinking water to A/J mice markedly inhibited N-nitrosodiethylamine (NDEA)-induced tumorigenesis. It decreased lung tumor incidence and multiplicity and forestomach tumor multiplicity. Oral administration of decaffeinated green tea and black tea extracts (0.6%) also inhibited 4-(methylnitrosamino)-1-(3-pyridyl)-1-butanone (NNK)-induced lung tumorigenesis in A/J mice. When given during the NNK treatment period, both tea extracts decreased tumor multiplicity by about 65%. When given during the post-initiation period, decaffeinated green tea was more effective than decaffeinated black tea in the reduction of tumor multiplicity (85% vs. 63%) and tumor incidence (30% vs. 7%). Inhibitory effects of tea on esophageal tumorigenesis induced by N-nitrosomethylbenzylamine (NMBzA) or its precursors are also discussed.

Tea (*Camellia sinensis*) is one of the most popular beverages worldwide. Green tea is mainly consumed in Asian countries where tea is the major beverage, and black tea is mainly consumed in the Western nations and some Asian countries. Green tea is made by steaming and drying fresh tea leaves at elevated temperatures. Its composition is similar to that of fresh tea leaves and contains polyphenols (up to 30% of the dry weight), most of which are flavan-3-ols, commonly known as catechins. Some major components are (-)-epigallocatechin-3-gallate (EGCG), (-)-epigallocatechin (ECG), (-)-epicatechin-3-gallate (EGC), (-)-epicatechin (EC), (+)-gallocatechin, (+)-catechin, flavonols, anthocyanidins, caffeine, and gallic acid. The flavan-3-ols possess nucleophilic centers at positions 6 and 8. They have antioxidative properties due to their free radical trapping ability. In the manufacture of black tea, the monomeric flavan-3-ols undergo polyphenol oxidase-dependent oxidative polymerization via C-O or C-C bond formation leading to the production of theaflavins, thearubigins, and other oligomers which give the characteristic color and taste of black tea.

[1]Corresponding author

0097-6156/92/0507-0292$06.00/0
© 1992 American Chemical Society

Previous Studies on Tea and Cancer

As summarized by the Committee on Diet and Health of the National Research Council (1) and a recent monograph by the International Agency for Research on Cancer (2), epidemiological studies on tea and cancer so far have yielded inconsistent and inconclusive results. For example, Kinlen *et al.* (3) reported a significant positive correlation between tea consumption and stomach cancer in a case-control study in London. However, in an earlier correlative study, there was a significant negative association between tea consumption and stomach cancer in both sexes in an investigation covering 20 countries (4). A case-control study in Nagoya, Japan indicated that black tea or green tea consumption did not increase the risk for stomach cancer (5). However, a case-control study in Kyushu, Japan showed that individuals consuming green tea more frequently or in larger quantities tended to have a lower risk for gastric cancer (6). Studies in Shizuoka Prefecture, Japan indicated that the cancer death rate in this tea production area, especially from stomach cancer, was lower than the national average, and inhabitants of towns having lower incidence rates tended to drink green tea more frequently than did inhabitants in other areas (7).

Early reports indicated that tea extracts may have enhanced carcinogenesis in animals (8, 9). For example, repeated subcutaneous injection of black tea extracts to rats once weekly for 45 to 77 weeks resulted in tumor formation at the injection site (8). Topical application of black tea infusion after initiation with benzo(a)pyrene resulted in weak tumor promotion (9). Recently, Nagabhushan *et al.* (10) reported that oral feeding of 60°C tea infusions did not increase the tumor incidence, and subcutaneous injection of tea infusions failed to produce tumors in Swiss mice.

Many recent studies have shown an inhibitory action of tea or tea components on carcinogenesis in animals. Application of polyphenolic extracts of green tea to mouse skin inhibited the tumorigenicity of polycyclic aromatic hydrocarbons and tumor promotion by 12-*O*-tetradecanoylphorbol-13-acetate (TPA), teleocidin, and okadaic acid (11-14). Oral administration of a polyphenolic extract of green tea also significantly inhibited skin tumor initiation by 7,12-dimethylbenz(a)anthracene (12) and skin tumor promotion by TPA in mice (15). Oral feeding of EGCG in the drinking water during the post-initiation stage inhibited N-ethyl-N'-nitro-N-nitrosoguanidine-induced tumor formation in the duodenum of mice (16). In additional studies, oral administration of a green tea polyphenol fraction (17) or green tea infusion (15) inhibited ultraviolet light – induced skin carcinogenesis in SKH-1 mice. However, the active compounds responsible for the inhibitory effects observed and the mechanisms of the inhibition remain to be investigated.

In this communication, the effects of tea on NDEA-, NNK-, and NMBzA-induced carcinogenesis are discussed. NDEA, a commonly used carcinogen in animal studies, has been found in the environment. NNK, a potent tobacco carcinogen, is believed to be an important etiological factor in human carcinogenesis (18). Although it remains to be confirmed, the presence of NMBzA, a potent esophageal carcinogen in rats, in the human diet in a high esophageal cancer incidence area in northern China has been reported (19).

Preparation and Compositions of Green Tea Infusion and Decaffeinated Tea Extracts

Green tea leaves (12.5 g) were placed in 500 ml of freshly boiled deionized water for 15 min and then filtered. The tea leaves were steeped a second time with 500 ml of freshly boiled water and filtered. The combined filtrate is referred to as 1.25%

infusion. A 0.63% infusion was prepared from the 1.25% infusion via a 1 to 2 dilution. The amount of solids present in the 1.25% infusion was 4.69 mg/ml. EGCG, caffeine, EGC, ECG, and EC were the major components, accounting for 15.1%, 8.1%, 6.9%, 3.0%, and 1.8% of the dry solid, respectively. Decaffeinated green tea powder and black tea powder were dehydrated water extracts from decaffeinated tea leaves, and were supplied by the Thomas J. Lipton Company (Englewood Cliffs, NJ). The decaffeinated tea leaves were prepared by extracting tea leaves with supercritical CO_2. EGCG, EGC, ECG and EC were the major catechins in decaffeinated green tea powder accounting for 11.1%, 11.0%, 2.6% and 3.1% of the dry weight, respectively, and caffeine was decreased to 0.3% of the dry weight. In the decaffeinated black tea powder, the amounts of EGCG, EGC, ECG, EC, and caffeine were 2.2%, 0.9%, 1.3%, 0.9%, and 0.4% of the dry weight, respectively. The amounts of theaflavins were 1.7% and thearubigins were about 12% of the dry weight.

Induction and Inhibition of Tumorigenesis

Female A/J mice (6-8 weeks old) were fed AIN-76A diet and given NDEA (10 mg/kg body weight) by gavage once weekly for 8 weeks or a single dose of NNK (103 mg/kg in saline, i.p.). The mice were given tea infusions or extracts as the sole source of drinking water during the "initiation" period (starting two weeks before and until one week after the carcinogen treatment), the post-initiation period (starting one week after carcinogen treatment and continuing until the end of the experiment), or the untire experimental period. The mice were killed 16 weeks after the last dose of carcinogen treatment; tumors in the lung and forestomach were counted.

Effects of Oral Administration of Green Tea on NDEA-Induced Lung and Forestomach Tumorigenesis

Treatment of A/J mice with NDEA caused lung tumors in more than 90% of the animals with an average of 8.3 ± 1.0 tumors per mouse 16 weeks after the last dose of NDEA (Table I). The administration of 0.63% or 1.25% of green tea infusion as the sole source of drinking water during the NDEA-treatment (initiation) period, post-initiation period, or the entire experimental period significantly decreased the forestomach tumor multiplicity (by 31% to 63%) and lung tumor multiplicity (by 36% to 60%). The 1.25% green tea infusion also decreased lung tumor incidence in all three protocols (by 36 to 44%) and significantly decreased forestomach tumor incidence when given during the initiation period or the entire experimental period. Inhibitory effects on lung and forestomach tumorigenesis by green tea infusion were also observed in a similar experiment using a high dosage of NDEA (20 mg/kg). Histopathological examination showed that almost all lung tumors were pulmonary adenomas. Most of the forestomach lesions were hyperplasia or papilloma; carcinoma-in-situ and squamous cell carcinoma were also observed. In all the tea treatment groups, the incidence of papilloma was lower than the positive control group. It was observed that the tumor size was reduced in the tea-drinking animals.

Effects of Oral Administration of Decaffeinated Green Tea and Black Tea on NNK-Induced Lung Tumorigenesis

Treatment of A/J mice with a single dose of NNK (103 mg/kg) resulted in 96% of mice bearing lung tumors and an average multiplicity of 9.3 ± 1.3 tumors per mouse after 16 weeks (Table II). When 0.6% decaffeinated green tea or black tea was given during the NNK-treatment (initiation) period as the sole source of drinking water, tumor multiplicity was reduced by 67% or 65%, respectively. When the tea extract

TABLE I. Effects of oral administration of green tea infusion (GT) on NDEA-induced lung and forestomach tumorigenesis in A/J mice

Group	Number of mice	Lung tumors		Forestomach tumors	
		% Animals with tumors	Tumors per mouse	% Animals with tumors	Tumors per mouse
Control groups					
1) Water alone	30	10	0.1 ± 0.1	0	0
2) GT (1.25%) alone	30	0	0	0	0
3) NDEA alone	47	93.6	2.5 ± 0.3	97.9	8.3 ± 1.0
GT given during the initiation period					
4) GT (0.63%)	33	81.8 (12.6%)	1.6 ± 0.2^b (36.0%)	97.0 (0.9%)	5.7 ± 0.7^b (31.3%)
5) GT (1.25%)	30	56.7^a (39.4%)	1.1 ± 0.3^b (56.0%)	83.3^a (14.9%)	3.5 ± 0.5^b (57.8%)
GT given after the initiation period					
6) GT (0.63%)	28	46.4^a (50.4%)	1.2 ± 0.3^b (52.0%)	96.4 (1.5%)	5.4 ± 0.6^b (34.9%)
7) GT (1.25%)	30	60.0^a (35.9%)	1.4 ± 0.3^b (44.0%)	90.0 (8.1%)	4.4 ± 0.6^b (47.0%)
GT given during the entire experimental period					
8) GT (0.63%)	30	76.7^a (18.1%)	1.6 ± 0.3^b (36.1%)	80.0^a (18.3%)	3.4 ± 0.6^b (59.0%)
9) GT (1.25%)	36	52.8^a (43.6%)	1.0 ± 0.3^b (60.0%)	72.2^a (26.3%)	3.1 ± 0.5^b (62.7%)

The number of tumors is expressed as the mean ± S.E. and the % inhibition is shown in parentheses.
[a] $p < 0.05$, significantly different from Group 3 using the χ^2 test.
[b] $p < 0.05$, significantly different from Group 3 using Newman-Keuls multiple comparison test.

TABLE II. Effects of decaffeinated black tea and green tea extracts on NNK-induced lung tumorigenesis in A/J mice

Group	Number of mice	% Animals bearing Tumors		Number of Tumors per mouse	
Control groups					
1) Water alone	25	4.0		0.1 ± 0.1	
2) Black tea extract (0.6%) alone	24	4.2		0.1 ± 0.1	
3) Green tea extract (0.6%) alone	25	4.0		0.1 ± 0.1	
4) NNK alone	27	96.3		9.3 ± 1.3	
Tea given during the initiation (NNK treatment) period					
5) Green tea extract (0.6%)	29	79.3	(16.9%)	3.1 ± 0.5^b	(66.7%)
6) Black tea extract (0.6%)	30	83.3	(13.5%)	3.3 ± 0.5^b	(64.5%)
Tea given after the initiation period					
7) Green tea extract (0.3%)	30	83.3	(13.5%)	2.4 ± 0.4^b	(74.2%)
8) Green tea extract (0.6%)	29	65.5^a	(30.4%)	1.4 ± 0.3^b	(85.0%)
9) Black tea extract (0.3%)	29	93.1	(3.3%)	3.5 ± 0.4^b	(62.4%)
10) Black tea extract (0.6%)	28	89.3	(7.3%)	3.4 ± 0.2^b	(63.4%)

The number of tumors is expressed as the mean $\pm$ S.E. and the % inhibition is shown in parentheses.
[a] $p<0.05$, significantly different from Group 4 using the χ^2 test.
[b] $p<0.05$, significantly different from Group 4 using Newman-Keuls multiple comparison test.

was given after the NNK-treatment period until the end of the experiment, 0.3% and 0.6% decaffeinated green tea reduced the tumor multiplicity by 74% and 85% and the tumor incidence by 14% and 30%, respectively. Under the same conditions, the decaffeinated black tea extracts decreased tumor multiplicity by about 63%, but did not significantly reduce the tumor incidence.

Effects of Tea Consumption on Esophageal Cancer

Han and Xu (20) reported that oral adminstration of infusions made from five different brands of green tea and black tea as the sole source of drinking water during the entire experimental period inhibited NMBzA -induced esophageal tumorigenesis in rats. For example, the positive control group (2.5 mg NMBzA/kg administered, p.o., once weekly for 2 weeks followed by administration twice weekly for another 10 weeks) had 90% tumor incidence with an average of 4.7 esophageal tumors per rat. Oral feeding of 2% infusions of black tea, jasmine tea, oolong tea, green tea from Hainan, and green tea from Zheijiang in the drinking water decreased tumor incidence by 26%, 51%, 53%, 28%, and 41%, respectively, and reduced tumor multiplity by 68%, 75%, 75%, 58, and 75%, respectively. Tumor size was also significantly reduced in all tea treatment groups.

Oral feeding of these tea infusions also markedly inhibited esophageal tumorigenesis caused by NMBzA precursors (methylbenzylamine plus sodium nitrite) in rats, reducing tumor incidence by 80% to 95% (21). Gao *et al* (22) also reported that oral feeding of green tea inhibited forestomach tumor formation induced by precursors of NMBzA in mice. These results suggested that blocking nitrosamine formation may be an important mechanism of reducing the incidence of carcinogenesis by tea. The inhibition of nitrosation by both green tea and black tea have been demonstrated in animals and in human volunteers (23). It was estimated that the consumption of 3-5 g of tea (leaves) per day can effectively block the endogenous nitrosation reactions.

It was suggested by Morton (24) that excessive consumption of tea may be a causative factor for the high incidence of esophageal cancer in many populations. However, several case-control studies showed that there was no association between tea drinking and esophageal cancer (25-27). Nevertheless, ingestion of very hot tea was associated with a two- to three-fold increase in risk for esophageal cancer (25-27). It appears that the hot temperature may be an important etiology factor in this cancer.

Discussion

Our present results demonstrate inhibitory effects of orally administered tea infusions on NDEA- and NNK-induced tumorigenesis in A/J mice. In the NDEA model, oral administration of the green tea infusion markedly decreased lung tumor multiplicity, lung tumor incidence, and forestomach tumor multiplicity, but there was a weaker inhibitory effect on forestomach tumor incidence. In the NNK model, oral administration of decaffeinated green tea and black tea extracts during the initiation period was effective in decreasing tumor multiplicity, but was not effective in reducing tumor incidence. When given in the post-initiation period, decaffeinated green tea was more effective than decaffeinated black tea in the reduction of both tumor incidence and multiplicity.

In studies with the green tea infusion, the body weights of the mice in the tea drinking group were lower than those in the control group (by 5% to 13%). A question may be raised as to whether the observed inhibitory effects are due to reduced growth rate. The studies on NNK-induced lung tumorigenesis showed that decaffeinated tea extract inhibited tumorigenesis without effecting the body weight. If we assume that after the carcinogen treatment period the NDEA- and NNK-induced tumorigenesis models are similar, then the results from Table II may be used to strengthen those in Table I.

The components in tea extracts responsible for the presently observed inhibitory action are not known. The results in Table II indicated that the effect was not due to caffeine. In studying the inhibitory effect of tea, many investigators have focused on green tea and green tea polyphenols, especially EGCG. The results in Table II demonstrate that black tea, which contained much lower amount of catechins than green tea, was just as effective as green tea in inhibiting tumorigenesis when given to the mice during the carcinogen treatment period. Inhibition of the metabolic activation of NNK may be one of the mechanisms of inhibition. Our preliminary results indicated that the ethyl acetate extractable polyphenol fractions from green tea and black tea inhibited the oxidative metabolic conversion of NNK to its N-oxide, keto alcohol, and keto aldehyde metabolites in mouse lung microsomes *in vitro*. Previously Wang *et al.* (28) reported that flavan-3-ols, green tea polyphenols, and black tea polyphenols inhibited benzo(a)pyrene metabolism in rat liver microsomes. Other mechanisms such as the induction of phase II enzymes (29) and the trapping of ultimate carcinogens (30) have also been suggested for the anticarcinogenic activities of tea. However, the importance of these mechanisms for the inhibition of NDEA- and NNK-induced carcinogenesis remain to be investigated.

The antioxidant properties of tea components may be important in the protection against cellular oxidative damage such as lipid peroxidation, DNA single strand breakage, or the formation of 8-hydroxydeoxyguanosine. These insults may contribute to carcinogenesis at both the initiation and post-initiation stages. (+)-Catechin was reported to inhibit NNK-induced DNA single strand breaks in rat hepatocytes *in vitro* and *in vivo* (31). EGCG was shown to inhibit TPA-induced 8-hydroxydeoxyguanosine formation in Hela cells (32).

The present work clearly demonstrates the protective effect of tea against lung tumorigenesis in the mouse model. However, such a protective effect has not been conclusively demonstrated in human populations in epidemiological studies. Because of the wide consumption of tea by many populations, the relationship between tea consumption and cancer warrants more laboratory and epidemiological studies.

Acknowledgments

This work was supported by NIEHS Center Grant ES-05022. The authors wish to thank Ms. Dorothy Wong for her excellent secretarial assistance.

Literature Cited

1. Committee on Diet and Health, Food and Nutrition Board, Commission on Life Sciences, National Research Council. *Diet and Health*; National Academy Press: Washington, D.C., 1989; pp.465-508.
2. World Health Organization, International Agency for Research on Cancer, *Coffee, Tea, Mate, Methylxanthines and Methylglyoxal*; IARC Monographs on the Evaluation of Carcinogenic Risks to Humans. IARC Press: Lyon, France, 1991; Vol. 51, pp 207-271.
3. Kinlen, L. J.; Willows, A. N.; Goldblatt., P. and Yudkin, J. *Br. J. Cancer* 1988, *58*, 397-401,.
4. Stocks, P. *Br. J. Cancer* 1970, *24*, 215-225.

5. Tajima. K. and Tominaga, S. *Jpn. J. Cancer Res. (Gann)* 1985, 76, 705-716.
6. Kono, S.; Ikeda, M.; Tokudome, S. and Kuratsune, M. *Jpn. J. Cancer Res. (Gann)* 1988, 79, 1067-1074.
7. Oguni, I.; Nasu, K.; Yamamoto, S. and Nomura, T. *Agric. Biol. Chem.* 1988, 52, 1879-1880.
8. Kapadia, G. J.; Paul, B. D.; Chung, E. B.; Ghosh, B. and Pradhan, S. N. *J. Natl Cancer Inst.* 1976, 57, 207-209.
9. Bogovski, P.; Day, N.; Chvedoff, M. and Lafaverges, F. *Cancer Letters* 1977, 3, 9-13.
10. Nagabhushan, M.; Sarode, A. V.; Nair, J.; Amonkar, A. J.; D'Souza, A. V. and Bhide, S. V. *Indian J. Exp. Biol.* 1991, 29, 401-4061.
11. Khan, W. A.; Wang, Z. Y.; Athar, M.; Bickers, D. R. and Mukhtar, H. *Cancer Letters* 1988, 42, 7-12.
12. Wang, Z. Y.; Khan, W. A.; Bickers, D. R. and Mukhtar, H. *Carcinogenesis* (Lond.) 1989, 10, 411-415.
13. Huang, M-T.; Wang, Z. Y.; Ho, C-T.; Ferraro, T.; Newmark, H.; Mitchell, J. M.; Laskin, J. D. and Conney, A. H. *Proc. Amer. Assoc. Cancer Res.* 1991, 32, 129.
14. Fujiki, H. In *Proc. of the Fourth Chemical Congress of North America*; New York, 1991, Abstract No. 66.
15. Wang, Z. Y.; Huang, M-T.; Ferraro, T.; Wong, C-Q.; Newmark, H.; Yang, C. S. and Conney, A. H. *Proc. Am. Assoc. Cancer Res.* 1991, 32, 129.
16. Fujita, Y.; Yamane, T.; Tanaka, M.; Kuwata, K.; Okuzumi, J.; Takahashi, T.; Fujiki, H. and Okuda, T. *Jpn. J. Cancer Res.* 1989, 80, 503-505.
17. Wang, Z. Y.; Agarwal, R.; Bickers, D. R. and Mukhtar, H. *Carcinogenesis* (Lond.) 1991, 12, 1527-1530.
18. Hecht, S. S. and Hoffmann, D. *Carcinogenesis* (Lond.) 1988, 9, 1665-1668.
19. Lu. S. H.; Yang, W. X.; Guo, L. P.; Li, F. M.; Wang. G. L.; Zhang, J. S. and Li, P. Z. In *Relevance of N-Nitro Compounds to Human Cancer: Exposures and Mechanisms*; Bartsch, H.; O'Neill, I. K.; and Schultz-Hermann, R., Ed.; IARC Publications: Lyon, France, 1987, pp. 538-543.
20. Han, C. and Xu, Y. *Biomed. Envir. Sci.* 1990, 3, 35-42.
21. Xu, Y. and Han, C. *Biomed. Envir. Sci.* 1990, 3, 406-412.
22. Gao, G. D.; Zhou, L. F.; Qi, G.; Wang, H. S. and Li, T. S. *Tumor (Chinese)* 1990, 10, 42-43.
23. Wang, H-Z. and Wu, Y. In *Proc. Tenth International Meeting on N-nitroso Compounds, Mycotoxins and Tobacco Smoke: Relevance to Human Cancer.* Lyon, France, 1989, Abstract, pp. 70.
24. Morton, J. F. In *Chemistry and Significance of Condensed Tannins;* Hemingway, R.W. and Karchesy, J.J., Ed.; Plenum Press: New York, 1989; pp.403-416.
25. Victora, C. C.; Muñoz, N.; Day, N. E.; Barcelos, L. B.; Peccin, D. A. and Braga, N. M. *Int. J. Cancer* 1987, 39, 710-716.
26. Cook-Mozaffari, P. J.; Azordegan, F.; Day, N. E.; Ressicaud, A.; Sabai, C. and Aramesh, B. *Br. J. Cancer* 1979, 39, 293-309.
27. de Jong, U. W.; Breslow, N.; Hong, J. G. E.; Sridharan, M. and Shanmugaratnam, K. *Int. J. Cancer* 1974, 13, 291-303.
28. Wang, Z. Y.; Bickers, D.R. and Mukhtar, H. *Drug Metab. Dispos.* 1988, 16, 93-103.
29. Sparnins, V. L.; Venegas, P. L. and Wattenberg, L. W. *J. Natl. Cancer. Inst.* 1982, 68, 493-496.
30. Wang, Z. Y.; Cheng, S. J.; Zhou, Z. C.; Athar, M.; Khan, W. A.; Bickers, D. R. and Mukhtar, H. *Mutation Res.* 1989, 223, 273-285.
31. Liu, L. and Castonguay, A. *Carcinogenesis* (Lond.) 1991, 12, 1203-1208.
32. Bhimani, R. and Frenkel, K. *Proc. Am. Assoc. Cancer Res.* 1991, 32, 126.

RECEIVED April 1, 1992

Chapter 23

Protection Against Tobacco-Specific, Nitrosamine-Induced Lung Tumorigenesis by Green Tea and Its Components

Fung-Lung Chung[1], Yong Xu[1,2], Chi-Tang Ho[3], Dhimant Desai[1], and Chi Han[2]

[1]Division of Chemical Carcinogenesis, American Health Foundation, One Dana Road, Valhalla, NY 10595
[2]Institute of Nutrition and Food Hygiene, Chinese Academy of Preventive Medicine, Nan Wei Road 29, Beijing 100050, People's Republic of China
[3]Department of Food Science, Rutgers, The State University of New Jersey, P.O. Box 231, New Brunswick, NJ 08903

The effects of Chinese green tea, (-)-epigallocatechin gallate (EGCG) and caffeine on 4-(methylnitrosamino)1-(3-pyridyl)-1-butanone (NNK) induced lung tumorigenesis in A/J mice were studied. Mice were given 2% tea infusion as drinking water, 560 ppm EGCG or 1120 ppm caffeine in water for 13 weeks. Two weeks later, NNK (11.65 mg/kg b.w.) was administered by gavage 3 times weekly for 10 weeks. The bioassay was terminated 6 weeks after the last NNK gavage. All mice treated with NNK only developed lung adenomas with a multiplicity of 22.5 tumors per mouse. NNK treated mice that drank tea or water containing EGCG developed only 12.2 and 16.1 tumors per mouse, respectively. This corresponds to a 45% and 30% reduction in lung tumor multiplicity. Little difference in water and dietary consumption was found among all the groups. However, mice that drank tea or caffeine solution showed consistently lower body weight gains than mice given water of EGCG solution. We also observed small but significant reduction of lung tumor multiplicity in the mice that were treated with caffeine solution. These results suggest that green teas can inhibit NNK-induced lung tumors in mice and that the effect appears primarily due to its major polyphenol, EGCG.

Cigarette smoking is a major cause of cancer in developed countries. It is estimated that 25-40% of all cancers in the U.S. can be attributed to tobacco use (1). In the U.S. alone, there are 50 million people who smoke cigarettes (2). Many smokers will not quit and are expected to continue the habit for the rest of their lives. Among tobacco-caused cancer, lung cancer has been the leading cause of death in males in the U.S., and recently it has surpassed breast cancer in cancer mortality in females (3). This tragedy is expected to be repeated in many developing countries such as China due to a sharp increase in the prevalence of cigarette smoking in recent years. Of course the best way to eradicate lung cancer is to stop smoking. However, it is difficult to effectively modify the life style in a large population. Therefore, there is an imminent need to identify chemopreventive agents, both natural and synthetic, to counteract the carcinogenic action of smoking. Numerous agents in cigarette smoke are found to be carcinogenic (4). The most intriguing

0097-6156/92/0507-0300$06.00/0

ones are the tobacco-specific alkaloid-derived nitrosamines. They are abundant, and some of them are potent carcinogens in laboratory animals (*5*). The most potent among them in 4-(methylnitrosamino) 1-(3-pyridyl)-1-butanone (NNK), a nicotine-derived nitrosamine (Figure 1), which is highly specific for the induction of lung tumor in various animal species (*5*). The organospecificity of NNK toward lung is independent of the route of administration. Due to its potency and organotropism, it is believed that NNK may be one of the compounds in tobacco, if not the major one, responsible for the development of lung cancers in smokers. In this chapter we will describe our recent work regarding the effects of green tea and its components on the lung tumor formation induced by NNK.

Modification of NNK Lung Tumorigenesis by Diet

The carcinogenic effect of NNK has been shown to be influenced by dietary components. We have observed a different response in the formation of lung tumor induced by NNK depending on the types of diet used in the A/J mouse bioassay. Mice fed a crude cereal-based diet (NIH-07) developed an average of 2.5 lung tumors per mouse as compared to an average of 8.3 tumors per mouse in mice fed a semipurified diet (AIN-76A) (*6*). Consistent with this finding, our studies showed that compounds found in cruciferous vegetables such as aromatic isothiocyanates and indole-3-carbinol inhibited lung tumor formation in mice and/or rats treated with NNK (*7,8*). Other dietary compounds which are protective against NNK-induced lung tumorigenesis in laboratory animal include phenolic compounds such as ellagic acid and butylated hydroxyanisole (BHA) and the citrus component D-limonene (*9,10*). All these studies support the notion that the incidence of smoking-caused lung cancer may be modulated by dietary factors.

Rationale for Selection of Green Tea

Epidemiological studies indicate that lung cancer mortality among males in Japan is much lower than in the U.S., although the average consumption of cigarette among males in Japan is considerably higher than in the U.S. (*11*). Many factors such as race, smoking pattern, or diet could account for the difference in risk between these two countries. Studies by Wynder *et al.* suggested that the lower amount of fat intake in Japan compared to the U.S. might contribute to this difference in the risk of lung cancers (*11*). Another such dietary factor which may be important is the prevalence of green tea consumption in Japan. Green tea is composed of at least 10 to 20% of polyphenols. These compounds are powerful antioxidants, capable of scavenging H_2O_2 and superoxide anion, thus preventing H_2O_2 and oxygen free radical-induced cytotoxicity (*12*). Green tea polyphenol fractions have been shown to be protective in mice against skin tumor induction by polycyclic aromatic hydrocarbons (*13*). It also inhibits aflatoxin B_1-induced hepatocarcinogenesis (*14*). Giving rats a variety of Chinese tea as drinking water resulted in a significant reduction in esophageal tumors induced by N-nitrosobenzylmethylamine (*15*). Since either tea or total tea polyphenol fraction was used in these studies, it is not possible to identify which is the active compound responsible for the inhibitory effect. (-)-Epigallocatechin gallate (EGCG), the main polyphenol in green tea, is an antimutagen (*16*). Topical application of EGCG inhibited teleocidine promoted DMBA induced skin tumors in mice (*17*). More recently, the antipromoting activity is EGCG was demonstrated in mouse duodenum carcinogenesis induced by N-ethyl-N-nitro-N-nitrosoguanidine (*18*). These results prompted us to test the hypothesis in laboratory animal that green tea and its major component EGCG or caffeine protect against lung carcinogenesis of NNK. The structures of EGCG and caffeine are shown in Figure 2.

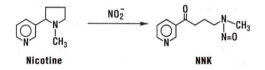

Nicotine   **NNK**

Fig. 1. Nicotine-derived carcinogenic nitrosamines.

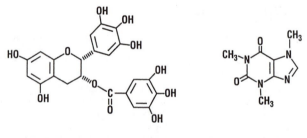

(-) Epigallocatechin-3-gallate (EGCG)   Caffeine

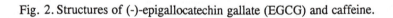

Fig. 2. Structures of (-)-epigallocatechin gallate (EGCG) and caffeine.

Inhibition of Lung Tumor Induction by Green Tea and its EGCG

Six weeks old female A/J mice fed AIN-76 diet were divided into 7 groups as shown in Table I. Two percent green tea infusion, and 560 ppm EGCG or 1120 mg caffeine in drinking water were consumed as sources of drinking water. The concentrations of EGCG and caffeine in solutions were determined on the basis of those found in the 2% tea infusion. The tea infusion was prepared by adding 50 ml of boiling water to 1 g of green tea leaves and the tea solution was obtained by filtration after standing at room temperature for 30 min. Two weeks after consumption of test substances in drinking water, mice were administered NNK (11.65 mg/kg b.w. in corn oil) by gavage 3 times weekly for 10 weeks. One week after the last NNK treatment, mice were switched back to tap water. Mice were sacrificed 6 weeks after last NNK treatment. Figure 3 shows body weight curves of different groups during the period of bioassay. Independent of NNK treatment, groups that drank tea or water containing caffeine showed consistently lower body weight gains compared to groups that drank water containing EGCG of only water. Comparable body weight curves were observed in tea and caffeine groups, suggesting caffeine is responsible for the decrease in weight gain in the tea group. In contrast to the caffeine and tea groups, mice treated with EGCG had normal growth. The weekly water consumption, 0.6-0.7 ml/g b.w. was similar among most of the groups, with the only exception of the tea group which consumed an average of 0.81 ml/g b.w. and no difference in diet consumption among groups was found (Figure 4).

Table I. Effect of Tea, EGCG and caffeine on NNK-induced lung adenomas in A/J mice

Groups	Number of Animals	Tumor/Mouse ($\pm$ S.D.)	% of Mice with Tumors
NNK	30	22.5$\pm$4.7	100
TEA+NNK	25	12.2$\pm$4.3**	100
EGCG+NNK	25	16.1$\pm$5.3**	100
Caffeine+NNK	15	19.2$\pm$4.8*	100
TEA	15	0.1$\pm$0.2	7
EGCG	15	0.3$\pm$0.6	20
Caffeine	15	0.3$\pm$0.6	20

* compared with NNK group, $P < 0.5$, using student's t test
** compared with NNK group, $P < 0.001$, using student's t test

In contrast to the single dose protocol used in our other studies, in this study NNK was administered by gavage as 30 doses over a period of 10 week. Each dose was 11.65 mg/k.g. b.w. which equals to a total dose of 350 mg/k.g. b.w. This dosing regimen induced an average of 22.5 lung adenomas per mouse at week 16 after NNK treatment began. Table I shows the effects of tea, EGCG and caffeine on the formation of lung adenomas in mice treated with NNK. While the tumor incidences are the same in all groups, mice that drank tea or water containing EGCG developed only 12.2 tumors or 16.1 tumors per mouse. These tumor multiplicities correspond to a 45% and 30% reduction of lung tumors in each group. Interestingly, the caffeine group also showed a marginal inhibition. Mice given only green tea or water containing EGCG or caffeine developed tumor multiplicity and incidence which are normally found in untreated mice.

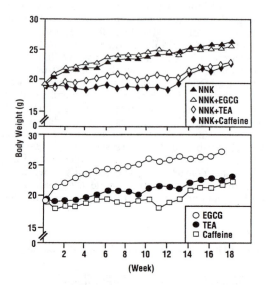

Fig. 3. Body weight curves of A/J mice during bioassay.

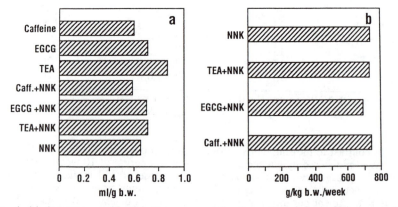

Fig. 4. (a) Average weekly drinking amount for the entire period of bioassay; (b) Diet consumption measured at weeks 6 and 7 following NNK treatment began.

The most encouraging finding in this study is that green tea appears to have a protective effect against NNK-induced lung tumorigenesis in mice. The results support the hypothesis that green tea consumption in Japan may be partially responsible for the reduced lung cancer mortality rate. Our results showed that EGCG, given in the concentration identical to that found in tea infusion, not only inhibited NNK in mice, but also appears to account for most of the activity in tea. Furthermore, EGCG, unlike caffeine, exerts little adverse effect on growth. Therefore, it might be an ideal candidate for consideration as a chemopreventive agent.

The Role of Caffeine in the Inhibition by Green Tea

In the tea preparation used in this study, caffeine constitutes about 5.6% by weight. It is a major component in tea and has been shown to inhibit chemical carcinogenesis (*19-21*). The exact mechanism by which caffeine inhibits tumorigenesis is not clear. Recently, Lagopoulos *et al.* found that the reduced body weight gain by caffeine treatment correlates with decreased hepatocarcinogenesis by diethylnitrosamine (*22*). Similarly, we found that, although diet consumption was unchanged by caffeine treatment, the body weight gains in caffeine groups were considerably lower than those in EGCG and NNK control groups. Therefore, the slight but significant reduction in lung tumor multiplicity by caffeine treatment could be related to its negative effect on body weight. A recent study by Welch *et al* suggested that the inhibition of metabolic activation of 7,2-dimethylbenz[a]anthracene by caffeine is involved in its suppression of the mammary gland tumor induction by this carcinogen in rats (*23*). Regardless of mechanism, the potential protective effect of caffeine should not be ignored because of the widespread consumption of this compound by man.

Possible Mechanism for the Inhibition

Several lines of evidence indicate that methylation of DNA by NNK following its metabolic activation by α-hydroxylation at the methylene carbon adjacent to the nitroso group is a critical step in its lung carcinogenicity in A/J mice (*7,24*). While green tea and EGCG treatment reduced the lung tumor formation induced by NNK, they had little effect on lung DNA methylation (Chung, F.-L., unpublished results). These results suggest a mechanism of NNK tumorigenesis in addition to that mediated by DNA methylation. Furthermore, in this study mice were treated with multiple small doses of NNK for a period of 10 weeks during which time tea or EGCG was also given. This dose regimen, mimicking more closely to human situation, is different from the pretreatment protocol used in our earlier studies on the inhibition of NNK tumorigenesis by aromatic isothiocyanates (*6*). In a separate study, we did not find any inhibition when mice were pretreated with green tea polyphenol fraction (5 mg/dose/day) for 4 consecutive doses followed by a single dose of 2 mg NNK. One possible explanation is that green tea and its polyphenol may exert protective effects during post-initiation stage. Ellagic acid, a polyphenol, and the antioxidant butylated hydroxyanisole (BHA) given in the diet inhibited lung adenoma formation in A/J mice treated with NNK in drinking water (*9*). These results suggest that the inhibition of NNK-induced lung carcinogenesis by green tea and EGCG may be due to the antioxidant property. Clearly, more studies are needed to clarify the mechanism of inhibition. The mechanism of inhibition by green tea and EGCG is being investigated in our laboratory.

Conclusion

Green tea infusion given in concentration comparable to that consumed by humans can inhibit lung tumor formation in mice caused by tobacco-specific nitrosamine NNK. Whereas the total amount of daily consumption in humans on a body weight basis is considerably less than that in mice, the NNK dose to which human is exposed is also much less than that used in the present study. The result also shows that the inhibitory effect of green tea appears primarily due to its major polyphenol EGCG. Studies are in progress to further delineate the protective effect of tea in mouse and other animals models and to understand the underlying mechanism of tumor inhibition by green tea.

Acknowledgments

This is paper 15 in the series "Dietary Inhibitors of Chemical Carcinogenesis." We thank Dr. Junshi Chen of the Institute of Nutrition and Food Hygiene, Beijing, China for the encouragement and support. We also thank Dr. Qikun Chen of Tea Research Institute, Hangzhou, China for supplying the green tea. This work was supported by a grant from the National Cancer Institute CA-46535.

Literature Cited

1. Doll, R.; Peto, R. *J. Natl. Cancer Inst.*, **1987**, *66*, 1191-1308.
2. U.S. Department of Commerce, Bureau of the Census. *Statistical Abstracts of the United States*, 107th Edition, Table No. 175, 1987.
3. Silverberg, E.; Boring, C. C.; Squires, T. S. *CA-A Cancer Journal for Clinicians*, **1990**, *40*, 9-26
4. Hoffmann, D.; Hecht S. S. *Advances in tobacco carcinogenesis in Handbook of Experimental Pharmacology, Vol. 94/I* Cooper,C. S. and Grover, P. L., Eds., *Springer-Verlag*, **1990**, 63-95.
5. IARC, In *Tobacco smoking, IARC Monographs on the Evaluation of the Carcinogenic Risk of Chemicals to Humans*; International Agency for Research on Cancer: Lyon, France, **1986**, 38.
6. Hecht, S. S.; Morse, M. A.; Amin, S.; Stoner, G. D.; Jordan, K. G.; Choi, C.-I.; Chung, F.-L. *Carcinogenesis, 1989, 10*, 1901-1904.
7. Morse, M. A.; Amin, S. G.; Hecht, S. S.; Chung, F.-L. *Cancer Res.*, **1989**, *49*, 2894-2897.
8. Morse, M. A.; Wang, C.-X.; Stoner, G. D.; Mandal, S.; Conran P. B.; Amin, S. G.; Hecht, S. S.; Chung, F.-L. *Cancer Res.*, **1989**, *49*, 549-553.
9. Pepin, P.; Rossignol, G.; Castonguay, A. *The Cancer J.*, **1990**, *3*, 266-273.
10. Wattenberg, L. W.; Coccia, J. B. *Carcinogenesis*, **1991**, *12*, 115-117.
11. Wynder, E. L.; Fujita, Y.; Harris, R. E.; Hiraguma, T.; Hiyarnor, T. University of Nagoya Press, **1989**, 103-127.
12. Rush, R. J.; Cheng, S. J.; Klaunig, J. E. *Carcinogenesis*, **1989**, *10*,1003-1008.
13. Wang, Z. Y.; Khan, W. A.; Bickers, D. R.; Mukhtar, H. *Carcinogenesis*, **1989**, *10*, 411-415.
14. Chen, Z. Y.; Yan, P. Q.; Qin, G. Z.; Qin, L. L. *Chinese J. Oncology*, **1987**, *9*, 109-111.
15. Han, C.; Xu, Y. *Biomed. and Environ. Sci.*, **1990**, *3*, 35-42.
16. Kada, T.; Kaneko, K.; Matsuzaki, S.; Matsuzaki, T.; Hara, Y. *Mutation Res.*, **1985**, *150*, 127-132.
17. Yoshizawa, S.; Horiuchi, T.; Fujiki, H.; Yoshida, T.;, Okuda, T.; Kugimura, T. *Phytother. Res.*, **1987**, *1*, 44-47.

18. Fujita, Y.; Yamane, T.; Tanaka, M.; Kuwata, K.; Osuzumi, J.; Takahashi, T.; Fujiki, H. *Japanese J. Cancer Res.*, **1989**, *80*, 503-505.
19. Rothwell, K. *Nature*, **1974**, *252*, 69-70.
20. Nomura, T. *Nature*, **1976**, *260*, 547-549.
21. Kacunaga, T. *Nature*, **1975**, *258*, 248-250.
22. Lagopoulos, L.; Sunahara, G. I.; Wurzner, H.; Fliesen, T.; Stalder, R. *Carcinogenesis*, **1991**, *12*, 211-215.
23. VanderPloeg, L. V. C.; Wolfrom, D. M.; Welsh, C. W. *Cancer Res.*, **1991**, *51*, 3399-3404.
24. Morse, M. A.; Eklind, K. I.; Hecht, S. S.; Jordan, K. G.; Choi, C.-I.; Desai, D. H.; Amin, S. G.; Chung, F.-L. *Cancer Res.*, **1991**, *51*, 1846-1850.

RECEIVED December 17, 1991

Chapter 24

Inhibitory Effect of a Green Tea Polyphenol Fraction on 12-O-Tetradecanoylphorbol-13-acetate-Induced Hydrogen Peroxide Formation in Mouse Epidermis

Jeffrey D. Laskin[1], Diane E. Heck[1], Debra L. Laskin[2], John M. Mitchell[1], Mou-Tuan Huang[3], Zhi Yuan Wang[3], Chung S. Yang[3], Chi-Tang Ho[4], and Allan H. Conney[3]

[1]Department of Environmental and Community Medicine, University of Medicine and Dentistry of New Jersey, Robert Wood Johnson Medical School, Piscataway, NJ 08854
Departments of [2]Pharmacology and Toxicology, [3]Chemical Biology and Pharmacognosy, and [4]Food Science, Rutgers, The State University of New Jersey, Piscataway, NJ 08855

Recent evidence suggests that reactive oxygen intermediates including superoxide anion, hydrogen peroxide and hydroxyl radicals are important in chemical carcinogenesis and tumor promotion. Using the two-stage mouse model of initiation and promotion, we found that topical treatment of skin with the tumor promoter, 12-O-tetradecanoyl-phorbol-13-acetate (TPA), results in a dramatic increase in epidermal production of hydrogen peroxide. A green tea polyphenol fraction (GTP), which is a potent inhibitor of TPA-induced skin tumor promotion, was found to inhibit TPA-induced hydrogen peroxide in mouse epidermis. This inhibition was observed after a single administration as well as following twice weekly treatments for periods up to 22 weeks. Our data suggest that the ability of GTP to inhibit tumor promotion may be related to its ability to decrease free radical formation in the skin.

Over the past several years, evidence has accumulated that strongly implicates active oxygen in the process of tumor promotion and carcinogenesis (see refs. *1-8* for reviews). Reactive oxygen intermediates including superoxide anion, hydrogen peroxide, and hydroxyl radicals are known to induce biological damage by reacting with cellular macromolecules such as nucleic acids and proteins, as well as various membrane components (*9-10*). DNA damage induced by reactive oxygen intermediates has been reported to cause chromosomal aberrations, sister chromatid exchanges and mutations, all processes that have been linked to the development of neoplasia (*11-12*).

In the two-stage mouse skin model of tumor initiation and promotion, it has generally been assumed that free radical species are derived from resident macrophages (Langerhans cells) and/or leukocytes that have infiltrated into the dermis and epidermis following tumor promoter treatment (*13*). These cells are known to produce reactive oxygen intermediates *in vitro* and have been postulated to play a role in tumor promotion (*8*) and in the metabolic activation of carcinogens

0097–6156/92/0507–0308$06.00/0

(*14*). Recent studies have suggested that keratinocytes also have the capacity to produce reactive oxygen intermediates. For example, using electron spin resonance, it has been shown that hydroxyl radicals are generated by epidermal cells following treatment with organic hydroperoxide tumor promoters (*3,7*). In addition, by quantifying chemiluminescence, Fisher *et al.* (*15,16*) demonstrated that phorbol ester tumor promoters induce oxidant production in isolated mouse epidermal cells. Chemiluminescence was inhibited by superoxide dismutase, but not by catalase or scavengers of hydroxyl radicals, suggesting that the response was due to the generation of superoxide anion.

In previous studies, we used techniques in flow cytometry to quantify hydrogen peroxide production in keratinocytes isolated from mice treated with the phorbol ester tumor promoter, TPA (*17*). This technique has the advantage of permitting us to simultaneously analyze different cell types within the epidermis. For measurement of hydrogen peroxide production, the fluorescent indicator dye, 2′,7′-dichlorofluorescin diacetate (DCFH-DA), was used. DCFH-DA is a non-polar compound that readily diffuses into cells where it is hydrolyzed by intracellular esterases to the non-fluorescent derivative, 2′,7′-dichlorofluorescin (DCFH). Hydrogen peroxide produced by the cells oxidizes DCFH to the highly fluorescent compound, dichlorofluorescein (DCF). Thus, DCF fluorescence, which can be detected by flow cytometry, is directly proportional to hydrogen peroxide content of the cells. In our studies we reported that TPA treatment caused significant increases in levels of hydrogen peroxide produced by specific subpopulations of keratinocytes (*17*). In addition, we demonstrated that TPA stimulated hydrogen peroxide production in a cloned mouse keratinocyte cell line. In both instances, hydrogen peroxide production by the cells was inhibitable by catalase. These studies provided the first direct evidence that epidermal keratinocytes produced hydrogen peroxide and suggested that this cell type may be a significant source of reactive oxygen intermediates during TPA-mediated mouse skin tumor promotion.

One of the most popular beverages in the world is tea (*Camellia sinensis*). Although epidemiological studies relating tea consumption and the development of cancer have not been conclusive (*18-24*), animal studies have shown an inhibitory action of tea, or tea components on chemical carcinogenesis (*25-31*). Using the two stage initiation-promotion model of mouse skin carcinogenesis, topical application of polyphenolic extracts of green tea have been shown to inhibit tumorigenesis (*25-28*). Similarly, skin tumor initiation by 7,12-dimethylbenz[*a*]anthracene (DMBA) (*26*) and skin tumor promotion induced by TPA or ultraviolet light, have been reported to be inhibited by oral administration of green tea (*29*). A number of studies have demonstrated that compounds with antioxidant activity inhibit TPA-induced tumor promotion in mouse skin (*32-39*). Green tea as well as some of the constituents of green tea possess antioxidant activity (*40-42*). It is likely, therefore, that the ability of GTP to inhibit the formation of free radicals in the skin may contribute to its antipromoting activity.

GTP Suppresses the Formation of Hydrogen Peroxide in Mouse Epidermis.

The effects of GTP on the formation of hydrogen peroxide in epidermal cells was monitored in female CD-1 mice (20 grams) following a single topical application of 5 nmol TPA in 200 μl of acetone. GTP (3.6 mg in 200 μl of acetone), prepared as previously described (*22*), was applied topically immediately following TPA. Control animals were treated with 200 μl of acetone alone. Twenty-four hr after treatment, the mice were sacrificed and epidermal cells were isolated as previously described (*17*). Briefly, treated areas of the skin were removed and floated, epidermis side down, in Hank's balanced salt solution containing 2.5% trypsin. After 45 min at 37°C, the epidermis was scraped off and washed in phosphate buffered saline. Cells (1 x 10^6) were then incubated at 37°C with 5 μM DCFH-DA. After 30 min, cell associated fluorescence was measured using a Coulter EPICS Profile flow cytometer. For each analysis, 10,000 events were accumulated.

Figure 1 shows the results obtained with cells from the epidermis of mice following various treatment protocols. The upper panel compares hydrogen peroxide production by epidermal cells isolated from acetone-treated and TPA-treated animals. Over 95% of the epidermal cells from acetone treated mice were found to produce relatively low basal levels of hydrogen peroxide (upper panel, solid line). A small subpopulation of cells from acetone treated animals produced elevated levels of hydrogen peroxide. Twenty-four hr after treatment with TPA, a marked increase in hydrogen peroxide production by the epidermal cells was observed (upper panel, broken line). The lower panel compares hydrogen peroxide production by epidermal cells from TPA-treated animals and from animals treated with TPA and GTP. From these data, it is apparent that GTP caused a dramatic decline in hydrogen peroxide production by the epidermal cells from TPA-treated animals to levels observed in cells isolated from control acetone-treated animals (lower panel, solid line). In further experiments, we found that when initiated animals were treated with TPA twice weekly for 16 weeks, there was an increase in hydrogen peroxide production by a subpopulation of epidermal cells. As observed in short term studies, topical application of GTP and TPA resulted in a marked reduction in TPA-stimulated levels of hydrogen peroxide (not shown).

Discussion and Conclusions

The results reported here and in an earlier publication (17) demonstrate that application of TPA to mouse skin stimulates epidermal cells to produce hydrogen peroxide. At the present time, the mechanisms by which TPA increases intracellular levels of hydrogen peroxide are not clear. Oxidants are produced by enzymes during intermediary metabolism. To protect the cell against these oxidants, cells contain chemicals that serve as antioxidants such as glutathione (43,44). Cells also possess enzymes which degrade oxidants (7,12). The balance between the production and detoxification of oxidants in cells of the epidermis is known to be altered following treatment with TPA (43-50). For example, TPA increases the activity of xanthine oxidase, an enzyme which produces both superoxide and hydrogen peroxide (49,50). Catalase, an enzyme which degrades hydrogen peroxide, has been reported to decrease in the epidermis following TPA-treatment (48). These data indicate that epidermal cells have the capacity to produce reactive oxygen intermediates and that TPA can modulate enzymes which regulate their levels in the epidermis. The fact that epidermal cells produce markedly elevated levels of hydrogen peroxide following TPA treatment supports the idea that oxidant-induced damage may contribute to the tumor promotion process.

Mouse epidermis is a stratified squamous epithelium consisting predominantly of keratinocytes. Cells that are actively dividing reside in the basal layer. These cells differentiate, migrating outward to form the stratum corneum. Topical application of TPA to mouse skin induces rapid alterations in tissue structure including increased basal cell proliferation along with alterations in the normal process of keratinocyte maturation (51-54). There is an increase in the thickness of the epidermis as well as the number of nucleated cell layers (53,54). This is

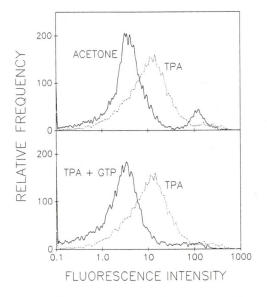

Figure 1. Inhibitory effect of topical application of GTP on TPA-induced hydrogen peroxide production by mouse epidermal cells. CD-1 female mice (n = 3 per treatment) were treated topically with 200 μl of acetone or TPA (5 nmol) in acetone (top panel) or TPA together with green tea polyphenol fraction (GTP, 3.6 mg) in acetone (bottom panel). Twenty-four hr later, the mice were sacrificed and epidermal cells isolated and analyzed for hydrogen peroxide content by flow cytometry as indicated in the text. Note the marked increase in hydrogen peroxide content, measured by fluorescence intensity of DCF, in epidermal cells isolated from animals treated with TPA (top panel). This increase was inhibited by treatment of mouse skin with GTP (bottom panel).

associated with a number of biochemical changes in the skin including alterations in epidermal DNA, RNA and protein biosynthesis (53-56), phospholipid metabolism (57) and keratinization (58,59). Previous studies have shown that GTP is anti-inflammatory and inhibits TPA-induced edema, epidermal ornithine decarboxylase as well as epidermal hyperplasia (27). Each of these events are early biochemical markers of the tumor promotion process (27,53,54,58).

As indicated above, green tea and green tea extracts are known to possess antioxidant activity (40-42). Antioxidant activity has also been described for several of the constituents of green tea including (-)-epigallocatechin gallate, (-)-epicatechin gallate, (-)-epigallocatechin, (-)-epicatechin and (+)-catechin (41). At the present time it is uncertain whether or not the ability of green tea extracts or its components to inhibit the biochemical and morphological changes in the epidermis induced by TPA as well as tumor promotion is dependent on its antioxidant activity. The fact that a large number of diverse chemicals that are scavengers of reactive oxygen intermediates or antioxidants inhibit tumor promotion or biochemical events associated with tumor promotion support the idea that this may be the mechanism of action of green tea extracts. Antioxidants that inhibit tumor promotion include ascorbic acid (38), α-tocopherol (37), ascorbyl palmitate (38). curcumin (39), butylated hydroxyanisole (32,34), sodium selenite (33), quercetin (35), butylated hydroxytoluene (34) and nordihydroguaiaretic acid (36). Earlier studies have shown that (-)-epigallocatechin gallate (a major component of green tea) inhibits the binding of TPA to its PKC receptor (25), and we cannot exclude the possibility that there are alternative sites of action for GTP in the epidermis. Further studies are necessary to determine the mechanisms of action of green tea extracts.

In summary, our data demonstrates that the tumor promoter TPA stimulates production of hydrogen peroxide by epidermal cells from mouse skin. The topical application of green tea polyphenol fraction, which is a potent inhibitor of mouse skin tumor promotion, was found to inhibit TPA-induced increases in epidermal hydrogen peroxide. This occurred in both short (24 hr) and long term (16 week) experiments. Many antioxidants are known to inhibit skin tumor promotion. Thus, the antioxidant activity of green tea extracts and its components may play a role in its biological effects on the skin.

Acknowledgements

This work was supported in part by NIEHS grant ES 03647 awarded to Jeffrey D. Laskin, NIEHS Center grant ES 05022 and grant CA49756 from the National Cancer Institute.

Literature Cited

1. Cerutti, P.A. Science 1985, 227, 375-381.
2. Troll, W. and Wiesner, R. Ann. Rev. Pharmacol. Toxicol. 1985, 25, 509-528.
3. Kensler, T.W. and Taffe, B.G. Adv. in Free Radic. Biol. Med. 1986, 2, 347-387.
4. Marnett, L. Carcinogenesis (Lond.), 1987, 8, 1365-1373.
5. O'Brien, P.J. Free Radic. Biol. Med. 1988, 4, 169-183.
6. Perchellet, J.P. and Perchellet, E.M. Free Radic. Biol. Med. 1989, 7, 377-408.

7. Trush, M.A. and Kensler, T.W. *Free Radic. Biol. and Med.* **1991**, *10*, 201-209.

8. Witz, G. *Proc. Soc. Expt'l Biol. and Med.* **1991**, *198*, 675-682.

9. Frenkel, K.; Chrzan, K.; Troll, W.; Teebor, G.W.; and Steinberg, J.J. *Cancer Res.* **1986**, *46*, 5533-5540.

10. Frenkel, K. *Environ. Health Perspect.* **1989**, *81*, 45-54.

11. Nagasawa, H. and Little, J.B. *Proc. Nat'l Acad. Sci., USA*, **1979**, *76*, 1943-1947.

12. Birnboim, H.C. *Science*, **1982**, 215, 1247-1249.

13. Lewis, J.G. and Adams, D.O. *Carcinogenesis (Lond.)*, **1987**, *7*, 889-898.

14. Trush, M.A.; Seed, J.L. and Kensler, T.W. *Proc. Nat'l Acad. Sci., USA*, **1985**, *82*, 5194-5198.

15. Fischer, S.M. and Adams, L.M. *Cancer Res.*, **1985**, 45, 3130-3136.

16. Fischer, S.M.; Baldwin, J.K. and Adams, L.M. *Carcinogenesis (Lond.)*, **1986**, *7*, 915-918.

17. Robertson, F.M.; Beavis, A.J.; Oberyszyn, T.M.; O'Connell, S.M.; Dokidos, A.; Laskin, D.L.; Laskin, J.D. and Reiners, J.J. *Cancer Res.* **1990**, *50*, 6062-6067.

18. Committee on Diet and Health, Food and Nutrition Board, Commission on Life Sciences, National Research Council. *Diet and Health;* National Academy Press: Washington, D.C., 1989; pp. 465-508.

19. World Health Organization, International Agency for Research on Cancer, *Coffee, Tea, Mate, Methylxanthines and Methylglyoxal*; IARC Monographs on the Evaluation of Carcinogenic Risks to Humans, IARC Press: Lyon, France, 1991, vol. 51; pp. 207-271.

20. Kinlen, L.J.; Willows, A.N.; Goldblatt, P. and Yudkin, J. *Br. J. Cancer* **1988**, *58*, 397-401.

21. Stocks, P. *Br. J. Cancer* **1970**, *24*, 215-225.

22. Tajima, K. and Tominaga, S. *Jpn. J. Cancer Res. (Gann)* **1985**, *76*, 705-716.

23. Kono, S.; Ikeda, M.; Tokudome, S. and Kuratsume, M. *Jpn. J. Cancer Res. (Gann)* **1988**, *79*, 1067-1074.

24. Oguni, I.; Nasu, K.; Yamamoto, S. and Nomura, T. *Agric. Biol. Chem.* **1988**, *52*, 1879-1880.

25. Yoshizawa, S.; Horiuchi, T.; Fujiki, H.; Yoshida, T.; Okuda, T.; Sugimura, T. *Phytother. Res.*, **1987**, *1*, 44-47.

26. Wang, Z.Y.; Khan, W.A.; Bickers, D.R. and Mukhtar, H. *Carcinogenesis (Lond.)* **1989**, *10*, 411-415.

27. Huang, M.-T.; Ho, C.-T.; Wang, Z.-Y.; Ferraro, T.; Finnegan-Olive, T.; Lou, Y.-R.; Mitchell, J. M.; Laskin, J. D.; Newmark, H.; Yang, C. S.; Conney, A. H. *Carcinogenesis* **1992**, (in press).

28. Fujiki, H. In *Proc. of the Fourth Chemical Congress of North America*; New York, **1991**, Abstract No. 66.

29. Wang, Z.-Y.; Huang, M.-T.; Ferraro, T.; Wong, C.-Q.; Lou, Y.-R.; Reuhl, K.; Iatropoulos, M.; Yang, C. S. and Conney, A. H. *Cancer Res.* **1992**, *52*(5), 1162-1170.

30. Fujita, Y.; Yamane, T.; Tanaka, M.; Kuwata, K.; Okuzumi, J.; Takahashi, T.; Fujiki, H. and Okuda, T. *Jpn. J. Cancer Res. (Gann)* **1989**, *80*, 503-505.

31. Wang, Z.Y.; Agarwal, R.; Bickers, D.R. and Mukhtar, H. *Carcinogenesis (Lond.)* **1991**, *12*, 1527-1530.

32. Slaga, T.J. *Acta Pharmacol. Toxicol. Copenh.* **1984**, Supp 2, 107-124.

33. Griffin, A.C. In *Molecular Interelations, Nutrition and Cancer;* M.S. Amott, J. Van Eys, and Wang, Y.M., Ed.; New York, Raven Press, 1982, pp. 401-408.
34. Kozombo, W.J.; Seed, J.L. and Kensler, T.W. *Cancer Res.*, **1983**, *43*, 2555-2559.
35. Kato, R.; Nakadate, T.; Yamamoto, S. and Sugimura, T. *Carcinogenesis (Lond.)*, **1984**, *4*, 1301-1305.
36. Nakadate, T.; Yamamoto, S.; Ishii, M. and Kato, R. *Carcinogenesis (Lond.)*, **1982**, *3*, 1411-1414.
37. Perchellet, J-P.; Owen, M.D.; Posney, T.D.; Orten, D.K. and Schneider, B.A. *Carcinogenesis (Lond.)*, **1985**, *6*, 567-573.
38. Smart, R.C.; Huang, M-T.; Han, Z.T.; Kaplan, M.C.; Focella, A. and Conney, A.H. *Cancer Res.*, **1987**, *47*, 6633-6638.
39. Huang, M-T.; Smart, R.C.; Wong, C-Q. and Conney, A.H. *Cancer Res.*, **1988**, *48*, 5941-5946.
40. Zhao, B.; Li, X.J.; He, R.; Cheng, S. and Wenjuan, X. *Cell Biophys.*, **1989**, *14*, 175-185.
41. Osawa, T., Namki, M. and Kawakishi, S., In *Basic Life Sciences: Antimutagenesis and Anticarcinogenesis Mechanisms*; Kuroda, Y., Shankel, D.M. and Waters, M.D., Ed.; Plenum Press, New York, 1990, vol. 2, pp. 139-153.
42. Ruch, R.J.; Cheng, S-J. and Klaunig, J.E. *Carcinogenesis (Lond.)*, **1989**, *10*, 1003-1008.
43. Solanki, V.; Rana, R.S. and Slaga, T. *Carcinogenesis (Lond.)* **1981**, *2*, 1141-1146.
44. Perchellet, J.P.; Perchellet, E.M.; Orten, D.K. and Schneider, *Carcinogenesis (Lond.)*, **1986**, *7*, 503-506.
45. Perchellet, J.P.; Orten, D.K.; Schneider, B.A. and Perchellet, E.M. *Proc. Am. Assoc. Cancer Res.* **1985**, 26, 133.
46. Perchellet, J.-P.; Abney, N.L.; Thomas, R.M.; Guislain, Y.L. and Perchellet, E.M. *Cancer Res.*, **1987**, *47*, 477-485.
47. Perchellet, J.P.; Perchellet, E.M., Orten, D.K. and Schneider, B.A. *Cancer Lett.* **1985**, *26*, 283-293.
48. Reiners, J.J.; Hale, M.A. and Cantu, A.R. *Carcinogenesis (Lond.)*, **1988**, *9*, 1259-1263.
49. Reiners, J.J.; Pence, B.C.; Barcus, M.C.S. and Cantu, A.R. *Cancer Res.*, **1987**, *47*, 1775-1779.
50. Pence, B.C. and Reiners, J.J. *Cancer Res.*, **1987**, *47*, 6388-6392.
51. Astrup, E.G. and Iverson, O.H. *Virchows Arch. B. Cell Path.*, **1983**, *42*, 1.
52. Argyris, T.S. *CRC Crit. Rev. Toxicol.*, **1981**, *9*, 151-200.
53. Raick, A.N.; Thumm, K. and Chivers, B.R. *Cancer Res.*, **1972**, *32*, 1562-1568.
54. Raick, A.N. *Cancer Res.*, **1973**, *33*, 269-286.
55. Balmain, A.; Alonso, A. and Fischer, J. *Cancer Res.* **1977**, *37*, 1548-1555.
56. Krieg, L.; Kuhlman, I. and Marks, F. *Cancer Res.*, **1974**, *34*, 3135-3146.
57. Rohrschneider, L.R.; O'Brien, D.H., and Boutwell, R.K. *Biochim. Biophys. Acta* **1972**, *280*, 57-70.
58. Molloy, C.J. and Laskin, J.D., *Cancer Res.*, **1987**, 47, 4674-4680.
59. Molloy, C.J.; Gallo, M.A. and Laskin, J.D. *Carcinogenesis (Lond.)*, **1987**, *8*, 1193-1199.

RECEIVED April 20, 1992

OTHER PHENOLIC COMPOUNDS
AND CANCER PREVENTION

Chapter 25

Penta-*O*-Galloyl-β-D-Glucose and (−)-Epigallocatechin Gallate

Cancer Preventive Agents

S. Yoshizawa[1], T. Horiuchi[1,5], M. Suganuma[1], S. Nishiwaki[1], J. Yatsunami[1], S. Okabe[1], T. Okuda[2], Y. Muto[3], K. Frenkel[4], W. Troll[4], and H. Fujiki[1,6]

[1]Cancer Prevention Division, National Cancer Center Research Institute, Tsukiji 5−1−1, Chuo-ku, Tokyo 104, Japan
[2]Faculty of Pharmaceutical Sciences, Okayama University, Tsushima, Okayama 700, Japan
[3]Faculty of Medicine, Gifu University School of Medicine, Tsukasa-machi, Gifu 500, Japan
[4]Department of Environmental Medicine, New York University Medical Center, New York, NY 10016

Penta-O-galloyl-β-D-glucose (5GG), obtained by methanolysis of tannic acid and (-)-epigallocatechin gallate (EGCG), the main constituent of Japanese green tea, were polyphenols which strongly inhibited the specific ^{3}H-12-O-tetradecanoylphorbol-13-acetate (TPA) binding to the phorbol ester receptors in a particulate fraction of mouse skin. Based on the evidence, these polyphenols were thought to inhibit tumor promotion of TPA on mouse skin. 5GG and EGCG inhibited tumor promotion of teleocidin, one of the TPA-type tumor promoters on mouse skin initiated with 7,12-dimethylbenz(a)anthracene (DMBA) in two-stage carcinogenesis experiments. Moreover, EGCG inhibited tumor promotion of okadaic acid, which acts differently on cells than the TPA-type tumor promoters. The mechanisms of action of EGCG were studied in relation to the reduction of specific binding of the tumor promoters, ^{3}H-TPA and ^{3}H-okadaic acid to their receptors in cell membrane. A single application of 5 mg EGCG decreased immediately to the minimum level the specific binding of ^{3}H-tumor promoters. The effect was understood to be a sealing of the membrane by EGCG, due to EGCG-protein interaction, resulting in reduction of protein phosphorylation in cells. In addition, EGCG inhibited H_2O_2 formation by TPA-activated human polymorpho-nuclear leukocytes. Since EGCG is a non-toxic compound, ingested in green tea in every day life in Japan, we extended the study of EGCG as a cancer preventive agent. This paper also reviews inhibitory effects of EGCG on duodenal carcinogenesis of male C57BL/6 mice and develop-ment of spontaneous hepatoma in male C3H/HeN mice. We think EGCG is a practical cancer chemopreventive agent to be implemented in every day life.

[5]Current address: Ehime University School of Medicine, Ehime 791−02, Japan
[6]Corresponding author

0097−6156/92/0507−0316$06.00/0

Polyphenols and Phorbol Ester Receptor Binding

Our study of EGCG for the purpose of cancer prevention started with the anticarcinogenic study of polyphenols derived from medicinal plants and drugs. About 30 polyphenols were first tested to find whether they shared the same phorbol ester receptor in a particulate fraction of mouse skin as TPA. Most of the polyphenols inhibited the specific ^{3}H-TPA binding dose-dependently, but some did not. Thirty polyphenols were roughly classified into four classes according to the ED_{50} values for inhibition (Fig. 1). The ED_{50} values of the Class 1 polyphenols were about one thousand times more than that of TPA (1). Table 1 lists some representatives of the 30 polyphenols, according to classification. 5GG , EGCG, pedunculagin, chebulinic acid and buddledin A were included in Class 1 with ED_{50} values of 1.7 μM. Since the structures of these five compounds are unrelated to that of TPA (Fig. 2), we thought that they might interact with the phorbol ester receptor in the cell membrane differently from TPA. Polyphenols have the ability to precipitate water-soluble proteins, such as a hemoglobin solution. The precipitability of hemoglobin by geraniin was taken as a standard and the potencies of various polyphenols were compared with their result with geraniin and these values were called the relative astringency to geraniin. The relative astringency to geraniin was used to estimate the biochemical effect of polyphenols that was caused by polyphenol-protein interaction (2). Some polyphenols showed similar values in two parameters, the relative astringency and inhibition of specific binding of ^{3}H-TPA to the receptor (Fig. 3). From these results, 5GG from hydrolyzable tannins and EGCG from condensed tannins were selected for studying inhibition of tumor promotion in two-stage carcinogenesis experiments. Polyphenols of Class 4, such as (+)-catechin, ellagic acid and methyl gallate, which did not bind to the receptor, were deleted from further experiments.

5GG and Its Inhibition of Tumor Promotion of Teleocidin

We first had to obtain 5GG on an order of gram amounts sufficient to carry out the inhibition of tumor promotion experiment. 25 g of tannic acid, which was isolated from a gall, *Schisandrae fructus*, was incubated in 500 ml of methanol:acetate buffer (9:1) at pH 6.0 at 40 °C for 6 hours. This methanolysis cleaved the depside linkage between 5GG and the galloyl group and released 5GG and the galloyl residue. According to the procedure summarized in Table 2, 8 g of 5GG were precipitated, after the methanol layer was diluted to 2% methanol solution and placed at 4 °C. The purity of 5GG was determined by HPLC on a TSK gel silica 60 column with n-hexane: methanol: tetrahydrofurane:formic acid (55:33:11:1) containing 450 mg/L oxalic acid as a solvent. UV absorption was monitored at 280 nm and resulted in 95% purity. This method was quick and applicable to large scale purification.

Inhibition by 5GG of tumor promotion was studied in a two-stage carcinogenesis experiment on skin of 8 week-old female CD-1 mouse. Initiation was carried out by a single application of 100 μg DMBA and tumor promotion was achieved by repeated applications of 2.5 μg teleocidin, twice a week (3). In the experimental group, 5 mg of 5GG were applied topically 15 min before each treatment with teleocidin. The 5GG treatment reduced the percentages of tumor-bearing mice from 100% to 53% and the average numbers of tumors per mouse from 3.3 to 0.9 in week 20 (4). The evidence encouraged us next to pursue an experiment with EGCG.

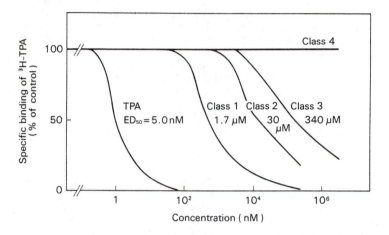

Fig. 1.   Classification of polyphenols according to their ED_{50} values of inhibition
of specific ^{3}H-TPA binding to a particulate fraction of mouse skin.

Table 1   Classification of Various Polyphenols by Phorbol Ester Receptor Binding

Class 1   ($ED_{50} = 1.7$ μM)
Penta-O-galloyl-β-D-glucose
(-)-Epigallocatechin gallate
Pedunculagin
Chebulinic acid
Buddledin A

Class 2   ($ED_{50} = 30$ μM)
Rugosins D and E
Coriariin A
Cornusiin A
Nobotaniin C

Class 3   ($ED_{50} = 340$ μM)
Tellimagrandin I
Tellimagrandin II

Class 4   (no inhibition)
(+)-Catechin
Ellagic acid
Methyl gallate

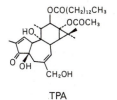

TPA

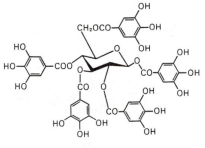

Penta-O-galloyl-β-D-glucose

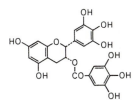

(−)-Epigallocatechin gallate

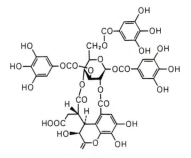

Chebulinic acid

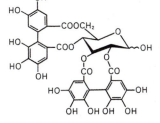

Pedunculagin

Buddledin A

Fig. 2. Structures of the Class 1 polyphenols compared with that of TPA.

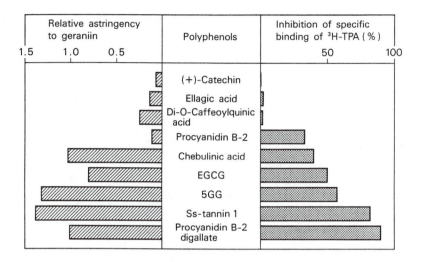

Fig. 3. Comparative values of biochemical effects of polyphenols on relative astringency to geraniin and inhibition of specific binding of ³H-TPA to a mouse skin particulate fraction. The precipitability of hemoglobin by geraniin was taken as a standard.

Table 2   Purification of Penta-O-galloyl-β-D-glucose

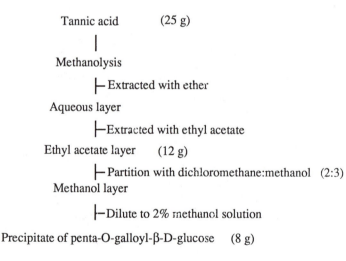

EGCG and Its Inhibition of Tumor Promotion of Okadaic Acid

In 1987, we first reported that EGCG inhibited tumor promotion of teleocidin on mouse skin (1). This paper first reports the inhibitory effects of EGCG on tumor promotion of okadaic acid, which acts differently than teleocidin (5). The EGCG used in the experiment was isolated from Japanese green tea leaves and was made up of EGCG (85%), (-)-epicatechin (10%) and (-)-epicatechin gallate (5%), as described previously (1).

According to the two-stage carcinogenesis experiment, initiation was carried out by a single application of 100 μg DMBA on skin of female CD-1 mouse and tumor promotion was achieved by repeated applications of 1 μg okadaic acid on the same area of the initiated skin, twice a week (5). In the experimental group, 5 mg EGCG was applied topically before each treatment with okadaic acid. As Figure 4 shows, EGCG treatment completely inhibited tumor promotion of okadaic acid up to week 20 of tumor promotion, with regard to the percentages of tumor-bearing mice and the average numbers of tumors per mouse. Our results indicated that EGCG inhibited tumor promotion of two tumor promoters, okadaic acid in this experiment, and teleocidin in the previous experiment, with different mechanisms of action (5-7). Okadaic acid inhibits dephosphorylation of phosphoserine and phosphothreonine through inhibition of protein phosphatase 1 and 2A activities (8), resulting in an increase of phosphoproteins (9), whereas teleocidin activates protein kinase C and produces phosphoproteins in larger amounts than in the usual state (3). In both cases, the increase of phosphoproteins might act as a signal for tumor promotion.

Since EGCG inhibits both tumor promotion pathways, we studied their receptor bindings using ^{3}H-okadaic acid and ^{3}H-TPA. Okadaic acid binds to protein phosphatases 1 and 2A (10, 11), whereas TPA and teleocidin bind to the phorbol ester receptor (3). These two different receptors are present in a membrane fraction of mouse skin. When mouse skin was treated with a single application of 5 mg EGCG, both the specific binding of ^{3}H-TPA and that of ^{3}H-okadaic acid decreased immediately and reached a minimum in 5 to 10 min (Fig. 5). The levels gradually returned to normal. In previous experiments of tumor promotion, EGCG was usually applied 15 min before each treatment with a tumor promoter. Therefore, the results of Figure 5 indicate that okadaic acid or teleocidin had been applied to the skin which was at the weakest condition for receptor binding. Based on the evidence, EGCG treatment possibly inhibited the interaction of tumor promoters with their receptors, resulting in reduction of protein kinase activity, as well as inhibiting the process of tumor promotion after tumor promoter receptor binding. Figure 6 is a schematic illustration of the suggested mechanisms of action of EGCG. We think EGCG first induces the sealing of the membrane, due to EGCG-protein interaction, which explains the reduction of specific binding of the tumor promoters to their receptors in the cell membrane. This sealing consequently causes inhibition of phosphorylation in cells, although phosphorylation originally acts as a signal for tumor promotion. In addition, it is generally accepted that a tumor promoter generates several active oxygen species. Since EGCG is a strong antioxidant, EGCG was expected to inhibit the effect of TPA. EGCG clearly inhibited H_2O_2 formation by TPA-activated human polymorphonuclear leukocytes, indicating that EGCG partly acts by suppressing oxyradical formation (12). In general, we assume that a membrane sealed by EGCG might additionally interrupt the interaction of various growth factors and hormones with their receptors in the membrane through autocrine and paracrine, subsequently resulting in a specific inhibition of the cell growth.

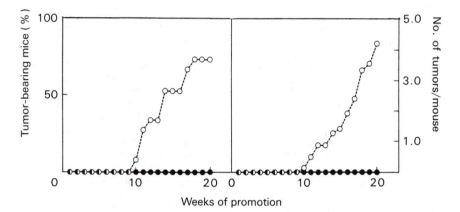

Fig. 4. Inhibition by EGCG on tumor promotion of the suggested okadaic acid. Groups treated with DMBA and okadaic acid (O) and with DMBA and okadaic acid plus EGCG (●).

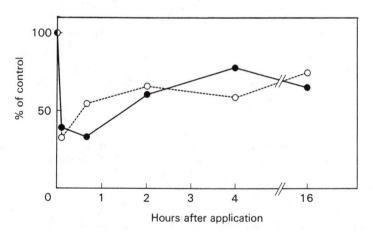

Fig. 5. Inhibition of specific binding of ^{3}H-TPA and ^{3}H-okadaic acid to a mouse skin particulate fraction after a single application of EGCG. ^{3}H-TPA (●) and ^{3}H-okadaic acid (O).

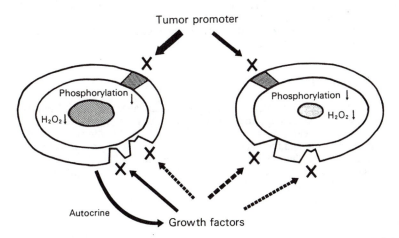

Fig. 6. A schematic illustration of the suggested mechanisms of action of EGCG, through a sealed cell membrane.

Anticarcinogenic Effects of EGCG

EGCG was effective in inhibition of carcinogenesis in other organs than the skin. The results are briefly reviewed. A solution of either 0.05% or 0.1% EGCG was significantly effective in inhibition of spontaneous hepatoma development in male C3H/HeN mice (Muto et al., manuscript in preparation). The effects might be related to the evidence that EGCG inhibits lipid peroxidation in liver mitochondria (13). As we previously reported, a solution of 0.005% EGCG significantly inhibited duodenal carcinogenesis of male C57BL/6 mice, which had been induced by N-ethyl-N'-nitro-N-nitrosoguanidine (14). However, EGCG in drinking water did not inhibit development of spontaneous thymic lymphoma in female AKR mice. Bladder cancer in female C3H/HeN mice induced by N-butyl-N-(4-hydroxyl-butyl)nitrosamine was also not inhibited (Ohtani et al., unpublished result). Moreover, the effects of EGCG on carcinogenesis in various organs, such as the colon and liver, are under investigation by scientists in Japan. Since inhibition of chemical carcinogenesis induced by large amounts of potent carcinogens is not a suitable test system to screen for a cancer preventive agent (15), we have to estimate the effects of EGCG based on various carcinogenesis experiments under different conditions. It is an important task to find out how we should apply EGCG for the prevention of human cancers, or what kinds of cancers are preventable by EGCG treatment. Drinking green tea should be evaluated as one of the most practical methods of cancer prevention available at present.

Acknowledgments

This work was supported in part by Grants-in-Aid for Cancer Research from the Ministry of Education, Science and Culture, a grant for the Program for a Comprehensive 10-Year Strategy for Cancer Control from the Ministry of Health and Welfare of Japan, and by grants from the Foundation for promotion of Cancer Research, the Uehara Memorial Life Science Foundation and the Princess Takamatsu Cancer Research Fund. We thank Dr. T. Sugimura at National Cancer Center for his encouragement of the work.

Literature Cited

1. Yoshizawa, S.; Horiuchi, T.; Fujiki, H.; Yoshida, T.; Okuda, T.; Sugimura, T. *Phytotherapy Res.* **1987**, 1, 44.
2. Okuda, T.; Mori, K.; Hatano, T. *Chem. Pharm. Bull.* **1985**, 33, 1424.
3. Fujiki, H.; Sugimura, T. *Adv. in Cancer Res.* **1987**, 49, 223.
4. Fujiki, H.; Yoshizawa, S.; Horiuchi, T.; Suganuma, M.; Yatsunami, J.; Nishiwaki, S.; Okabe, S.; Matsushima, R.N.; Okuda, T.; Sugimura, T. *Preventive Medicine,* **1991**, in press.
5. Fujiki, H.; Suganuma, M.; Sugimura, T. *Envir. Carcino. Revs.* **1989** C7(1), 1.
6. Sassa, T.; Richter, W.W.; Uda, N.; Suganuma, M.; Suguri, H.; Yoshizawa, S.; Hirota, M.; Fujiki, H. *Biochem. Biophys. Res. Commun.* **1989**, 159, 939.
7. Fujiki, H.; Tanaka, Y.; Miyake, R.; Kikkawa, U.; Nishizuka, Y.; Sugimura, T. *Biochem. Biophys. Res. Commun.* **1984**, 120, 339.

8. Takai, A.; Bialojan, C.; Troschka, M.; Rüegg, J. C. *FEBS Lett.* **1987**, 217, 81.
9. Yatsunami, J.; Fujiki, H.; Suganuma, M.; Yoshizawa, S.; Eriksson, J. E.; Olson, M.O.J.; Goldman, R.D. *Biochem. Biophys. Res. Commun.* **1991**, 177, 1165.
10. Suganuma, M.; Suttajit, M.; Suguri, H.; Ojika, M.; Yamada, K.; Fujiki, H. *FEBS Lett.* **1989**, 250, 615.
11. Nishiwaki, S.; Fujiki, H.; Suganuma, M.; Ojika, M.; Yamada, K.; Sugimura, T. *Biochem. Biophys. Res. Common.* **1990**, 170, 1359.
12. Zhong, Z.; Tius, M.; Troll, W.; Fujiki, H.; Frenkel, K. *Proceedings of the AACR*, **1991**, 32, 127.
13. Okuda, T.; Kimura, Y.; Yoshida, T.; Hatano, T.; Okuda, H.; Arichi, S. *Chem. Pharm. Bull.* **1983**, 31, 1625.
14. Fujita, Y.; Yamane, T.; Tanaka, M.; Kuwata, K.; Okuzumi, J.; Takahashi, T.; Fujiki, H.; Okuda, T. *Jpn. J. Cancer Res.* **1989**, 80, 503.
15. Ashby, J.; Morrod, R.S. *Nature* **1991**, 352, 185.

RECEIVED January 13, 1992

Chapter 26

Protective Effects Against Liver, Colon, and Tongue Carcinogenesis by Plant Phenols

T. Tanaka, N. Yoshimi, S. Sugie, and H. Mori

First Department of Pathology, Gifu University School of Medicine,
40 Tsukasa-machi, Gifu 500, Japan

Three different experiments were performed in order to examine
the potential inhibitory effects of the plant phenolic
compounds ellagic acid (EA) and chlorogenic acid (CA) on
chemical carcinogenesis. Experiment I: The effect of dietary
EA (400 ppm for 18 weeks) on the carcinogenicity to rats of
concurrently administered N-2-fluorenylacetamide (FAA, 200
ppm in diet for 16 weeks) was determined. Experiment II: The
effect of EA in diet (400 ppm for 10 weeks) on the rat tongue
carcinogenesis by 4-nitroquinoline 1-oxide (4-NQO, 10 ppm in
drinking water for 8 weeks) during the initiation stage was
investigated. Experiment III: The protective effect of dietary
CA (250 ppm for 24 weeks) on methylazoxymethanol (MAM) acetate
(a single i.v. injection, 20 mg./kg body weight)-induced colon
and liver carcinogenesis during the post-initiation stage was
examined in Syrian golden hamsters. In the rats receiving EA
together with FAA or 4-NQO, the incidence of neoplasm in the
liver (30%) or tongue (20%) was significantly decreased
compared with those of rats given the carcinogen alone (100%
in liver; 71% in tongue). Similarly, colon tumor (13%) and
liver cell foci (6.0 ± 2.7/cm²) incidence in hamsters given
MAM acetate and CA was significantly smaller than those of
hamsters given MAM acetate alone (50% and 10.8 ± 2.7/cm²).
These results suggest that the plant phenolics EA and CA are
potential chemopreventive agents in carcinogenesis. The number
and area of AgNORs in epithelial cell nuclei of liver, tongue,
and colon of animals were also quantified in order to evaluate
the proliferative activity of cells. The number and area of
AgNORs of nuclei of the target organs from the animals treated
with the carcinogens and test chemicals were significantly

0097–6156/92/0507–0326$06.00/0

lower than those of animals given the carcinogen alone, indicating that EA and CA suppressed the proliferative activity of targeted epithelial cells and AgNOR enumeration could be a useful method for estimating the chemopreventive potential of certain chemicals in carcinogenesis.

Chemical carcinogenesis is now generally recognized as a multistage process (initiation, promotion, and progression) (*1*). Environmental factors (mutagens, carcinogens, and promoters) are important for occurrence of human cancers (*2*). There is growing evidence for the involvement of genetic alterations of various types of cancers in mammals (*3, 4*). However, the process leading from the initiation stage to the overt cancer stage is less understood. Although recent advances in cancer diagnosis and therapy have improved the short-term prognosis and quality of life for many cancer patients, there has been little or no effects on the mortality rate for most cancers. Thus, it is considered that primary and secondary prevention of cancer development in humans is very important (*5*).

Recently, much attention has been given to chemopreventive agents that may act as protection against mutagens or carcinogens in the environment. Several compounds have been found to inhibit development of neoplasms induced by chemical carcinogens when administered prior to and/or simultaneously with the carcinogen (*6*). These include naturally occurring constituents of food, particularly edible plants, vegetables and fruits, and also synthetic compounds (*6-9*). We present here our recent studies (*10, 11*) on the inhibitory effects of the plant phenolics ellagic acid (EA) and chlorogenic acid (CA) on liver, colon, and tongue carcinogenesis.

Materials and Methods

Chemicals. Test chemicals, ellagic acid (EA, >97% pure) and chlorogenic acid (CA) were purchased from Tokyo Chem. Ind. Co., Ltd., Tokyo, Japan and Sigma Chem. Co., St. Louis, MO, U.S.A., respectively. N-2-Fluorenylacetamide (FAA, Nacalai Tesque, Inc., Kyoto, Japan), 4-nitroquinoline 1-oxide (4-NQO), and methylazoxymethanol (MAM) acetate (Ash Stevens, Milwaukee, WI, U.S.A.) were used as a chemical carcinogen. EA and CA were mixed with powdered basal diet, CE-2 (Japan CLEA Inc., Tokyo, Japan) at concentrations 400 ppm and 250 ppm, respectively. These diet were made every two weeks.

Animals. Male inbrd ACI/N rats (6 weeks of age), which were maintained in our laboratory, were used for examining the inhibitory effect of EA on liver carcinogenesis induced by FAA (Experiment I) and on tongue carcinogenesis induced by 4-NQO (Experiment II). Syrian golden hamsters of both sexes (Japan CLEA Inc., Tokyo, Japan), 2 months old were used for examining the suppressing effects of CA on large intestinal and liver carcinogenesis induced by MAM acetate (Experiment III).

Experiment I: Inhibitory Effect of EA on FAA-Induced Hepatocarcinogenesis.
A total of 70 male ACI/N rats were divided into 4 groups as shown in Figure
1. Group 1 (19 rats) was fed the diet containing 200 ppm FAA for 16 weeks
and maintained on the basal diet, CE-2 (Japan CLEA Inc., Tokyo, Japan), for
20 weeks. Group 2 (19 rats) was fed the diet containing 200 ppm FAA and 400
ppm EA for 16 weeks. Rats in this group were fed 400 ppm EA containing diet
for one week before and after the carcinogen exposure, and then maintained
on the basal diet for 19 weeks. Group 3 (16 rats) was fed the diet
containing 400 ppm EA begining one week before the start of the study and
continued for 18 weeks. Group 4 (16 rats) was fed only the basal diet during
the experiment. At weeks 17, 20, and 24, three rats from each group were
sacrificed to determine the inhibitory effect of EA on the occurrence of
γ-glutamyltranspeptidase (GGT)-positive liver cell foci. The remaining
animals were killed at the end of the study (week 36) in order to obtain the
final incidence of liver neoplasms. At autopsy, two slices were taken from
each sublobe of liver and fixed in 10% buffered formalin for hematoxylin and
eosin (H & E) staining and in 95% cold ethanol (4℃) for GGT reaction,
respectively. Liver cell focus, adenoma, and carcinoma were diagnosed
according to the criteria described by Stewart et al. (12). The incidence of
altered liver cell foci (no./cm²) was quantified on the H & E-stained and
GGT-reacted sections using a microscope.

Experiment II: Inhibitory Effect of EA on 4-NQO-Induced Tongue
Carcinogenesis. A total of 48 male ACI/N rats were divided into 4
experimental groups as shown in Figure 2. Groups 1 (17 rats) and 2 (15 rats)
were given 4-NQO in the drinking water (10 ppm) for 8 weeks. Group 2 was
given the diet containing 400 ppm EA, starting one week before the
commencement of the study until one week after the carcinogen exposure. Rats
were then switched to the basal diet, CE-2, and maintained on this diet for
27 weeks. Group 3 (8 rats) was fed EA diet for 10 weeks as for group 2.
Group 4 (8 rats) was fed the basal diet alone during the experiment and
served as an untreated control. All animals were sacrificed 36 weeks after
the start of the experiment in order to evaluate the incidence of
preneoplastic and neoplastic lesions of the tongue. After complete necropsy
of animals, all organs were fixed in 10% buffered formalin and all tissues
and gross lesions were processed for histology by the conventional methods.
Epithelial lesions (hyperplasia, dysplasia and neoplasia) of the tongue were
diagnosed on the H & E stained sections according to the criteria described
by Rubio (13).

Experiment III: Inhibitory Effect of CA on MAM Acetate-Induced Large Bowel
and Liver Carcinogenesis. A total of 94 Syrian golden hamsters of both
sexes (47 each) were used and divided into 4 groups as shown in Figure 3.

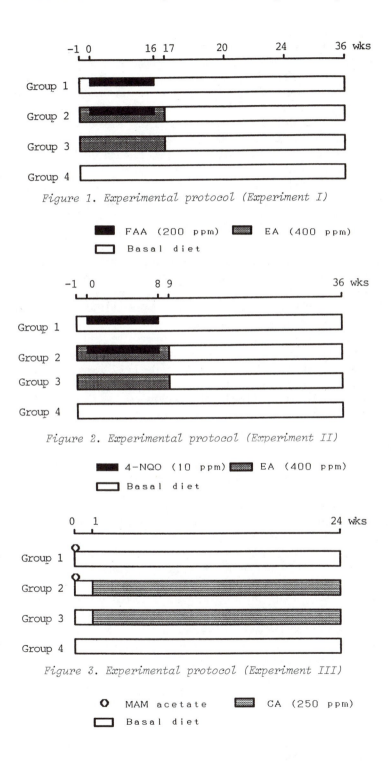

Figure 1. *Experimental protocol (Experiment I)*

FAA (200 ppm)   EA (400 ppm)
Basal diet

Figure 2. *Experimental protocol (Experiment II)*

4-NQO (10 ppm)   EA (400 ppm)
Basal diet

Figure 3. *Experimental protocol (Experiment III)*

MAM acetate   CA (250 ppm)
Basal diet

Groups 1 (10 males and 10 females) and 2 (12 males and 12 females) were given a single i.v. injection of MAM acetate (20 mg/kg body weight). After the injection, animals in group 1 were kept on the basal diet, CE-2, and those in group 2 were fed the diet containing 250 ppm CA for 24 weeks. Group 3 (10 males and 10 females) was fed the diet containing 250 ppm CA for 24 weeks. Group 4 (15 males and 15 females) was fed the basal diet alone throughout the experiment. On termination of the study, all animals were sacrificed and autopsied. At autospy, the gastrointestinal tract was removed and a longitudinal incision was made from the stomach to the rectum. After gross examination of all organs, all tissues, including the digestive tract and liver, were fixed in 10% buffered formalin, stained with H & E, and examined histologically according to the criteria described by Ward for colon neoplasms (*14*) and those described by Stewart *et al.* for liver lesions (*12*).

Enumeration of Silver-Stained Nucleolar Organizer Regions (AgNORs). For assessment of proliferative activity of epithelial cells, number and area of AgNORs were quatified in three animals from each group in Experiments I, II, and III. One-step silver colloid method for AgNOR staining was carried out on the liver (Experiment I), tongue (Experiment II), and colon (Experiment III) specimens from the animals sacrificed at the end of the study. Computer-assisted image analysis quantitation of the number and area of AgNORs in 300 interphase cells from each nonlesional areas of these organs was performed using the image analysis system SPICCA II (Japan Abionics, Co., Tokyo, Japan) with a Olympus BH-2 microscope and a color CCD camera (Hamamatsu Photonics, Co., Hamamatsu City, Japan).

Results

Experiment I.   The results are summarized in Table I. Administration of EA to rats in group 3 did not affect the body weights compared with the controls in group 4. Treatment with FAA increased the liver weight in group 1 due to the development of tumors. The liver weight of rats in group 2 was significantly greater than that in group 3, but was slightly smaller than that in group 1 without statistical significance. Altered liver cell foci and hepatocellular neoplasms developed in rats of groups 1 and 2. The number of foci in group 2 was almost half of that in group 1 during and at the end of the experiment. In rats of group 1, the incidence and multiplicity were 100% and 4.0 neoplasms/rat, respectively. Feeding of EA together with FAA (group 2) reduced the incidence (30%) and multiplicity (0.6/rat) of liver cell neoplasms. In other organs, no preneoplastic and neoplastic lesions were found in any group.

Experiment II.   The results are given in Table II. The average body weights, liver weights, and relative liver weights of rats in all groups

Table I. Results of Experiment I

	Group 1 (FAA)	Group 2 (FAA + EA)
Body wt (g)	246 ± 24[a]	269 ± 18
Liver wt (g)	11.7 ± 1.3	10.9 ± 0.4
Relative liver wt	4.75 ± 0.31	4.06 ± 0.34
Liver cell tumors		
Incidence: Total	10/10 (100%)	3/10 (30%)[b]
Adenoma	7/10 (70%)	2/10 (20%)[c]
Carcinoma	8/10 (80%)	2/10 (20%)[d]
Multiplicity	4.00 ± 2.41	0.60 ± 1.02[e]
GGT-positive liver cell foci (/cm²)		
Week 16	45.7 ± 5.4	21.5 ± 5.4[f]
Week 20	54.0 ± 0.8	20.0 ± 4.1[e]
Week 24	55.7 ± 3.3	23.7 ± 4.2[e]
Week 36	59.8 ± 6.3	24.3 ± 0.9[e]

[a]Mean ± SD.
[b-d]Significantly diffrent from group 1 by Fisher's exact probability test ([b]P=0.02, [c]P=0.004, [d]P=0.01).
[e, f]Significantly different from group 1 by Student's t-test ([e]P<0.001, [f]P<0.05).

Table II. Results of Experiment II

	Group 1 (4-NQO)	Group 2 (4-NQO + EA)
Body wt (g)	318 ± 19[a]	310 ± 17
Liver wt (g)	10.0 ± 0.8	10.4 ± 1.0
Relative liver wt	3.14 ± 0.12	3.35 ± 0.25
Incidence of tongue tumors		
Total	12/17 (71%)	3/15 (20%)[b]
Papilloma	5/17 (29%)	1/15 (7%)
Squamous cell carcinoma	12/17 (71%)	3/15 (20%)[b]
Incidence of preneoplastic lesions of tongue		
Total	17/17 (100%)	10/15 (67%)[c]
Hyperplasia	12/17 (71%)	7/15 (47%)
Dysplasia	10/17 (59%)	5/15 (33%)

[a]Mean ± SD.
[b, c]Significantly different from group 1 by Fisher's exact probability test ([b]P=0.005, [c]P=0.02).

were similar. Preneoplastic and neoplastic lesions were observed in rats of groups 1 and 2. The combined incidence (20%, 3/15) of tongue neoplasms (papilloma and squamous cell carcinoma) and the incidence of squamous cell carcinoma (20%, 3/15) in group 2 significantly lower than those in group 1 (71%, 12/17; 71%, 12/17). The incidence of preneoplastic lesions (hyperplasia and dysplasia) of the tongue in group 2 was also smaller than that in group 1. No other leaions were noted in any group.

Experiment III. The results obtained are summarized in Table III. Large intestinal tumors developed in the cecum and the upper-third of the colon of rats in groups 1 and 2. Histologically, they were adenomas or adenocarcinomas. In rats of group 2, no adenocarcinomas of the large intestine were noted. Statistically, the combined incidence of total neoplasms from males and females or males of group 2 was significantly smaller than that of group 1. Similarly, the combined incidence of large bowel adenocarcinomas in males and females and the incidence of carcinomas of males or females of group 2 were significantly lower than those of group 1. Besides large intestinal tumors, liver cell lesions (neoplasms and liver cell foci) were found in groups 1 and 2. The frequencies of liver cell foci (/cm^2) in males and females of group 2 were significantly smaller than those of group 1. No significant differences in the incidences of liver neoplasms between groups 1 and 2 were noted.

Enumeration of AgNORs. The number and total area/nuclei of liver cells (2.04±0.55; 2.84±1.64 μm^2), tongue squamous cells (1.87±0.76; 2.81±1.55 μm^2) or colonic crypt cells (2.85±1.21; 4.09±2.10 μm^2) from animals given the carcinogen and test chemical were smaller than that from animals give the carcinogen alone (liver: 2.94±0.91, 3.30±1.69 μm^2; tongue: 2.36±1.03, 3.02± 2.01 μm^2; colon: 3.59±1.95, 5.57±3.53 μm^2). Significant differences in the number of AgNORs were obtained between these two groups in Experiments I-III and in the area of AgNORs in Experiment III.

Discussion

It is well known that dietary factors influence the process of carcinogenesis. Some naturally occurring products have been described as chemopreventive agents that inhibit the tumorigenic effects of several chemical carcinogens (6-9). EA and CA are naturally occurring plant phenolics (15-17). EA is present in grapes, strawberries, raspberries, and certain nuts (15) and CA in coffee beans, blue berries, apples, and peaches (17). These are normally consumed by humans. EA and CA have been shown to inhibit the mutagenicity of benzo[a]pyrene-7,8-diol-9,10-epoxide in both *S. typhimurium* TA100 and in Chinese hamster V79 (18) and of the nitrosation products of methylurea (19). They also suppress the carcinogenicity of ultimate carcinogens or carcinogens that require metabolic activation in

Table III. Results of Experiment III

Incidence of lesions		Group 1 (MAM acetate)	Group 2 (MAM acetate + CA)
Large bowel tumors			
Total	♂ + ♀	10/20 (50%)	3/24 (13%)[a]
	♂	6/10 (60%)	2/12 (17%)[b]
	♀	4/10 (40%)	1/12 (8%)
Adenoma	♂ + ♀	6/20 (30%)	3/24 (13%)
	♂	4/10 (40%)	2/12 (17%)
	♀	2/10 (20%)	1/12 (8%)
Carcinoma	♂ + ♀	8/20 (40%)	0/24 (0%)[c]
	♂	4/10 (40%)	0/12 (0%)[d]
	♀	4/10 (40%)	0/12 (0%)[d]
Liver cell foci (/cm^2)			
	♂ + ♀	10.8 ± 2.7[e]	6.0 ± 2.7[f]
	♂	10.0 ± 3.8	5.3 ± 2.3[f]
	♀	11.2 ± 4.7	6.4 ± 3.0[f]

[a-d]Significantly different from group 1 by Fisher's exact probability test ([a]P=0.007, [b]P=0.04, [c]P=0.0007, [d]P=0.03).
[e]Mean ± SD.
[f]Significantly different from group 1 by Student's t-test (P<0.01).

order to exert their carcinogenic activity in the mouse skin (20, 21) and lung (22). However, no studies on the modifying effects of these chemicals on liver and tongue carcinogenesis have been reported. Also, the carcinogens used in previously reported studies on inhibitory effects of EA and CA on carcinogenesis are polycyclic aromatic hydrocarbons (20-22). In the present study, EA clearly inhibited the development of both preneoplastic and neoplastic lesions in the liver or tongue carcinogenesis induced by FAA or 4-NQO, when it was orally given during the initiation stage. It is also clearly demonstrated that dietary administration of CA in the post-initiation phase produced a significant decrease in the incidence of large bowel tumors and preneoplastic hepatocellular lesions induced by MAM acetate. These results indicate that plant phenols, EA and CA have the ability to suppress the tumorigenic potential of nonaromatic carcinogens in the liver and tongue. With respect to N-nitrosocompounds, EA inhibited N-methyl-N-nitrosourea-induced mutations in *S. typhimurium* (23). Mandal *et al.* recently reported that dietary exposure of EA inhibits methylbenzylnitrosamine-induced esophageal tumors in rat (24). The inhibitory effect of EA on the occurrence of lung tumors is greater when it was administered by i.p. injection than in the diet (22). In the present study, EA and CA were given to rats or hamsters in the diet, and this was effective. Moreover, dietary administration of EA and CA produced no toxic effects. Lesca reported in his study using mice that i.p. injection of EA resulted in severe toxicity, but dietary exposure of EA did not cause any toxicity (22).

Several potential mechanisms for the antimutagenic and anticarcinogenic effect of chemopreventive agents have been proposed (6, 9, 25): (a) by the formation of adducts with DNA, acting in a competitive manner to inhibit ultimate carcinogen:DNA adduct formation; (b) by the inhibition of the microsomal enzymes involved in the carcinogen formation; (c) by the stimulation of the detoxification enzymes; (d) by the antioxidative properties which could inactivate of the electrophilic radical forms of carcinogens. EA and CA are also considered to exert their protective effects on mutagenesis and carcinogenesis by any one or more of the above mechanisms (18, 26-30). Teel described that the highest amounts of radiographic activity after i.p. injections of [³H]EA are present in the liver and kidney of mice (27). Moreover, chronic administration of EA in drinking water enhances the detoxicifying enzyme glutathione S-transferase's activity in the mouse liver (28). Phenolic types of synthetic antioxidants such as butylated hydroxyanisole and butylated hydroxytoluene are reported to inhibit hepatocarcinogenesis (31-33) and tongue carcinogenesis (34) when they are administered together with the carcinogens. Thus, it seems likely that the protective effect of EA in the present study is due in major part to a change in the balance between enzyme activation and detoxification of the carcinogen. CA as well as EA has been known to have antioxidative effect and to be a potent inhibitor of many enzyme activities (35-37). Recently, Kasai *et al.* found that CA purified from the extracts of carrot, apricot, and prune inhibits an oxidative DNA damage, 8-hydroxyguanine induced by lipid peroxide (*38*). The protective effect of CA on large bowel and liver carcinogenesis may be related to such effects. It is reported that several plant phenolics inhibit arachidonic acid metabolism to prostaglandins and an increase of ornithine decarboxylase activity during promotion phase of carcinogenesis (*39, 40*). These mechanisms may also be involved in the effect of CA administered to hamsters after the initiation stage. Since high concentrations of EA and CA are present in human foods, such as fruits and vegetables eaten by humans (*15, 17*) and are well tolerated by experimental animals (*41*) and/or humans (*42*), these compounds are considered to offer considerable promise as chemopreventive agents in human carcinogenesis. However, their efficacy in chemoprevention of carcinogenesis and the precise mechanism(s) need to be investigated. In the present study, the number and area of AgNORs in epithelial cell nuclei of liver, tongue and colon were measured. AgNORs may be considered as a marker of rDNA transcriptional activity and/or of rDNA transcriptional potential (*43, 44*). The number of AgNORs is reported to reflect the proliferative activity of cells (*45-47*). The results of the AgNOR enumeration in the present study suggest that AgNORs could be useful as a biomarker (proliferative marker) in chemoprevention studies.

Literature Cited

1. Pitot, H.C.;Beer, D.;Hendrich, S. In *Theories of Carcinogenesis*;Iverson, O., Ed.;Hemispher Press: Washington, 1988;pp.159-177.
2. Sugimura, T. In *Environmental Mutagens and Carcinogens*, Sugimura, T.;Kondo, S., Takebe, H., Eds.;Univ. Tokyo Press and Alan R. Liss, Inc.: Tokyo and New York, 1982;pp.3-20.
3. Bishop, J.M. *Science* 1987, 235, 305-311.
4. Levine, A.J.;Momand, J.;Finlay, C.A. *Nature* 1991, 351, 453-456.
5. Sugimura, T. In *Antimutagenesis and Anticarcinogenesis Mechanisms II*; Kuroda, Y;Shankel, D.M.;Water, M.D., Eds.;Plenum Press: New York and London,1990;pp.23-34.
6. Wattenberg, L.W. *Cancer Res.* 1985, 45, 1-8.
7. Fiala, E.S.;Reddy, B.S.;Weisburger, J.H. *Ann. Rev. Nutr.* 1985, 5, 295-321.
8. Newmark, H.L. *Can. J. Physiol. Pharmacol.* 1987, 65, 461-466.
9. Hocman, G. *Comp. Biochem. Physiol. [B]* 1989, 93, 201-212.
10. Mori, H.;Tanaka, T;Shima, H.;Kuniyasu, T.;Takahashi, M. *Cancer Lett.* 1986, 30, 49-54.
11. Tanaka, T.;Iwata, H.;Niwa, K.;Mori, Y.;Mori, H. *Jpn. J. Cancer Res. (Gann)* 1988, 79, 1297-1303.
12. Stewart, H.L.;Williams, G.M.;Keysser, C.H.;Lombard, L.S.;Montali, R.J. *J. Natl. Cancer Inst.* 1980, 64, 177-207.

13. Rubio, C.A. *Pathol. Res. Pract.* 1983, 176, 269-275.
14. Ward, J.M. *Lab. Invst.* 1974, 30, 505-513.
15. Bate-Smith, E.C. In *The Pharmacology of Plant Phenolics*; Fairbarian, J.W., Ed.; Academic Press: New York, 1959; pp.133-147.
16. Conn, E.E. *The Biochemistry of Plants - A Comprehensive Treatise; Secondary Plant Products*; Academic Press: New York, 1981; Vol.7, pp.269-316 and pp.403-478.
17. Sondheimer, E. *Bot. Rev.* 1964, 30, 667-712.
18. Wood, A.W.; Huang, M.T.; Chang, R.L.; Newmark, H.L.; Lehr, R.E.; Yagi, H.; Sayer, D.M.; Jerina, D.M..; Conney, A.H. *Proc. Natl. Acad. Sci. U.S.A.* 1982, 79, 5513-5517.
19. Stich, H.F.; Rosin, M.P.; Bryson, L. *Mutat. Res.* 1982, 95, 119-128.
20. Mukhtar, H.; Das, M.; Del Tito, B.J.; Bickers, D.R. *Biochem. Biophys. Res. Commun.* 1984, 119, 751-757.
21. Mukhtar, H.; Das, M.; Bickers, D.R. *Cancer Res.* 1986, 46, 2262-2265.
22. Lesca, P. *Carcinogenesis* 1983, 4, 1651-1653.
23. Dixit, R.; Gold, B. *Proc. Natl. Acad. Sci. U.S.A.* 1986, 83, 8039-8043.
24. Mandal, S.; Shivapurkar, N.M.; Superczynski, M.; Stoner, G.D. *Proc. Am. Assoc. Cancer Res.* 1986, 27, 124.
25. Newmark, H.L. *Nutr. Cancer* 1984, 6, 58-70.
26. Mandal, S.; Ahuja, A.; Shivapurkar, N.M.; Cheng, S.-J.; Groopman, J.D.; Stoner, G.D. *Carcinogenesis* 1987, 8, 1651-1656.
27. Teel, R.W. *Cancer Lett.* 1987, 34, 165-171.
28. Das, M.; Bickers, D.R.; Mukhtar, H. *Carcinogenesis* 1985, 6, 1409-1413.
29. Del Tito, B.J.; Mukhtar, H.; Bickers, D.R. *Biochem. Biophys. Res. Commun.* 1983, 114, 388-394.
30. Teel, R.W. *Cancer Lett.* 1986, 30, 329-336.
31. Williams, G.M.; Maeura, Y.; Weisburger, J.H. *Cancer Lett.* 1983, 19, 55-60.
32. Williams, G.M.; Tanaka, T.; Maeura, Y. *Carcinogenesis* 1986, 7, 1043-1050.
33. Maeura, Y; Weisburger, J.H.; Williams, G.M. *Cancer Res.* 1984, 44, 1604-1610.
34. Tanaka, T.; Iwata, H.; Kanai, N.; Nishikawa, A.; Mori, H. *J. Nutr. Growth Cancer* 1987, 4, 239-248.
35. Hayase, F.; Kato, H. *J. Nutr. Sci. Vitaminol.* 1984, 30, 37-46.
36. Namiki, M. *CRC Crit. Rev. Food Sci. Nutr.* 1990, 29, 273-300.
37. Osawa, T.; Namiki, M.; Kawakishi, S. In *Antimutagenesis and Anticarcinogenesis Mechanisms II*; Kuroda, Y.; Shankel, D.M.; Water, M.D., Eds.; Plenum Press: New York and London, 1990; pp.139-153.
38. Kasai, H.; Chung, M.-H.; Yamamoto, F.; Ohtsuka, E.; Laval, J.; Grollman, A.P.; Nishimura, S. *Third International Conference of Mechanisms of Antimutagenesis and Anticarcinogenesis*; Italy, 1991; pp.26.
39. Baumann, J.; von Bruchhausen, F.; Wurm, G. *Prostaglandins* 1980, 20, 627-639.
40. Kato, R.; Nakadate, T.; Yamamoto, S.; Sugimura, T. *Carcinogenesis* 1983, 4, 1301-1305.

41. Doyle, B.;Griffiths, L.A. *Xenobiotica* 1980, 10, 247-256.
42. Giirolami, A.;Cliffton, E.E. *Thromb. Diath. Haemorrh.* 1967, 17, 165-175.
43. Dimova, R.N.;Markov, D.V.;Gajdardjieva, K.C.;Dabeva, M.D.;Hadjiolov, A.A. *Eur. J. Cell Biol.* 1982, 28, 272-277.
44. Busch, H. In *Chromosomal Non-Histone Proteins*;Hnilica, L.S., Ed.;CRC Press: Boca Raton, FL, 1984;pp.233-286.
45. Tanaka, T.;Takeuchi, T.;Nishikawa, A.;Takami, T.;Mori, H. *Jpn. J. Cancer Res.* 1989, 80, 1047-1051.
46. Takeuchi, T.;Tanaka, T.;Ohno, T.;Yamamoto, N.;Kobayashi, S.;Kuriyama, M.;Kawada, Y.;Mori, H. *Virch. Arch. B Cell Pathol.* 1990, 58, 383-387.
47. Tanaka, T.;Kojima, T.;Okumura, A.;Yoshimi, N.;Mori, H. *Carcinogenesis* 1991, 12, 329-333.

RECEIVED December 2, 1991

Chapter 27

Inhibitory Effects of Curcumin on Carcinogenesis in Mouse Epidermis

Mou-Tuan Huang[1], Fredika M. Robertson[2], Thomas Lysz[3],
Thomas Ferraro[1], Zhi Yuan Wang[1], Constantine A. Georgiadis[4],
Jeffrey D. Laskin[4], and Allan H. Conney[1]

[1]Laboratory for Cancer Research, College of Pharmacy, Rutgers, The State
University of New Jersey, Piscataway, NJ 08855
[2]Department of Surgery, Ohio State University, Columbus, OH 43210
[3]Department of Surgery, University of Medicine and Dentistry of New
Jersey, Robert Wood Johnson Medical School, Newark, NJ 07103
[4]Department of Environmental and Community Medicine, University
of Medicine and Dentistry of New Jersey, Robert Wood Johnson Medical
School, Piscataway, NJ 08854

Curcumin (diferuloylmethane) inhibits tumor initiation by
benzo[a]pyrene and tumor promotion by 12-O-tetradecanoylphorbol-
13-acetate (TPA) in female CD-1 mice. Topical application of
curcumin also inhibits benzo[a]pyrene-mediated DNA adduct
formation in the epidermis. Co-application of curcumin with TPA
inhibits TPA-induced skin inflammation (mouse ear edema), increased
epidermal ornithine decarboxylase activity, increased epidermal DNA
synthesis, increased epidermal thickness, increased number of
epidermal cell layers and leukocyte infiltration. Curcumin also inhibits
arachidonic acid-induced edema of mouse ears in vivo and epidermal
cyclooxygenase and lipoxygenase activities in vitro. Curcumin and
demethoxycurcumin are equipotent as inhibitors of TPA-induced ear
inflammation, and bisdemethoxycurcumin and tetrahydrocurcumin are
less active. The structurally related dietary compounds chlorogenic
acid, caffeic acid and ferulic acid are weak antioxidants, weak
inhibitors of TPA-induced increases in epidermal ornithine
decarboxylase activity, weak inhibitors of TPA-induced mouse ear
inflammation, weak inhibitors of cyclooxygenase and lipoxygenase
activity, and weak inhibitors of TPA-induced tumor promotion in
mouse skin.

Carcinogenesis is a multistage process that can be divided into initiation, promotion
and progression. Initiation occurs relatively rapidly (within a few hours to one or
two days), while promotion requires long periods of time (up to a year or longer in
some rodent models (1). It is estimated that the promotion stage may require up to 30
years or even longer in some people. Although efforts in cancer chemotherapy have
intensified over the years, many cancers still remain very difficult to cure and cancer
prevention could become an increasingly useful strategy in our fight against cancer.

Human epidemiology and animal studies have indicated that cancer risk may
be modified by changes in dietary habits or dietary components (1-4). Humans
ingest large numbers of naturally occurring antimutagens and anticarcinogens in food

0097–6156/92/0507–0338$06.00/0

(*1, 4-6*), and these antimutagens and anticarcinogens may inhibit one or more stages of carcinogenesis and delay or prevent the formation of cancer. Recent studies have indicated that compounds with antioxidant and/or anti-inflammatory properties can inhibit tumor initiation and tumor promotion in mouse skin (*7-14*). Epidemiological studies indicate that dietary factors play an important role in the development of human cancer (*2, 15*), and attempts to identify naturally occurring dietary anticarcinogens may lead to new strategies for cancer prevention.

Curcumin (Figure 1) is the major pigment in turmeric (the ground rhizome of *Curcuma longa* Linn) which is widely used as a spice and coloring agent in curry, mustard and other foods. Curcumin has been reported to possess antioxidant and anti-inflammatory activity (*16-19*). This paper describes the inhibitory effect of curcumin and some structurally related compounds on chemical carcinogenesis in mouse skin. The effects of curcumin on certain biochemical and morphological markers of carcinogenesis are also described.

Effect of Curcumin on Benzo[*a*]pyrene-Mediated DNA Adduct Formation and Benzo[*a*]pyrene-Induced Tumor Initiation

Inhibitory Effect of Curcumin on the Formation of Benzo[*a*]pyrene-Mediated DNA Adducts in Mouse Epidermis. Fifteen hours after the application of 20 nmol ^{3}H-benzo[*a*]pyrene (BP) to the backs of CD-1 mice, there was an average of 0.74 pmol of covalently bound adducts per mg of epidermal DNA. Topical application of 3 or 10 μmol curcumin 5 minutes before the application of ^{3}H-BP inhibited the formation of BP metabolite DNA-bound adducts by 30 or 61%, respectively (Figure 2).

Inhibitory Effect of Curcumin on the Initiation of Skin Tumors by BP. Female CD-1 mice (30 per group) that were initiated with 20 nmol of BP once weekly for 10 weeks and then promoted with 15 nmol TPA twice weekly for 21 weeks developed an average of 7.1 skin tumors per mouse. Topical application of 3 or 10 μmol curcumin 5 minutes before each application of BP inhibited the number of skin tumors per mouse by 58 or 62%, respectively, after 21 weeks of TPA promotion (Figure 3).

Similar results were obtained using a single, high dose of BP. Mice that were initiated with 200 nmol BP and then promoted with 15 nmol TPA twice weekly for 21 weeks developed an average of 6.9 skin tumors per mouse; 77% of the mice had tumors. Topical application of 10 μmol curcumin to the skin at 120, 60 and 5 minutes before the application of 200 nmol BP decreased the average number of skin tumors per mouse by 58%, and the percent of mice with tumors was decreased by 23%.

In another experiment studying curcumin's effect on BP-induced skin tumorigenesis, the topical application of 10 μmol curcumin to the backs of mice once daily for 4 days prior to a single application of 200 nmol BP (24 hours after the last dose of curcumin) neither induced nor inhibited the number of skin tumors per mouse or the number of mice with tumors. These observations suggest that curcumin does not induce enzymes that metabolically activate or detoxify BP.

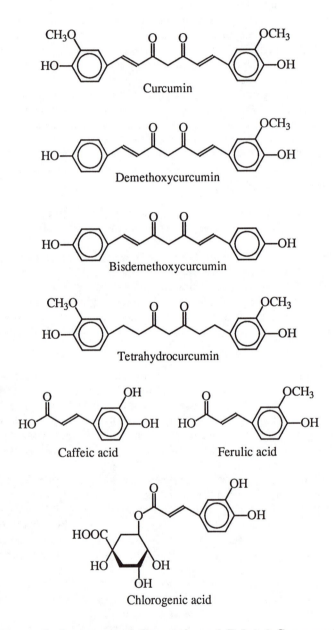

Figure 1. Structures of Curcumin and Related Compounds

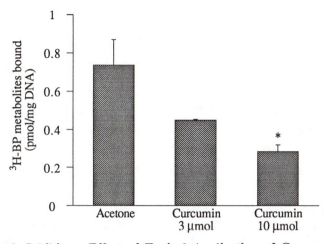

Figure 2. Inhibitory Effect of Topical Application of Curcumin on the Formation of ^{3}H-BP-DNA Adducts in Mouse Epidermis

Mice were treated topically with 200 μl acetone or curcumin in acetone 5 minutes before the application of ^{3}H-BP (100 μCi, 20 nmol) in 200 μl acetone. Fifteen hours later, the mice were sacrificed and epidermal DNA was isolated. Data are expressed as the mean ± SE from 3 mice.
*Statistically different from acetone alone as determined by Student's t test ($p < 0.05$).

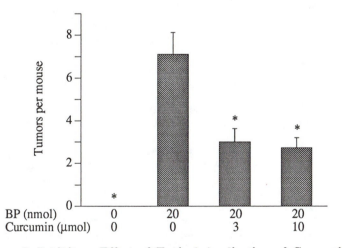

Figure 3. Inhibitory Effect of Topical Application of Curcumin on BP-Induced Skin Tumorigenesis

Female CD-1 mice were treated topically with acetone alone or curcumin in acetone 5 minutes before the application of 20 nmol BP in acetone once weekly for 10 weeks. One week later, the mice were treated with 15 nmol TPA twice weekly for 21 weeks. Data are expressed as the mean ± SE from 30 mice.
*Statistically different from BP alone as determined by Student's t test ($p < 0.01$).

Inhibitory Effect of Curcumin on TPA-Induced Biochemical and Morphological Changes in Mouse Epidermis

Table I: Inhibitory Effects of Curcumin on TPA- and Arachidonic Acid-Induced Biochemical and Morphological Changes in Mouse Skin

Effect observed	Inhibition
TPA-induced biochemical changes	
a. skin inflammation (edema of mouse ears)	+++
b. ornithine decarboxylase activity	+++
c. thymidine incorporation into DNA	+
TPA-induced morphological changes	
a. epidermal thickening	+++
b. increased epidermal layers	+++
c. leukocyte infiltration	++
d. intercellular edema	++
e. increased proportion of cells in S phase	++
Arachidonic acid-induced changes	
a. skin inflammation (edema of mouse ears)	++
b. epidermal lipoxygenase activity	+++
c. epidermal cyclooxygenase activity	+++

The amount of inhibition is indicated by + (weak), ++ (moderate) or +++ (strong).

Curcumin inhibits many TPA-induced biochemical and morphological changes in the skin that are associated with epidermal carcinogenesis (Table I). Some or all of these effects may be important for the inhibitory action of curcumin on TPA-induced tumor promotion in mouse skin.

Inhibitory Effect of Curcumin on TPA-Induced Edema of Mouse Ears. The anti-inflammatory activity of curcumin *in vivo* was evaluated by determining its effect on TPA-induced edema of mouse ears (*14*). Five hours after the topical application of 0.5 nmol TPA to the ears of mice, the average weight of a 6 mm diameter punch increased from an average value of 6.6 mg for the control group to 12.7 mg for the TPA group. Co-application of 0.1, 0.3 or 1.0 µmol curcumin with the TPA inhibited this increase by 67, 94 or 98%, respectively. During the course of these studies, we found that curcumin also inhibited TPA-induced ear reddening (erythema). Curcumin is a stronger inhibitor of TPA-induced edema of mouse ears than either quercetin or nordihydroguaiaretic acid (inhibitors of lipoxygenase) or indomethacin (an inhibitor of cyclooxygenase) (*20*).

Inhibitory Effect of Curcumin on TPA-Induced Increases in Epidermal Ornithine Decarboxylase Activity. The application of 5 nmol TPA to the dorsal surface of the skin of CD-1 mice resulted in more than a 100-fold increase in epidermal ornithine decarboxylase activity at 5 hours. When 1, 3, or 10 µmol curcumin was applied topically with TPA, the increase in epidermal ornithine decarboxylase activity was inhibited by 46, 84, or 98%, respectively.

An additional study indicated that i.p. injection of curcumin can inhibit the induction of ornithine decarboxylase activity in mouse epidermis that occurs after the topical application of TPA. The i.p. injection of 40 or 120 µmol curcumin (dissolved in dimethyl sulfoxide) into mice one hour before the topical application of 5 nmol TPA inhibited the induction of epidermal ornithine decarboxylase activity by 46 or 86%, respectively. It was noted during necropsy that curcumin precipitated out of solution in the abdominal cavity and thus may not have been completely bioavailable.

Curcumin could be seen as yellow flecks in the abdominal cavity throughout the experimental period of 6 hours (*14*).

Inhibitory Effect of Curcumin on TPA-Induced Increases in the Incorporation of ^{3}H-Thymidine into Epidermal DNA. The i.p. injection of ^{3}H-thymidine into mice resulted in its incorporation into epidermal DNA, and the amount incorporated was used as a measure of DNA synthesis. Topical application of 2 nmol TPA to CD-1 mice induced a 3- to 4-fold increase in the incorporation of ^{3}H-thymidine into epidermal DNA (*14*). Co-application of 1, 3, or 10 μmol of curcumin with the TPA inhibited this increase by 4, 14, or 49%, respectively. When 5 nmol TPA was applied to mouse skin, co-application of 1, 3, or 10 μmol curcumin inhibited the TPA-induced increase in the incorporation of ^{3}H-thymidine into epidermal DNA by 3, 4, or 29%, respectively. The topical application of 2 nmol fluocinolone acetonide, a known inhibitor of TPA-induced epidermal DNA synthesis, inhibited by 86% the incorporation of ^{3}H-thymidine into epidermal DNA.

Inhibitory Effect of Curcumin on TPA-Induced Morphological Changes in Mouse Epidermis. Topical application of TPA caused an increase in epidermal thickness, an increased number of epidermal cell layers, inflammatory cell infiltration and intercellular edema. Topical application of 1 nmol TPA in acetone twice daily for 4 days increased the epidermal thickness 5-fold. Co-application of 10 μmol curcumin with the TPA completely prevented the TPA-induced epidermal thickening and markedly inhibited the TPA-induced increases in the number of epidermal cell layers, inflammatory cell infiltration and intercellular edema. Topical application of 10 μmol curcumin alone twice daily for 4 days did not affect cellular histology.

Effect of Curcumin on TPA-Induced Changes in the Cell Cycle of Mouse Epidermis. The percent of cells in different phases of the cell cycle, as measured by DNA content, has been used as a marker of proliferation. Cells that are actively proliferating have an amount of DNA between 2N and 4N and are in the synthetic (S) phase of the cell cycle. Cell cycle analysis provides a quantitative measure of the fraction of cells that are proliferating (*21, 22*), and this is correlated with ornithine decarboxylase activity (*23*) and the incorporation of ^{3}H-thymidine into DNA (*24*).

    Flow cytometry was used in the present study to evaluate the effect of curcumin on the distribution of epidermal cells in different phases of the cell cycle after 22 weeks of tumor promotion with TPA. The detailed methodology is described elsewhere (*23*). In brief, cells were isolated from mice, fixed in 70% ethanol, treated with 0.1% Triton X-100 and RNase, stained with propidium iodide (a DNA specific fluorescent dye) and examined by flow cytometry. As shown in Table II, epidermis from female CD-1 mice treated once with 200 μl acetone or 7,12-dimethyl-benz[*a*]anthracene (DMBA) in acetone and after one week treated with 200 μl acetone twice weekly for 22 weeks had 16 - 17% of the cells in the S phase, whereas epidermis from mice treated with DMBA and then promoted with 5 nmol TPA twice weekly for 22 weeks had 34% of the cells in the S phase. In papillomas, 31% of the cells were in the S phase of the cell cycle. Mice initiated with DMBA and promoted with 5 nmol TPA together with 10 μmol curcumin twice a week for 22 weeks had a significantly decreased percentage of epidermal cells (23%) in the S phase (Table II). The results indicate that curcumin inhibits TPA-induced epidermal cell proliferation.

Table II: Effect of Curcumin on the Epidermal Cell Cycle

Treatment of mice	Percent of cells in:		
	G_0/G_1	S	G_2/Mitosis
DMBA/acetone	79	17	4
Acetone/TPA	67	28	5
DMBA/TPA	62	34	4
DMBA/TPA + curcumin	74	23	3
Papillomas	64	31	5

Mice were treated once with acetone alone or 200 nmol DMBA in acetone then one week later with acetone alone, 5 nmol TPA or 5 nmol TPA with 10 µmol curcumin twice weekly for 22 weeks. Cell cycle analysis was performed as previously described (23). Each value represents the mean of 2 experiments (cells pooled from 7 or 8 mice).

Inhibitory Effect of Curcumin on TPA-Induced Tumor Promotion in Mouse Skin

Female CD-1 mice that were initiated with a single topical application of 200 nmol DMBA and then promoted with 5 nmol TPA twice weekly for 18 weeks developed an average of 17.9 tumors per mouse, and 93% of the mice had tumors. Topical application of 1, 3, or 10 µmol curcumin together with TPA inhibited the number of skin tumors formed by 46, 79 or 98%, respectively, and the percent of tumor-bearing mice was inhibited by 24, 66 or 82%, respectively (14, Figure 4).

Not only was the number of tumors reduced, but the appearance of tumors was delayed. In DMBA-initiated animals that were treated with TPA, the first tumor was observed during the sixth week of promotion, whereas animals that received 10 µmol curcumin with the TPA did not develop any skin tumors until the fourteenth week of promotion (14). These results, as well as others (25, 26), indicate that curcumin can cause a delay in the appearance of tumors.

Inhibitory Effects of Curcumin on the Action and Metabolism of Arachidonic Acid

Inhibitory Effect of Curcumin on Arachidonic Acid-Induced Edema of Mouse Ears. Topical application of arachidonic acid to the ears of mice resulted in the rapid induction of edema and in an increased weight of ear punches (a measure of inflammation). The time course and dose-response relationship for arachidonic acid-induced edema of mouse ears have been described (14). Topical application of 1 µmol arachidonic acid to the ear of a mouse caused a rapid and substantial inflammatory response within 1 hour. The topical application of 3 or 10 µmol curcumin 30 minutes before the application of arachidonic acid caused a 30 or 80% inhibition, respectively, of the inflammatory response. During the course of this study, we noted that curcumin also inhibited arachidonic acid-induced ear reddening.

Inhibitory Effect of Curcumin on Epidermal Lipoxygenase and Cyclooxygenase Activities. Curcumin had a dose-dependent inhibitory effect on the *in vitro* metabolism of arachidonic acid to hydroxyeicosatetraenoic acids (HETEs) by mouse epidermal lipoxygenase (20). Curcumin at 3, 10, 30, and 100 µM inhibited the *in vitro* metabolism of arachidonic acid to 5-HETE by 40, 60, 66, and 83%, respectively, and the formation of 8-HETE was inhibited by 40, 51, 77, and 85%, respectively. These results indicate that the ID_{50} for the inhibitory effect of curcumin on the metabolism of arachidonic acid to 5- and 8-HETE is 5 - 10 µM.

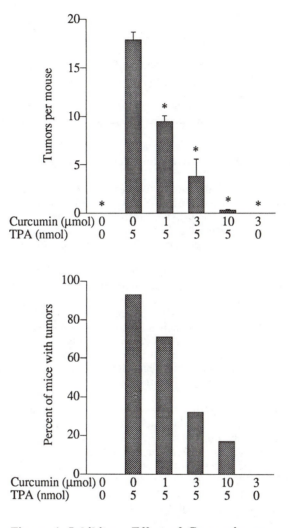

**Figure 4. Inhibitory Effect of Curcumin on
TPA-Induced Tumor Promotion**

Female CD-1 mice were initiated with 200 nmol DMBA in 200 μl acetone. One week later, the mice were treated with acetone alone, TPA or curcumin alone, or TPA and curcumin in 200 μl acetone twice weekly for 21 weeks. Data are expressed as the mean ± SE from 30 mice.
*Statistically different from TPA alone as determined by Student's *t* test (p < 0.01).

Curcumin had a dose-dependent inhibitory effect on the *in vitro* metabolism of arachidonic acid to prostaglandins (PGE_2, $PGF_{2\alpha}$ and PGD_2) by mouse epidermal cyclooxygenase. Curcumin at 1, 10, or 100 μM inhibited the metabolism of arachidonic acid to PGE_2 by 42, 86 and 99%, respectively, to $PGF_{2\alpha}$ by 2, 64, and 85%, respectively, and to PGD_2 by 22, 73, and 94%, respectively. These results indicate that the ID_{50} for the inhibitory effect of curcumin on epidermal cyclooxygenase activity is 5 - 10 μM.

These results suggest that curcumin is different from the cyclooxygenase inhibitor indomethacin and the lipoxygenase inhibitor nordihydroguaiaretic acid in that curcumin is an equally potent inhibitor of both lipoxygenase and cyclooxygenase activities.

Effects of Curcumin Derivatives and Related Compounds

Effects of Demethoxycurcumin, Bisdemethoxycurcumin and Tetrahydrocurcumin on TPA-Induced Edema of Mouse Ears.   Food grade curcumin prepared by extraction from the rhizome of the plant *Curcuma longa* Linn consists of a mixture of three naturally occurring curcuminoids. Approximately 70 - 80% is curcumin, 17 - 20% is demethoxycurcumin, and 0.3 - 6% is bisdemethoxycurcumin.

The anti-inflammatory activities of food grade curcumin, 98% pure curcumin, demethoxycurcumin, bisdemethoxycurcumin (all obtained from Kalsec Inc., Kalamazoo MI), and tetrahydrocurcumin (a synthetic analog obtained from Hoffmann-La Roche Inc., Nutley NJ) were compared in dose response studies by determining the effect of topical application of these compounds on TPA-induced edema of mouse ears. The structures of these compounds are shown in Figure 1. The results of these studies indicate that food grade curcumin, pure curcumin, and demethoxycurcumin have equal anti-inflammatory activity and are more active than bisdemethoxycurcumin and tetrahydrocurcumin.

Studies With Chlorogenic Acid, Caffeic Acid and Ferulic Acid. Chlorogenic acid, caffeic acid and ferulic acid (Figure 1) are structurally related to curcumin and are widely found in many edible fruits and vegetables. For example, chlorogenic acid is about 6 - 8% of the dry weight of the coffee bean (*27*) and substantial amounts are present in brewed coffee. The effects of these compounds on TPA-induced increases in epidermal ornithine decarboxylase activity, epidermal DNA synthesis and tumor promotion in mouse skin have been studied (*14*), as well as their effect on *in vitro* epidermal lipoxygenase and cyclooxygenase activities (*20*). The results shown in Table III indicate that the topical application of chlorogenic acid, caffeic acid or ferulic acid has an inhibitory effect on TPA-induced ornithine decarboxylase activity and tumor promotion, but these compounds are less active than curcumin.

Wattenberg, *et al.* (*28*) reported an inhibitory effect of caffeic acid and ferulic acid on BP-induced forestomach carcinogenesis in mice, while Tanaka, *et al.* (*29*) have shown that 0.025% chlorogenic acid in the diet inhibits methyloxymethanol acetate-induced colon cancer in hamsters. It is of interest that epidemiological studies suggest that heavy coffee drinkers may have a decreased risk of colon cancer (*30, 31*). Since chlorogenic acid is a major constituent of coffee, and caffeic acid is readily formed from chlorogenic acid by hydrolysis, it is possible that these compounds contribute to the reported protective effects of coffee.

Table III: Inhibitory Effects of Curcumin, Chlorogenic Acid, Ferulic Acid and Caffeic Acid on TPA-Induced Ornithine Decarboxylase Activity and Tumor Promotion in Mouse Epidermis

Treatment	Ornithine decarboxylase (pmol CO_2/mg protein/hr)	Tumors per mouse	Percent of mice with tumors
Acetone : DMSO (90:10)	17 ± 11*	0*	0
TPA	2683 ± 250	6.18 ± 1.84	68
TPA + chlorogenic acid (10 µmol)	2022 ± 509	2.50 ± 0.78	47
TPA + chlorogenic acid (20 µmol)	1215 ± 118*	0.53 ± 0.05*	27
TPA + caffeic acid (10 µmol)	1553 ± 439*	4.43 ± 1.89	47
TPA + caffeic acid (20 µmol)	1068 ± 159*	1.12 ± 0.67*	39
TPA + ferulic acid (10 µmol)	1470 ± 190*	4.00 ± 1.19	64
TPA + ferulic acid (20 µmol)	1032 ± 171*	—	—
TPA + ferulic acid (50 µmol)	901 ± 244*	0.17 ± 0.08*	13
TPA + curcumin (1 µmol)	1636 ± 440*	—	—
TPA + curcumin (10 µmol)	255 ± 130*	0*	0

For ornithine decarboxylase activity determination, female CD-1 mice were treated topically with 5 nmol TPA alone or with appropriate inhibitor in 200 µl acetone: DMSO (90:10). Mice were sacrificed 5 hours later, epidermis was isolated and ornithine decarboxylase activity was determined. Data are expressed as the mean ± SE of duplicate assays from at least 6 mice.

For the tumor study, female CD-1 mice (30 per group) were treated topically with 200 nmol DMBA in 200 µl acetone. One week later the mice were treated topically with 5 nmol TPA alone or with appropriate compound in 200 µl acetone twice weekly for 19 weeks. Tumors per mouse are expressed as the mean ± SE from 30 mice.

*Statistically different from TPA alone as determined by Student's *t* test (p < 0.05).

Other Effects of Curcumin

Recent studies in our laboratory have shown that 2% curcumin (food grade) in the diet inhibits azoxymethane-induced foci in the colon of CF-1 mice and inhibits duodenal tumorigenesis in C57BL/6J mice previously initiated with N-ethyl-N'-nitro-N-nitrosoguanidine (unpublished studies). Curcumin has also been reported to inhibit the metabolism-mediated mutagenicity of BP and DMBA (*32, 33*), to inhibit the growth of 3-methylcholanthrene-induced sarcomas in mice (*26*), to inhibit the growth of tumor cells *in vitro* and to increase the survival of animals with lymphomas (*34*). In additional studies, curcumin was also reported to inhibit lipid peroxidation (*17, 35*), lipid peroxide-induced DNA damage (*36*), and smoke-induced DNA damage (*37*).

In conclusion, many studies have demonstrated inhibitory effects of curcumin on tumorigenesis and tumor growth *in vitro* and *in vivo*. These observations suggest a need for further research on the chemopreventive and anticancer activities of curcumin and its derivatives in animals and in humans.

Acknowledgments

This work was supported in part by grant CA 49756 from the National Cancer Institute, United States Public Health Services.

Literature Cited

1. Bertram, J. S.; Kolonel, L. N.; Meyskens, F. L., Jr. *Cancer Res.* **1987**, *47*, pp 3012-3031.
2. Doll, R.; Peto, R. *J. Natl. Cancer Inst.* **1981**, *66*, pp 1191-1308.
3. Peto, R.; Doll, R.; Buckley, J. D.; Sporn, M. B. *Nature* **1981**, *290*, pp 201-208.
4. Wattenberg, L. W. *Cancer Res.* **1985**, *45*, pp 1-8.
5. Ames, B. *Science* **1983**, *221*, pp 1256-1264.
6. Ames, B. N. In *Genetic Toxicology of the Diet*; Knudson, I. B., Ed.; Alan R. Liss, Inc.: New York, 1986, pp 3-32.
7. Slaga, T.; Scribner, J. *J Natl. Cancer Inst.* **1973**, *51*, pp 1723-1725.
8. Schwartz, J. A.; Viaje, A.; Slaga, T. J.; Yuspa, S. H.; Hennings, H.; Lichi, U. *Chem.-Biol. Interact.* **1977**, *17*, pp 331-347.
9. Slaga, T. J.; Brack, W. M. *Cancer Res.* **1977**, *37*, pp 1631-1635.
10. Viaje, A.; Slaga, T.; Wigler, M.; Weinstein, I. *Cancer Res.* **1977**, *37*, pp 1530-1536.
11. Verma, A. K.; Shapas, B. G.; Rice, H. M.; Boutwell, R. K. *Cancer Res.* **1979**, *39*, pp 1339-1346.
12. Kozumbo, W. J.; Seed, J. L.; Kensler, T. W. *Cancer Res.* **1983**, *43*, pp 2555-2559.
13. Nakadate, T.; Yamamoto, S.; Aizu, E.; Kato, R. *Gann* **1984**, *75*, pp 214-222.
14. Huang, M.-T.; Smart, R.; Wong, C.-Q.; Conney, A. *Cancer Res.* **1988**, *48*(21), pp 5941-5946.
15. Committee on Diet, Nutrition and Cancer of the National Research Council. *Diet, Nutrition and Cancer*; National Academy Press: Washington DC, 1982.
16. Srimal, R. C.; Dhawan, B. N. *J. Pharm. Pharmacol.* **1973**, *25*, pp 447-452.
17. Sharma, O. P. *Biochem. Pharmacol.* **1976**, *25*, pp 1811-1812.
18. Mukhopadhyay, A.; Basu, N.; Gatak, N.; Gujral, P. K. *Agents Actions* **1982**, *12*, pp 508-512.
19. Toda, S.; Miyase, T.; Arichi, H.; Tanizawa, H.; Takino, Y. *Chem. Pharm. Bull.* **1985**, *33*, pp 1725-1728.
20. Huang, M.-T.; Lysz, T.; Ferraro, T.; Abidi, T. F.; Laskin, J. D.; Conney, A. H. *Cancer Res.* **1991**, *51*, pp 813-819.
21. Dean, P. N. In *Techniques in Cell Cycle Analysis*; Gray, J. W.; Darzynkiewicz, Z., Eds.; Humana Press: Clifton NJ, 1987, pp 207-253.
22. Gray, J. W.; Dolbeare, F.; Pallavicini, M. G.; Vanderlaan, M. In *Techniques in Cell Cycle Analysis*; Gray, J. W.; Darzynkiewicz, Z., Eds.; Humana Press: Clifton NJ, 1987, pp 93-137.
23. Robertson, F. M.; Gilmour, S. K.; Conney, A. H.; Huang, M.-T.; Beavis, A. J.; Laskin, J. D.; Hietala, O.; O'Brien, T. G. *Cancer Res.* **1990**, *50*(15), pp 4741-4746.
24. Wilson, G. D.; McNally, N. J.; Dunphy, E.; Kärcher, H.; Pfragner, R. *Cytometry* **1985**, *6*, pp 641-647.

25. Kuttan, R.; Bhanumathy, P.; Nirmala, K.; George, M. C. *Cancer Lett.* **1985,** *29,* pp 197-202.
26. Soudamini, K. K.; Kuttan, R. *J. Ethnopharmacol.* **1989,** *27*(1-2), pp 227-233.
27. Weiss, L. C. *J. Assoc. Official Agricul. Chem.* **1957,** *40,* pp 350-354.
28. Wattenberg, L. W.; Coccia, J. B.; Lam, L. K. T. *Cancer Res.* **1980,** *40,* pp 2820-2823.
29. Tanaka, T.; Nishikawa, A.; Shima, H.; Sugie, S.; Shinoda, T.; Yoshimi, N.; Iwata, H.; Mori, H. *Basic Life Sci.* **1990,** *52,* pp 429-440.
30. La Vecchia, C.; Ferraroni, M.; Negri, E.; D'Avanzo, B.; Decarli, A.; Levi, F.; Francesci, S. *Cancer Res.* **1989,** *49*(4), pp 1049-1051.
31. Rosenberg, L. *Cancer Lett.* **1990,** *52,* pp 163-171.
32. Nagabhushan, M.; Bhide, S. V. *Nutr. Cancer* **1986,** *8,* pp 201-210.
33. Nagabhushan, M.; Amonkar, A. J.; Bhide, S. V. *Food Chem. Toxicol.* **1987,** *25,* pp 545-547.
34. Kuttan, R.; Bhanumathy, P.; Nirmala, K.; George, M. C. *Cancer Lett.* **1985,** *29,* pp 197-202.
35. Donatus, I. A.; Sardjoko; Vermeulen, N. P. *Biochem. Pharmacol.* **1990,** *39*(12), pp 1869-1875.
36. Shalini, V. K.; Srinivas, L. *Mol. Cell. Biochem.* **1987,** *77*(1), pp 3-10.
37. Shalini, V. K.; Srinivas, L. *Mol. Cell. Biochem.* **1990,** *95*(1), pp 21-30.

RECEIVED February 18, 1992

Chapter 28

Tocopherol

Natural Phenolic Inhibitor of Nitrosation

William J. Mergens

Roche Vitamins and Fine Chemicals, Vitamins Research and Development Department, Hoffmann–La Roche, Inc., Nutley, NJ 07110

N-Nitroso compounds, once formed and present *in vivo*, generally do not easily or readily revert back to precursors. Instead, many of them are metabolized to, or are converted to alkylating agents as terminal or proximate carcinogens. One effective method of controlling the formation of these compounds in foods and *in vivo* has been to use agents that inhibit their formation from precursors (e.g., amines, amides, plus nitrosating agents). The formation of N-nitroso compounds is directly dependent on the source of nitrosating agent which recent surveys indicate are numerous, ranging from atmospheric exposure to oxides of nitrogen through bacterial reduction of nitrate in the gastrointestinal tract. The generally accepted mechanism of blocking these reactions is one of competitive kinetics between the susceptible amine (amide) and potential blocking agent for the nitrosating specie. The ultimate effectiveness of any given blocking agent then depends on being able to deliver it in sufficient concentration to the site where nitrosation takes place. α-Tocopherol and its water soluble antioxidant counterpart, ascorbic acid, have been found to be extremely efficient inhibitors of nitrosation reactions in food products as well as *in vivo* in man. The mechanisms which apply to these systems will be presented along with the results of many recently reported investigations.

The concern about human exposure to N-nitroso compounds relates to their carcinogenicity. It is now clearly established that the N-nitroso compounds, as a class, are highly carcinogenic and many of the tested ones have been found to produce tumors in a large variety of organs and in many animal species including non-human primates. While the presence of N-nitroso compounds in the diet or production through endogenous synthesis has not been positively linked with cancer in man, substantial evidence has been assembled suggesting a possible health risk from exposure to these compounds.

0097–6156/92/0507–0350$06.00/0
© 1992 American Chemical Society

If *N*-nitroso compounds are toxicologically significant, then it is important to prevent their formation through elimination of one or both reactants or through prevention of the reactions that lead to their formation.

Many compounds have been studied in the latter role, but two in particular, α-tocopherol and ascorbic acid, have the unique properties of being both essential nutrients and effective blocking agents of nitrosation reactions. Before presenting the results of the many studies involving inhibition of nitrosamine formation in food and animal systems, it may be beneficial to first review some of the key aspects of *N*-nitroso compound formation, the level of man's exposure to nitrosating agents, plus some remarks regarding the mechanism of nitrosamine inhibition by α-tocopherol, related tocopherols and ascorbic acid.

Formation of *N*-Nitroso Compounds

Essentially all amines and amides must be considered capable of forming *N*-nitroso compounds to a lesser or greater degree. Excellent reviews of this subject have been published (1,2). Generally, the susceptibility of any particular compound towards the formation of a nitroso derivative reflects the mutual optimization of two conditions: (a) the presence of a potential nitrosating agent and its ultimate activation and (b) the chemical state in which the amine or amide exists in the reaction milieu.

For simple basic amines ($pK_a > 5$), the rate of nitrosation in water is maximum at pH 3.0-3.4, which is near the pK_a of nitrous acid, HNO_2. For a given amine, the nitrosation rate decreases as the pH increases above 3.4 because the concentration of nitrous acid decreases. When the pH decreases below 3.0, the rate again decreases because the concentration of unprotonated amine decreases. At a given pH, the rate of nitrosation increases as the basicity of the amine decreases because of a higher relative concentration of unprotonated amine present.

In moderately acidic aqueous solutions, the nitrosating agent is essentially nitrous anhydride, N_2O_3, formed from 2 mols of nitrous acid (HONO, pK_a 3.14 at 25°C), which is in turn formed from acidification of nitrite ions. At lower pH, the more reactive nitrosating agent, H_2ONO^+ is formed. Of significant interest is the catalytic effect on nitrosation that results when certain anions combine with nitrite to form alternate nitrosating agents (3-6). Thiocyanate is the most active of such catalysts, followed by the halides. This catalyzed nitrosation of amines competes most favorably with the N_2O_3 mechanism under conditions of high acidity and low nitrite ion concentration. The latter occurs because the reaction becomes first order in nitrite ion concentration through the formation of an NOX nitrosating intermediate (X = SCN or halide).

Secondary amines such as diethanolamine will react with a nitrosating agent according to the following reaction:

$$R_2NH + HONO \longrightarrow R_2N\text{--}NO + H_2O \qquad (1)$$

The formation of nitrosamines from secondary amines as described above is usually the most rapid of all the amine nitrosations. In fact, it should be observed that primary amines must be alkylated and tertiary amines dealkylated prior to formation of the ultimate nitroso compound through the nitrosation of the correspondingly generated secondary amine. Hence, the reaction described above is common to all the mechanisms discussed below.

Primary amines can react with a nitrosating agent leading to the formation of a nitrosamine by the following reaction mechanism (2):

$$RNH_2 + HONO \xrightarrow{-H_2O} RN{=}N{-}OH \xrightarrow{H^+} RN{\equiv}N^+ + H_2O \qquad (2)$$
$$\mathbf{I} \qquad\qquad\qquad\qquad\qquad \mathbf{II}$$

$$RN{\equiv}N^+ + RNH_2 \xrightarrow{-N_2} R_2NH_2^+ \xrightarrow{HONO} R_2N{-}NO + H_3O^+ \qquad (3)$$
$$\mathbf{II}$$

The initial attack of a nitrosating agent on a primary amine (**I**) leads to the formation of an alkyl diazonium ion (**II**) (Equation 2). The decomposition of this diazonium ion will normally lead to the formation of the corresponding alcohols and alkenes. These compounds are the classical deamination products. Under the proper conditions as demonstrated by Equation 3, however, this diazonium ion can react further with another mol of primary amine to form a secondary amine. This secondary amine can be nitrosated subsequently to form a nitrosamine. While the nitrosamine product is not favored for the reaction of nitrosating agent with a primary amine, the presence of certain catalysts like thiocyanate, formaldehyde, or other aldehydes can lead to at least some minor conversion of a primary amine to a secondary nitroso compound (7).

Tertiary amines can be converted to secondary nitrosamines, but the reaction occurs to a significant extent only at elevated temperatures in weakly acidic media. The rate of reaction is dependent on the pK_a of the amine. At 25°C and pH 3.4, the nitrosation of tertiary amines is about 10,000 times slower than that of related secondary amines. The following mechanism has been proposed for nitrosative dealkylation of tertiary amines and subsequent nitrosamine formation (7-9).

$$R_2N{-}CH_2R + 2\,HONO \rightleftharpoons R_2\overset{\displaystyle H}{\underset{\displaystyle NO}{\overset{|}{\underset{|}{N^+{-}CHR}}}} + H_2O + ONO^-$$

$$\downarrow {-}HNO$$

$$R_2N{-}\underset{+}{CHR} \xrightarrow{H_2O} RCHO + R_2NH_2^+ \qquad (4)$$

$$R_2NH + HONO \longrightarrow R_2N{-}NO + H_2O \qquad (1)$$

Quaternary amines and amine oxides also appear to react slowly with nitrite in acidic media to form nitrosamines (10). Compounds of this type are very similar to tertiary amines in their reactivity toward nitrosating agents. They also require a dealkylation step prior to further reaction to form a nitroso derivative.

The chemistry of N-alkylamide, as well as the related N-alkylurea and N-alkylcarbamate nitrosation, has not been studied as thoroughly as that of nitrosamines, mainly because suitable analytical methods have not been developed, particularly for their determination in complex systems like foods and body fluids.

The nitrosation of a typical amide is shown in Equation 5. In these reactions, the nitrosating agent is believed to be nitrous acid ion, H_2ONO^+ *(3,7,11)*.

$$RNHCOR + H_2ONO^+ \longrightarrow RN\!\!-\!\!COR + H_3O^+ \qquad (5)$$
$$\overset{|}{NO}$$

Like amides, many ureas and carbamates occur naturally or as drugs and some readily produce nitrosoureas or nitroso-carbamates. *N*-Alkyl ureas and carbamates are rapidly nitrosated at pH 1 - 2. The reaction rate increases about ten times for each pH unit drop from 3 to 1 and does not show a pH maximum as has been found for the formation of nitrosamines.

N-Nitroso compounds also can be rapidly obtained through the reaction of the parent compound with N_2O_3, or N_2O_4, in organic solvents *(12)*. The reactivity of amines and amides in a lipophilic environment, such as in a cosmetic preparation, is significantly greater than in the aqueous systems described above. Once the conditions are set for a nitrosating agent and susceptible amine, for example, to enter a lipid or nonpolar phase, the reaction is generally extremely rapid. Free amines readily dissolve in aprotic solvent systems as the unprotonated base, yielding a high proportion (if not all) of the more reactive form of the molecule. The nitrosating agents N_2O_3, and N_2O_4, are gases with appreciable solubility in lipid, nonpolar solvents *(13)*. This is in marked contrast to the poor lipid solubility of nitrite ion. Indeed, an aprotic solvent will be the first choice of an organic chemist for the nitrosation of amines by N_2O_3 or N_2O_4.

It is apparent, then, that the formation of nitrosamines and nitrosamides from the corresponding parent amine or amide can be demonstrated for all classes of these compounds. Generally, however, it is still the secondary amine or amide that is the most facile reactant. The other classes require either a preliminary alkylation or dealkylation.

Dietary Sources of Nitrate and Nitrite

For most Americans, food is the major source of exogenous exposure to NO_3^- and NO_2^-. NO_3^- occurs in many foods, but estimates indicate that the majority of American dietary NO_3^- comes from vegetables (87-97%), with a small portion coming from cured meats (2%) (Table I). Breads, dairy products and fruit and fruit juices contribute minor amounts (11%) *(14)*. NO_2^- consumption was estimated to be 0.002 and 0.004 mmol/person/day for average and high cured meat diets, respectively (Table II). Cured meats contributed from 39 to 71% of daily NO_2^- intake, with the balance coming from baked goods (15 to 34%), vegetables (7 to 16%), and fresh meats (3.5 to 7.7%). These estimates were based on published per capita consumption data and literature values of the NO_3^- and NO_2^- content of foods *(14)*.

Endogenous Sources of Nitrate and Nitrite

Estimates of the endogenous levels of NO_3^- and NO_2^- have been derived from studies on the pharmacology and metabolism of NO_3^-. These works have been reviewed *(15-17)*. The human studies of Wagner *et al.* *(18)*, confirmed by Leaf *et al.* *(19)*, have clearly indicated that NO_3^- is synthesized through a *de novo* pathway in

Table I. Dietary Sources of Nitrate (United States)

Source	μmol/person/day		% of total	
Vegetables	1050	(4200)[a]	86	(97)[a]
Fruits, juices	69		6	
Water	32	(2840)[b]	3	(68)[b]
Cured meat	26		2	
Baked goods, cereals	26		2	
Others	__13__		__1__	
Total	1216		100	

[a]Vegetarian diet
[b]Nitrate-rich water
SOURCE: Reprinted with permission from ref. 14. Copyright 1981 National Academy of Science.

Table II. Dietary Sources of Nitrite (United States)

Source	μmol/person/day		% of total	
Cured meat	6.5	(26)[a]	39	(71)[a]
Baked goods, cereals	5.7		34	
Vegetables	2.6	(10)[b]	15	(62)[b]
Fresh meats	1.3		8	
Others	__0.7__		__4__	
Total	16.8		100	

[a]High cured meat diet
[b]Vegetarian diet
SOURCE: Reprinted with permission from ref. 14. Copyright 1981 National Academy of Science.

humans in amounts of approximately 1.2 mmol/day. Thus, the quantity of endogenously formed NO_3^- is a significant portion of overall nitrate exposure, perhaps nearly equal in level to the estimates of ingested nitrate. Tannenbaum *et al.* (*17*) have shown that NO_3^- is concentrated and secreted in human saliva where a portion is reduced by oral microflora to NO_2^-. The NO_3^- and NO_2^- content of human saliva peaks approximately one hour after ingestion of NO_3^-, although there appears to be wide inter-individual variability.

Assuming a 5% reduction of NO_3^- to salivary NO_2^-, Hotchkiss (*20*) calculated that NO_2^- exposure through this path would be 0.12 mmol/person/day compared to 0.02 mmol/person/day through the diet.

Preventing *N*-Nitroso Compound Formation with α-Tocopherol and Ascorbic Acid

If *N*-nitroso compounds are toxicologically significant, then it is important to prevent their formation through elimination of one or both of the reactants or through prevention of the reactions leading to their formation. The formation of *N*-nitroso compounds can be reduced, minimized or even completely prevented by the presence of blocking agents when nitrosation potential exists. Blocking agents are essentially substances capable of rapidly reducing the nitrosating agent to the non-nitrosating nitric oxide (NO). They act as competitive substrates for the nitrosating species in the specific system (Figure 1). Both the absolute and relative concentrations of nitrosating agent, blocking agent and amine or amide substrate will determine the degree of "blocking" that will occur.

Many types of compounds can function as blocking agents, but two compounds in particular, ascorbic acid and α-tocopherol, have the unique properties of being both essential nutrients and effective blocking agents for nitrosation reactions. Together, the pair can act to inhibit nitrosation in multiphase systems, a desirable characteristic for foods and for the lumen of the gastrointestinal tract.

Ascorbic acid has been shown to be an excellent blocking agent against nitrosation in aqueous systems, particularly under weakly acidic conditions. The ascorbic acid competes for the nitrosating agent, forming dehydroascorbic acid and the non-nitrosating nitric oxide (NO) as indicated in Figure 2. The chemistry of α-tocopherol inhibition in lipid systems and possibly at the interface of micellar emulsions is quite analogous to that of ascorbic acid (Figure 2). The mechanism and kinetics of these reactions have been reported by a number of investigators (*21-25*). The effectiveness of α-tocopherol and ascorbic acid as inhibitors of nitrosation reactions can vary with the system under study. Simply, the interconversions that are possible among nitrogen oxide species (Figure 3) indicate that nitrosating agents can occur not only in an aqueous environment, but in the gaseous and lipid states as well. Hence, the combined use of both blocking agents is best suited for all but the simplest systems, where it can be clearly defined that nitrosations are taking place solely in the aqueous or lipophilic phase.

Isomers and Derivatives of Vitamin E as Inhibitors. Not all forms of "vitamin E" are equally effective as inhibitors of nitrosation. Of the naturally occurring forms of vitamin E, α-tocopherol has the highest biological activity as the vitamin. There are, however, other tocopherol and tocotrienols (tocopherols with unsaturated side chains) in foods of plant origin. The structures of the α-, β-, γ- and δ- tocopherols are shown in Figure 4.

I. Amine Nitrosation:

Amine + nitrosating agent ──────────────────────► Nitrosamine

II. Principle of Inhibition:

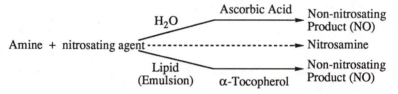

Figure 1. Principle of blocking nitrosation reactions using water soluble and fat soluble inhibitors. (Reproduced with permission from ref. 49. Copyright 1981 Cold Spring Harbor.)

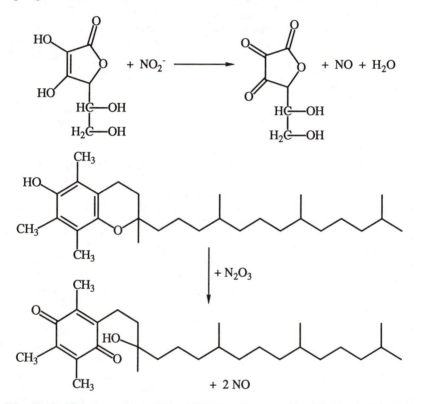

Figure 2. Reactions of ascorbic acid (top) and α-tocopherol (bottom) with nitrite. The actual nitrosating agent varies with pH, reaction conditions, and/or the presence of catalysts. (Reproduced with permission from ref. 50. Copyright 1980 Plenum Publishing.)

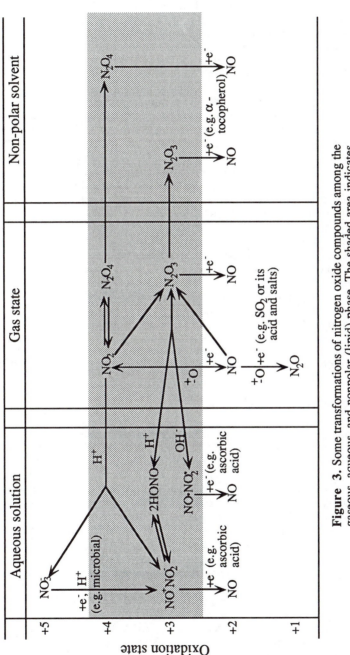

Figure 3. Some transformations of nitrogen oxide compounds among the gaseous, aqueous, and nonpolar (lipid) phase. The shaded area indicates potential nitrosating agents. Adapted from ref. 49. Copyright 1981 Cold Spring Harbor Laboratory.

From the viewpoint of blocking of nitrosation reactions, these structures are very interesting. Although all of the tocopherols can act as reactive phenols to form quinones in reducing a nitrosating agent to NO, only the α-tocopherol, being fully substituted on the aromatic ring, can completely avoid forming a C-nitroso derivative. The latter can potentially transnitrosate or even catalyze the nitrosation of a nitrosatable amine or amide (26). Kamm et al. (27) have shown that α-tocopherol is a very effective blocking agent against nitrosation reactions, whereas γ-tocopherol is somewhat less effective both *in vitro* and *in vivo*. This superiority of α-tocopherol over γ-tocopherol as a nitrosation inhibitor is in marked contrast to the superiority of γ- over α-tocopherol as an *in vitro* or food-type antioxidant. The difference in effectiveness of blocking nitrosation reactions, as compared with lipid oxidation, apparently is due to the lack of complete substitution on the aromatic ring in the γ isomer.

In addition to the difference in nitrosation blocking ability that occurs among tocopherol isomers, it should be noted that there are some forms of vitamin E that are not inhibitors. These are the ester deriviatives of tocopherol (Figure 5). The esters do not appear naturally in foods, but were developed as means of stabilizing the molecule against oxidation of the phenol group during processing and storage, particularly in pharmaceutical dosage forms such as tablets and capsules. These esters per se are inactive as antioxidants or N-nitroso blocking agents. They are hydrolyzed slowly in the acidic conditions of the stomach. Lower in the gastrointestinal tract hydrolysis is much more efficient. The acetate ester may be partly absorbed unhydrolyzed and is only very slowly hydrolyzed on parenteral administration (28). With this in mind, it is always better to utilize the unesterified α-tocopherol, d (natural), or dl (synthetic) as the preferred nitrosation inhibitor. This may be especially true of *in vivo* studies seeking activity of α-tocopherol in the upper portion of the gastrointestinal tract (stomach, duodenum, etc.) where little or no free active tocopherol can be expected to be formed when tocopheryl acetate or succinate is administered.

Inhibition of Nitrosation in Foods

Of the various foods monitored for nitrosamine content, cured meats (primarily bacon) are probably the most important, principally because these products are the main known source of nitrite in the Western diet. The two nitrosamines usually detected in such foods are N-nitrosopyrrolidine (NPYR) and N-nitrosodimethyl-amine (NDMA). In cured meats other than bacon, only extremely low concentrations (<1 µg/kg) of nitrosamines have been detected (29-32). Although raw bacon is also low in nitrosamines, significantly higher amounts are found in the fried product. The nitrosamines formed during frying are found especially in the cooked-out fat. In view of this, Bharucha et al. (30) have suggested that an effective nitrosation inhibitor in bacon should be lipid-soluble, nonvolatile, able to trap a nitrosating agent, and heat stabile up to 174°C (typical frying temperature). Ascorbic acid has been shown to reduce nitrosamine formation in cooked bacon while its lipid-soluble derivatives are more effective in reducing the nitrosamine content of both lean meat and cooked-out fat. As a result of these findings, it is now mandatory to add ascorbic acid or erythorbic acid at 550 mg/kg to all bacon manufactured in the U.S. The lipid-soluble α-tocopherol has also been affirmed as GRAS (Generally Recognized as Safe) by both the USDA and by the FDA as an effective inhibitor of nitrosamine formation in bacon (33) and can be added at the level of 500 ppm. The use of α-tocopherol in

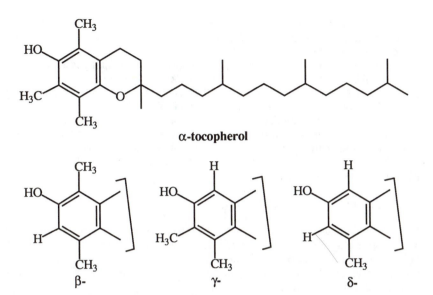

α-tocopherol

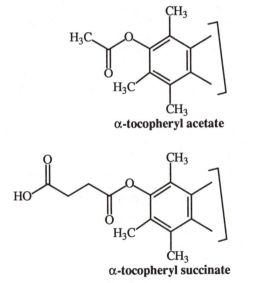

β-   γ-   δ-

Figure 4. Chemical structures of tocopherol isomers. The balance of the abbreviated β, γ and δ structures are identical to α. (Reproduced with permission from ref. 21. Copyright 1980, NY Acad. Sci.)

α-tocopheryl acetate

α-tocopheryl succinate

Figure 5. Two stabilized ester pharmaceutical forms of α-tocopherol, α-tocopheryl acetate and α-tocopheryl succinate. The balance of the abbreviated structures are identical to α-tocopherol (Figure 4).

foods and model food systems was recently summarized (34). In foods, its use has been focused almost exclusively on bacon where it has been found it can significantly reduce nitrosamine formation (Table III).

In a quite different application, but one that involves lipid phase nitrosation, Mergens and DeRitter (35) have shown that α-tocopherol can be used to inhibit the formation of N-nitrosodiethanolamine (NDELA) in thin films of cosmetic creams exposed to low levels of nitrogen dioxide (Table IV). While the cream was formulated using triethanolamine, it should be noted that the cosmetic grade triethanolamine employed in this study did contain diethanolanine (DELA) as an impurity. It is the latter which is believed to be nitrosated by nitrogen dioxide rather than NDELA formation occurring through the preliminary dealkylation of triethanolamine and subsequent nitrosation of DELA.

Inhibition of Nitrosation *In Vivo*

Endogenous nitrosation occurs through the interaction of nitrogenous compounds and nitrosating agents either generated in the body or taken in from the environment. Food, as well as natural body secretions, contain numerous nitrosatable substrates such as dimethylamine, citrulline, agmatine, methylguanidine, and, of course, amide linkage-containing peptides.

The advent of the nitrosoproline (NPRO) test (36) has allowed testing the effects of ascorbic acid and α-tocopherol on endogenous nitrosation in man. Ohshima and Bartsch (36) first showed that, following ingestion of 325 mg (5.9 mmol) nitrate and 250 mg (2.2 mmol) proline by a human volunteer, the amount of NPRO excreted in urine over the next 24 h increased from a background of 2.9 µg (20 nmol) to 14.9 µg (103 nmol). This increase was completely inhibited if 1 g (5.7 mmol) of ascorbic acid was taken at the same time as proline, clearly demonstrating the blocking effect of ascorbic acid on gastric nitrosation (Table V). Administration of ascorbic acid or α-tocopherol simultaneously with nitrite and amines or amides to experimental animals and the nitrate-proline model in man have been shown in many experiments to inhibit the endogenous formation of the corresponding N-nitroso compounds and the appearance of the toxic effects thereof. A number of these studies are summarized in Table VI.

Wagner et al. (37) have shown that 4 doses of ascorbic acid, 500 mg (2.8 mmol) each taken over 24 h, inhibited by 81% the endogenous synthesis of NPRO caused by the ingestion of 3.5 mmol nitrate and 4.3 mmol proline. Four hundred mg of α-tocopherol inhibited NPRO synthesis by 59%. An additional nitrosamino acid, nitrosothiazolidine-4-carboxylic acid (NTCA), was also found in the urine. Ascorbic acid and α-tocopherol blocked this nitrate-induced synthesis as well. Significantly, no inhibition of the background levels of NPRO (i.e., that occurring in the absence of ingestion of nitrate and proline) could be brought about by ascorbic acid. It appears that while intragastric nitrosation (resulting from ingested nitrate or nitrite) can be efficiently blocked by ascorbic acid, additional pathways of endogenous nitrosation exist (possibly related to the activity of the reticuloendothelial system) (38).

Garland et al. (39) developed a mass spectrometric analytical technique for NDMA in urine sensitive to 5 ng/l. Using this technique, these investigators suggested that humans excrete an average of 38 ng/day. The source of NDMA was not determined, but changes in the amounts excreted correlated positively with changes in the amount of atmospheric nitrogen dioxide (NO_2). This suggests that the NDMA could have been endogenously formed from inhaled NO_x compounds, but

Table III. Summary of Nitrite Scavenger and Inhibitory Effects
of α-Tocopherol on
Nitrosamine Formation in Foods and Food Model Systems

α-Tocopherol mM	(mg)	Food/Food Model	N-Nitrosamine	% Inhibition
5.00	(2150)	model system	DMN	92
0.93	(400)	fried lean bacon	NPYR	73
1.86	(800)	fried lean bacon	NYPR	87
0.93	(400)	fried lean bacon	DMN	76
1.86	(800)	fried lean bacon	DMN	65
1.16	(500)	model system	NYPR	57
1.16	(500)	fried bacon	NYPR	70
			DMN	—
0.60	(250)	dry cured bacon	NYPR	65
			DMN	50
1.16	(500)	dry cured bacon	NYPR	77
			DMN	77
1.74	(750)	dry cured bacon	NYPR	67
			DMN	28

SOURCE: Reprinted with permission from ref. 34. Copyright 1989 Hogref & Huber.

Table IV. Effect of α-Tocopherol on
the Lipid Phase Nitrosation of Diethanolamine

Triethanolamine[a] (μM)	dl-α-Tocopherol (μM)	NO_2 (μM)	Nitrosodiethanolamine found (μM)	% Reduction
67	0	7	0.127	—
67	7	7	0.037	71
67	21	7	0.010	92
67	28	7	0.007	96

Thin films of cosmetic cream were exposed to atmospheric nitrogen dioxide.
[a]Diethanolamine is a normal component of this cosmetic grade raw material.

Table V. Effects of Ascorbic Acid and α-Tocopherol on the Formation of Nitrosoproline *In Vivo* in a Human

Material Ingested	Nitrosoproline excreted in urine (mean) (μg/24 hours)
Beet juice (250 ml containing 325 mg nitrate)	1.69, 3.44, 3.66 (2.93)
Proline (250 mg)	1.05, 1.40, 3.42 (1.96)
Beet juice (250 ml containing 325 mg nitrate) + proline (250 mg)	14.0, 14.8, 15.9 (14.9)
Beet juice (250 ml containing 325 mg nitrate) + proline (250 mg) + ascorbic acid (1 g)	2.39, 2.96, 3.13 (2.83)
Beet juice (250 ml containing 325 mg nitrate) + proline (250 mg) + α-tocopherol (500 mg)	8.17, 7.45, 6.16 (7.26)

SOURCE: Reprinted with permission from ref. 36. Copyright 1981 American Association of Cancer Research.

Table VI. Summary of *In Vivo* N-Nitrosocompound Inhibition Studies Using Ascorbic Acid or α-Tocopherol

Ref.	Model	Nitrosating Agent	Amine	Ascorbic Acid (% Inhibition)	α-Tocopherol (% Inhibition)	N-Nitroso Compound
36	human	NO_3^- 325 mg (5.9 mmol)	proline 250 mg (2.2 mmol)	100 mg (100)	500 mg (60)	NPRO
37	human	^{15}N nitrate 190 mg (3.5 mmol)	proline 490 mg (4.3 mmol)	4 x 500 mg (81)	500 mg (59)	NPRO
39	human	baseline	baseline	1 g (0)	400 mg (0)	NDMA
40	mouse	NO_2 47 ppm x 2 hours	morpholine 2 mg	— —	5 μg/g body weight (70)	NMOR

direct evidence for endogenous formation was not presented. Neither ascorbic acid nor α-tocopherol influenced this excretion. Considering (a) the rapid rate at which NDMA is metabolized, (b) the fact that it equilibrates rapidly with total body water, and (c) urinary excretion levels may only reflect a small fraction of that water, it may be interesting to repeat this study using an inhibitor of NDMA metabolism.

The possibility that α-tocopherol could inhibit the formation of nitrosomorpholine (NMOR) *in vivo* was investigated in mice injected i.p. with 2 mg morpholine and subsequently exposed to 47 ppm of NO_2 in an inhalation chamber for 2 hours. Pretreatment of these animals with oral doses of α-tocopherol (2.5-100 mg/kg body weight) once daily for 6 days significantly decreased NMOR formation *in vivo* during exposure to nitrogen dioxide. Whole body α-tocopherol levels increased in mice over the 6-day period in a dose-dependent manner. In this system, a maximal 50-70% inhibition of nitrosamine formation occurred at total body levels of α-tocopherol of 5 μg/kg (*40*). One of the important observations from these studies, which represented extended work on earlier investigations with this model (*41-44*) was to demonstrate that body levels of α-tocopherol, rather than dietary intake, correlated more precisely with inhibition effects (see Figure 6).

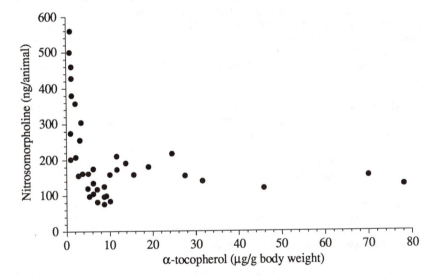

Figure 6. *N*-Nitrosomorpholine formation in the mouse as a function of α-tocopherol body levels. (Reproduced with permission from ref. 40. Copyright Oxford Univ. Press.)

Not all studies have shown that blocking agents can significantly inhibit *in vivo* nitrosamine formation. Stich *et al.* (*45*) found that the consumption of nitrite-cured meats led to an increase in urinary excretion of NPRO in humans. They suggested that the NPRO was preformed in the meat, and not endogenously formed, because the excretion could not be reduced by consuming large amounts of ascorbic acid. Later these same workers (*46*) reported that while the NPRO was not detected during direct analysis of the cured meat, analyses following enzymatic digestion yielded substantial amounts of NPRO. Kubacka *et al.* (*47*) have shown that peptides which are N-terminal in proline can be nitrosated.

Perciballi (*48*) has studied this excess urinary NPRO with ferrets consuming nitrite-cured meat diets. NPRO excretion increased 13-fold when ferrets were switched from a semi-purified diet to a cured-meat diet. Using stable isotopes of NO_2^- and proline, it was shown that a majority of the NPRO was formed in the cured meat before ingestion, and less than 17% was formed in the stomach. As found by Stich *et al.* (*46*) these data also show that the exposure to N-nitroso compounds due to consumption of cured meats may be higher than direct analysis of the food indicates and may be a partial explanation of the baseline contribution and individual variability of endogenous levels.

Clearly, however, further studies are needed to confirm these findings and to elucidate the factors which determine the effectiveness of α-tocopherol and ascorbic acid as nitrosation inhibitors in different individuals.

Conclusions

The substantial body of evidence which now exists regarding man's exposure to nitrosating agents and *N*-nitroso compounds, the mechanisms of *N*-nitroso compound formation and inhibition both *in vitro* and *in vivo* presents a compelling case that *N*-nitroso compounds are present in the diet and in addition can be formed from precursors both ingested and synthesized *de novo* in the body. It is also clear that *N*-nitroso compound formation can be inhibited by both α-tocopherol and ascorbic acid.

From the evidence presented in this paper, the following conclusions can be drawn:

• Man's total exposure to nitrate and nitrite is significant and of the same order of magnitude used in the Ohshima-Bartsch model through which *in vivo* nitrosation has been conclusively demonstrated.

• α-Tocopherol and ascorbic acid can inhibit the *in vivo* formation of several *N*-nitroso compounds and presumably others.

• There appears to be a refractory (baseline) level of *N*-nitroso compounds upon which α-tocopherol and ascorbic acid have no effect. It is not clear whether this is due to an alternate mechanism or site of *in vivo* nitrosation, or if this contribution is released from the diet during digestion.

• α-Tocopherol and ascorbic acid have, by and large, been employed independently as inhibitors of nitrosation reactions in most models. There appears to be an untapped potential to use these compounds in combination since one is a fat soluble and the other is a water soluble nitrosation inhibitor.

Nitrosamine Abbreviations

NDELA = nitrosodiethanolamine
NDMA = nitrosodimethylamine
NMOR = nitrosomorpholine
NPRO = nitrosoproline
NPYR = nitrosopyrrolidine
NTCA = nitrosothiozolidine-4-carboxylic acid

Literature Cited

1. Douglas, M.L.; Kabacoff, B.L.; Anderson, G.A.; Cheng, M.C. *J. Soc. Cosmet. Chem.* **1978**, *29*, 581.
2. Mirvish, S.S. *Tox. Appl. Pharmacol.* **1975**, *31*, 325.
3. Mirvish, S.S. *J. Natl. Cancer Inst.* **1971**, *46*, 1183.
4. Boyland, E.; Nice, E.; Williams, K. *Food Cosmet. Toxicol.* **1971**, *9*, 639.
5. Williams, D.L.H. *J. Chem. Soc. Perkin Trans.* **1977**, *2*, 128.
6. Fan, T.Y.; Tannenbaum, S.R. *J. Agric. Food Chem.* **1972**, *21*, 237.
7. Ridd, J.H. *Q. Rev. Chem. Soc.* **1961** *15*, 418.
8. Smith, P.A.S.; Pars, H.G. *J. Org. Chem.* **1959**, *24*, 1325.
9. Smith, P.A.S.; Leoppky, R.N. *J. Amer. Chem. Soc.* **1967**, *89*, 1147.
10. Fiddler, W.; Pensabene, J.W.; Doerr, R.C.; Wasserman, A.E. *Nature* **1972**, *236*, 307.
11. Mirvish, S.S. In *N-Nitroso Compounds: Analysis and Formation*; Bogovsky, P.; Preussmann, R.; Walker E.A., Eds., IARC Scientific Publications No. 3, International Agency for Research on Cancer: Lyon, France, 1974; pp. 132-136.
12. White, E.H. *J. Amer. Chem. Soc.* **1955**, *77*, 6008.
13. Lovejoy, D.J.; Vosper, A.J. *J. Chem. Soc.* **1968**, 2325.
14. National Academy of Science. *The Health Effects of Nitrate, Nitrite and N-Nitrosocompounds*; National Academy Press: Washington, D.C. 1981
15. Hartman, P.E. In *Chemical Mutagens*; deSerres F.J.; Hollander, A., Eds; Plenum Publishing: New York, NY, 1982, Vol. 7; p. 211.
16. Tannenbaum, S.R.; Young V.R. *J. Environ. Path. Toxicol.* **1980** *3*, 357.
17. Tannenbaum, S.R., Weisman, M.; Fett, D. *Food Cosmet. Toxicol.* **1976**, *14*, 549.
18. Wagner, D.A.; Schultz, D.S.; Deen, W.M.; Young, V.R.; Tannenbaum, S.R *Cancer Res.* **1983**, *43*, 1921.
19. Leaf, C.D.; Vecchio, A.J.; Roe, D.A; Hotchkiss, J.H. *Carcinogenesis* **1987**, *8* 791.
20. Hotchkiss, J. H. In *Food Toxicology: A Perspective on the Relative Risk*; Taylor, S.L.; Scanlan R.A., Eds.; Marcel Dekker: New York, NY, 1989; pp. 57.
21. Tannenbaum, S.R.; Mergens, W. J. *Annal. N.Y. Acad. Sci.* **1980**, *355*, 267.
22. Kim, Y.K.; Tannenbaum, S.R.; Wishnok, John S. In *Ascorbic Acid: Chemistry, Metabolism & Uses*; Seib, P.A.; Tolbert, B.M., Eds.; ACS Symposium Series 200; American Chemical Society: Washington, D.C., 1982; p. 571
23. Licht, W. R.; Tannenbaum, S. R.; Deen, W.M. *Carcinogenesis* **1988**, *9*, 365.
24. Licht, W.R.; Deen, W. M. *Carcinogenesis* **1988**, *9*, 2227
25. Licht, W.R; Fox, J.G.; Deen, W.M. *Carcinogenesis* **1988**, *9*, 373.
26. Walker, E.A.; Pignatelli, B; Castegnaro, M. *J. Agric. Food Chem.* **1979**, *27*, 393
27. Kamm, J.J.; Dashman T.; Newmark, H. L; Mergens, W. J. *Toxicol. Appl. Pharmacol* **1977**, *41*, 575
28. Newmark, H. L.; Pool, W. C; Bauernfeind, J.C.; DeRitter, E. *J. Pharm. Sci.* **1975**, *64*, 665
29. Nitrite Safety Council. *Food Technol.* **1980**, *34*, 45.

30. Bharucha, K.R.; Cross, C.K.; Rubin L. J. *J. Agric. Food. Chem.* **1979**, *27*, 63.
31. Gray, J. I.; Reddy, S.K.; Price, J.F.; Mandagere, A.; Wilkens, W.F. *Food Technol.* **1982**, *36*, 39
32. Gray, J. I. In *N-Nitrosocompounds*, Scanlan, R.A.; Tannenbaum, S.R., Eds.; ACS Symposium Series 174; American Chemical Society: Washington, D.C., 1981; p. 165
33. Code of Federal Regulations: 21 CFR 184.1890 and 9 CFR. 318.7 (c) (4).
34. Lothia D.; Blum, A. *Internat. J. Vit. Nutr. Res.* **1989**, *59*, 430.
35. Mergens, W. J.; DeRitter, E. *Cosmet. Technol.* January **1980**, 56.
36. Ohshima, H.; Bartsch, H. *Cancer Res.* **1981**, *41*, 3658.
37. Wagner, D.A.; Shuker, D.E.G.; Bilmazes, C.; Obiedzinski, M.; Baker, I.; Young, V. R.; Tannenbaum, S.R. *Cancer Res.* **1985**, *45*, 6519.
38. Wagner, D.A.; Shuker, D.E.G.; Bilmazes, C.; Obiedzinski, M.; Young, V.R.; Tannenbaum, S.R. In *N-Nitroso Compounds: Occurrence, Biological Effects and Relevance to Human Cancer*; O'Neil, I.K.; von Borstel, R.C.; Long, J.E.; Miller, C.T.; Bartsch, H., Eds., IARC Publication No. 57; International Agency for Research on Cancer: Lyon, France, 1984; p. 223.
39. Garland, W.A.; Kuenzig, W.; Rubio, F.; Kornychuk, H.; Norkus, E.P.; Conney, A.H. *Cancer Res.* **1986**, *46*, 5392.
40. Norkus, E.P.; Kuenzig, W.A.; Chau, J.; Mergens, W.J.; Conney, A.H. *Carcinogenesis* **1986**, *7*, 357.
41. Iqbal, Z.M.; Dahl, K.; Epstein, S.S. *Science* **1980**, *207*, 1475.
42. Epstein, S.S.; Iqbal, Z.M.; Johnson, M.D. In *N-Nitroso Compounds: Analysis, Formation and Occurrence*; Walker, E.A.; Griciute, L; Castegnaro, M.; Borzony, M., Eds.; International Agency for Research on Cancer: Lyon, France, 1980; p. 195.
43. Van Stee, E.W.; Sloane, R.A.; Simmons, J.E.; Brunnemann, K.D. *J. Nat. Cancer Inst.* **1983**, *70*, 375.
44. Norkus, E.P.; Boyle, S.; Kuenzig, W.; Mergens, W. J. *Carcinogenesis* **1984**, *5*, 549.
45. Stich, H.F.; Hornby, A.P.; Dunn, B.P. *Int. J. Cancer* **1984**, *33*, 625.
46. Dunn, B.P.; Stich, H.F. *Food Chem. Toxicol.* **1984**, *22*, 609.
47. Kubacka, W.; Libbey, L.M.; Scanlan, R.A. *J. Agric. Food Chem.* **1984**, *32*, 401.
48. Perciballi, M.; Conboy, J.J.; Hotchkiss, J.H. *Food Chem. Toxicol.* **1989**, *27*, 111.
49. Newmark, H.L.; Mergens, W.J. In *Gastrointestinal Cancer: Endogenous Factors*; Bruce, W.R.; Correa, P.; Lipkin, M.; Tannenbaum, S.R.; Wilkins, T.D., Eds.; Banbury Report 7; Cold Spring Harbor Laboratory, 1981; pp. 285-304.
50. Mergens, W.J.; Newmark, H.L. In *Autoxidation in Food and Biological Systems*; Simic M.G.; Karel, M., Eds.; Plenum Publishing: New York, NY, 1980; pp. 387-403.

RECEIVED April 20, 1992

Chapter 29

Plant Phenolic Compounds as Cytotoxic Antitumor Agents

Kuo-Hsiung Lee

Natural Products Laboratory, Division of Medical Chemistry and Natural Products, School of Pharmacy, University of North Carolina, Chapel Hill, NC 27599

The cytotoxic antitumor plant phenolics and their analogs are reviewed with emphasis on those discovered from the author's laboratory. The active compounds include, in addition to those which are in clinical use or trials as anticancer drugs, alkaloids, anthraquinones, coumarins, flavonoids, lignans, macrolides, naphtho-quinones, and polyphenols. The compounds are discussed briefly with respect to their *in vitro* cytotoxicity and *in vivo* antitumor activity as well as their structure-activity relationships and mechanism of action, when sufficient data are available.

Plant-derived natural products and their synthetic analogs have been a promising source of novel anticancer drugs. The clinical utility of the vinca alkaloids, vinblastine and vincristine, and their analog vindesine, as well as the podophyllotoxin-derived lignan glycosides, etoposide and teniposide, testify to the validity of this thesis (*1-5*). The progress in the discovery and development of plant antitumor agents and analogs during the past two decades has been reviewed (*6-16*). Numerous plant phenolics have been isolated as the cytotoxic antitumor principles involved in this study.

Plant Phenolics and Analogs in Clinical Use or Trials As Anticancer Drugs

Currently, there are at least six plant phenolics and their analogs in clinical use or trials as anti-cancer drugs. These include the above-mentioned etoposide (**1**) and teniposide (**2**), 10-hydroxycamptothecin (**3**), lycobetaine (**4**), gossypol (**5**), and 9-hydroxy-N-2-methylellipticinum acetate (**6**).

Etoposide (1) and teniposide (2) are semi-synthetic analogs of podophyllotoxin, a cytotoxic lignan isolated from *Podophyllum peltatum* and related species. Compounds **1** and **2** are effective anticancer drugs used clinically for the treatment of small cell lung cancer, testicular cancer, lymphoma, and leukemia (*17,18*). There exists ample evidence suggesting that these drugs block the catalytic activity of DNA

0097–6156/92/0507–0367$06.00/0
© 1992 American Chemical Society

topoisomerase II by stabilizing an enzyme-DNA complex in which the DNA is cleaved and covalently linked to the enzyme (*19-21*). The cytotoxicity of **1** and **2** might also be associated with the phenoxy free radical and its resulting *ortho*-quinone species formed by biological oxidation of the drugs (*22,23*). The phenoxy free radical and *ortho*-quinone derivatives of **1** were believed to be produced by metabolic activation of **1**, which could bind to critically important cellular macromolecules as alkylating species causing dysfunction and, subsequently, cell death (*24,25*). Recent studies involving the replacement of the 4β-sugar moiety with 4β-aryl-amino groups in the author's laboratory has yielded several compounds (**7-12**) which were 5- to 10-fold more potent than **1** as inhibitors of DNA topoisomerase II *in vitro*. All of these compounds (**7-12**) could generate the same amount or more protein-linked DNA breaks in cells than **1** at 1-20 μM. In addition, these new compounds were cytotoxic not only to KB cells but also to their **1**-resistant and vincristine-resistant variants which showed decreased cellular uptake of **1** and a decrease in DNA topoisomerase II content or over expression of MDR1 phenotype (*26-28*).

10-Hydroxycamptothecin (**3**) was isolated from *Camptotheca acuminata* a tree common to Southern China. *C. acuminata* also yielded camptothecin (**13**), which was first isolated, tested, and found to be extremely toxic in the United States (*29*). 10-Hydroxycamptothecin was found to be more potent and less toxic than **13**, and has been in clinical use in China in patients with head and neck malignancies and primary liver carcinoma. It is also useful for the treatment of gastric cancer and leukemia (*30*). It inhibits strongly the S phase of tumor cells, DNA synthesis, and in particular DNA polymerase (*31*). Recently, **3** and **13** were both found to be the potent inhibitors of DNA topoisomerase I (*32,33*). Three analogs and derivatives of **3** and **13**, including hycamp-tamine (**14**) (*34*), 9-aminocamptothecin (**15**) (*35*), and CPT-11 (**16**) (*36*), are currently in clinical trials as anticancer drugs for colon and other cancers in Europe, U.S.A., and Japan, respectively (*37*).

Lycobetaine (**4**) is a derivative of lycorine, an alkaloid common to many *Lycoris* and *Narcissus* species, such as *Lycoris radiata* and *Narcissus pseudonarcissus* (*38*). Compound **4** is used in patients having ovarian carcinoma and gastric cancer in China (*30,39*).

Gossypol (**5**), isolated from *Gossypium herbaceum*, is a well-known male contraceptive developed in China (*40*). It is also used as an anticancer drug for the treatment of stomach, esophageal, liver, mammary, and bladder cancers in China (*39*).

9-Hydroxy-N-2-methylellipticinum acetate (**6**) is an analog of ellipticine and 9-methoxyellipticine, which are alkaloids isolated from *Ochrosia elliptica* or *O. acuminata* (*9,41*). Compound **6** is a DNA topoisomerase II inhibitor and active against thyroid and renal cancers. Several of its analogs are in European clinical trials as anticancer drugs (*4,37*).

Plant Phenolics As Cytotoxic Antitumor Agents Discovered in the Author's Laboratory

Studies on plant cytotoxic antitumor agents including phenolic compounds are proceeding in many laboratories (*6-16*). However, to limit the scope of the present discussion, this review will only deal with the work carried out in the author's laboratory. In our program, bioassay-directed fractionation and isolation studies are employed. This includes mainly an *in vitro* KB (cell culture) and an *in vivo* P-388 murine lymphocytic leukemia screening system as routine procedure for detecting cytotoxic antileukemic substances according to a protocol of the U.S. National Cancer

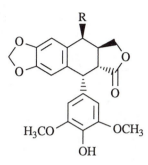

1, Etoposide : R =

2, Teniposide : R =

7 , NH—⟨benzene⟩—NH$_2$ • HCl

8 , NH—⟨benzene⟩—CN

9 , NH—⟨benzene⟩—NO$_2$

10, NH—⟨benzene⟩—F

11, NH—⟨benzene⟩—COOCH$_2$CH$_3$

12, NH—⟨benzodioxane⟩

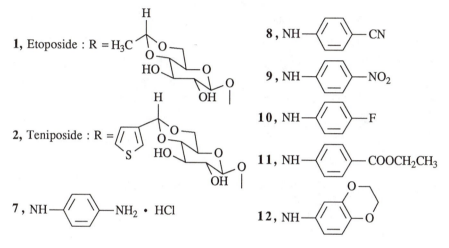

3, 10-Hydroxycamptothecin : R$_1$ = OH, R$_2$ = R$_3$ = H

13, Camptothecin : R$_1$ = R$_2$ = R$_3$ = H

14, Hycamptamine : R$_1$ = OH, R$_2$ = CH$_2$N(CH$_3$)$_2$, R$_3$ = H

15, 9-Aminocamptothecin : R$_1$ = OH, R$_2$ = NH$_2$, R$_3$ = H

16, CPT-11 : R$_1$ = OCO—N⟨piperidine⟩—N⟨piperidine⟩ • HCl, R$_2$ = H, R$_3$ = CH$_2$CH$_3$

Institute (NCI) (42). In keeping up with the new screening strategy developed by the NCI (43-46), we have recently also included several disease-oriented human tumor cell lines, such as A-549 (lung carcinoma), HCT-8 (colon carcinoma), and MCF-7 (breast adenocarcinoma), into our routine screening program in order to discover new compounds with potential specificity against slowly growing solid tumors. On the basis of this approach, many new cytotoxic antitumor phenolics have been isolated and characterized, and are briefly discussed below.

Alkaloids. 1.   (+)-Thalifarazine (**17**) was isolated from *Thalictrum sessile* (Ranunculaceae). It showed cytotoxicity with $ED_{50}^{\ddagger}$ (KB) = 1.50, (HCT) = 4.30, (A-549) = 6.53, (P-388) = 3.75, and (L-1210) = 5.69 µg/ml for the first time (47).

2.   Three canthin-6-one alkaloids, including 11-hydroxycanthin-6-one [**18**, ED_{50} (KB) = 1.0 µg/ml], 11-hydroxy-1-methoxycanthin-6-one [**19**, ED_{50} (KB) = 2.0 µg/ml], and 1-hydroxy-11-methoxycanthin-6-one [**20**, ED_{50} (KB) = 2.0 µg/ml] were first isolated as cytotoxic phenolic alkaloids from the stem of *Brucea antidysenterica* (Simaroubaceae) (48). A structure-activity relationships study of **18-20** and their related derivatives indicated that hydroxylation or methoxylation of canthin-6-one alkaloids at C-11 or C-10 are essential for the cytotoxicity in the KB cells (49).

3.   (-)-Norannuradhapurine • HBr [**21**, ED_{50} (HCT-8) = 3.32 µg/ml] isolated from *Fissistigma oldhamii* (Annonaceae) and (-)-armepavine N-oxide [**22**, ED_{50} (KB) = 4.00 µg/ml] were found to show significant cytotoxicity (50).

Anthraquinones. 1.   Five cytotoxic antileukemic phenolic anthraquinones were isolated from *Morinda parvifolia* (Rubiaceae). These include the new morindaparvin-B [**23**, ED_{50} (KB) = 4.0 µg/ml], and the known alizarin-1-methyl ether [**24**, T/C[†] (P-388) = 136% (10 mg/kg)], lucidin-ω-methyl ether [**25**, ED_{50} (KB) = 0.62 µg/ml], digiferruginol [**26**, ED_{50} (P-388) = 5.9 µg/ml, (KB) = 0.09 µg/ml], and 1-hydroxy-6 or 7-hydroxymethylanthraquinone [**27**, ED_{50} (KB) = 2.6 µg/ml, T/C (P-388) = 150% (10 mg/kg)]. Compounds **24-27** were isolated for the first time to demonstrate significant cytotoxic antileukemic activity (51,52). Compounds **15** and **18** were also synthesized by this laboratory (53).

Coumarins. 1.   The *Euphorbia splendens* (Euphorbiaceae) afforded a weak cytotoxic coumarin, 3,3',4-tri-O-methylellagic acid [**28**, ED_{50} (KB) = 7.7 µg/ml]. However, its demethylated polyphenolic ellagic acid [**29**, ED_{50} (KB) = 4.6 µg/ml] was reported to be able to inactivate carcinogen benzo[α] pyrene-7,8-diol-9,10-epoxide (Lee, K. H.; Hayashi, N.; Nozaki, H.; Okano, M. unpublished data.) (54).

[‡] ED_{50}: The drug concentration in µg/ml which inhibits the growth by 50% as compared with controls. KB (human epidermoid carcinoma of the nasopharynx) response of ≤ 4 µg/ml is considered as indicative of cytotoxicity (42).

[†] T/C: The ratio of mean survival rate of treated animals to the mean survival rate of control animals. According to NCI protocols, T/C $\geq 125\%$ is required for significant activity in the P-388 lymphocytic leukemia screen (42).

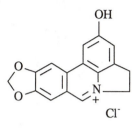

4, Lycobetaine

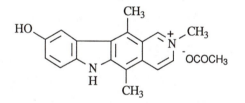

5, Gossypol

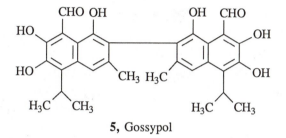

6, 9-Hydroxy-N-2-methylellipticinum Acetate

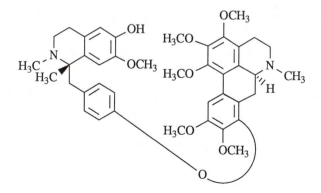

17, (+)-Thalifarazine

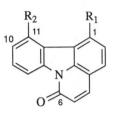

18, 11-Hydroxycanthin-6-one : R_1 = H, R_2 = OH

19, 11-Hydroxy-1-methoxycanthin-6-one : R_1 = OCH$_3$, R_2 = OH

20, 1- Hydroxy-11-methoxycanthin-6-one : R_1 = OH, R_2 = OCH$_3$

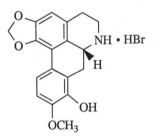

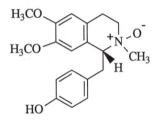

21, (-)-Norannuradhapurine • HBr   **22,** (-)-Armepavine N-oxide

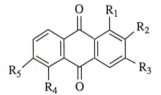

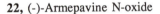

23, Morindaparvin-B : R_1 = OH, R_2 = CH$_2$OH, R_3 = R_5 = H, R_4 = OH

24, Alizarin-1-methyl ether : R_1 = OCH$_3$, R_2 = OH, R_3 = R_4 = R_5 = H

25, Lucidin-ω-methyl ether : R_1 = OH, R_2 = CH$_2$OCH$_3$, R_3 = OH,
R_4 = R_5 = H

26, Digiferruginol : R_1 = OH, R_2 = R_3 = R_4 = H, R_5 = CH$_2$OH

27, 1-Hydroxy-6 or 7-hydroxymethylanthraquinone : R_1 = OH,
R_2 = R_3 = R_4 = H, R_5 = CH$_2$OH

2. The chloroform extract of *Brucea javanica* (Simaroubaceae) furnished a phenolic coumarino-lignan cleomiscosin-A (3 0), which showed potent cytotoxicity [ED_{50} (P-388) = 0.4 µg/ml] (*55*).

3. The known phenolic dicoumaryl ether, daphnoretin (3 1), was isolated from the whole plant of *Wikstroemia indica* (Thymelaeaceae). Compound 3 1 showed significant inhibitory activity *in vivo* against the Ehrlich ascites carcinoma growth in mice (97% inhibition at 3 mg/kg) but did not demonstrate significant activity against P-388 lymphocytic leukemia growth (*56*). Compound 3 1 inhibits DNA and protein syntheses of Ehrlich ascites tumor cells (*57,58*).

Flavonoids. 1. The aforementioned *W. indica* also yielded two antileukemic flavonoids, tricin [3 2, T/C (P-388) = 133 (6 mg/kg) and 174% (12.5 mg/kg)] and kaempferol-3-*O*-β-glucopyranoside [3 3, T/C (P-388) = 130% (12.5 mg/kg)]. The good antileukemic activity demonstrated by 3 2 is noteworthy, as the cytotoxic flavonoids seldom show significant *in vivo* antileukemic (P-388) activity. It is also interesting to note that both 3 2 and 3 3 possess the same 7,5,4'-trihydroxylated pattern (*56*).

2. Hinokiflavone [3 4, ED_{50} (KB) = 2.0 µg/ml] was isolated as the cytotoxic principle from the drupes of *Rhus succedanea* (Anacardiaceae). A comparison of the cytotoxicity of 3 4 and other related biflavonoids indicates that an ether linkage between two units of apigenin as seen in 3 4 is structurally required for significant cytotoxicity (*59*).

Lignans. 1. The antitumor lignan isolated from *W. indica* mentioned above was identified as (+)-nortrachelogenin (i.e. wikstromol) (3 5) Compound 3 5 revealed activities of T/C = 122, 130, and 130% at doses of 4, 8, and 16 mg/kg, respectively, when tested against *in vivo* P-388 lymphocytic leukemia in BDF_1 mice (*56*). The reported data for 3 5 in the same *in vivo* screen were T/C = 130, 141, 137, 146, and 154% at doses of 1, 2, 4, 10, and 16 mg/kg in CDF_1 mice (*60*).

2. A bioassay-directed fractionation of an antileukemic (P-388) extract of *Justicia pro-cumbens* (Acanthaceae) has led to the isolation of diphyllin (3 6), whose potent cytotoxicity [ED_{50} (KB) < 1.0 µg/ml] was shown for the first time. A comparison of the cytotoxicity of 3 6 and its related compounds indicated that the γ-lactone ring carbonyl α to C-3 instead of C-2 is required for the potent cytotoxicity (*61*).

3. The known phenolic lignan, α-peltatin [3 7, ED_{50} (KB) = 0.09 µg/ml], was isolated as a potent cytotoxic principle from *Podophyllum peltatum* (Berberidaceae) (*62*).

Macrolides. Lasiodiplodin (3 8), a resorcylate macrolide, was first isolated as a potent antileukemic [T/C (P-388) > 140% (0.2 mg/kg)] agent from the stem, root, and latex of the aforementioned *E. splendens* (*63*). Compound 3 8 was previously reported as a fungal metabolite of *Lasiodiplodia theobromae* (*64*).

Naphthoquinones. The binaphthoquinone isodiospyrin [3 9; ED_{50} (HCT-8) = 4.9 µg/ml, (P-388) = 0.59 µg/ml], isolated from *Diospyros morrisiana* (Ebenaceae), was shown for the first time to be cytotoxic in the HCT-8 and P-388 screens (*65*).

Polyphenols. Acrovestone (4 0) was isolated from the stem and root bark of *Acronychia pedunculata* (Rutaceae). It demonstrated potent cytotoxicity for the first

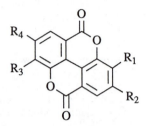

28, 3,3',4-Tri-O-methylellagic Acid : $R_1 = R_3 = R_4 = OCH_3$, $R_2 = OH$

29, Ellagic Acid : $R_1 = R_2 = R_3 = R_4 = OH$

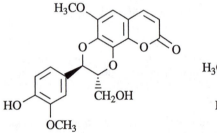
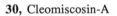

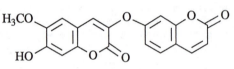

30, Cleomiscosin-A   **31,** Daphnoretin

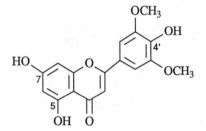

32, Tricin

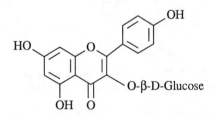

33, Kaempferol-3-O- β-D-glucopyranoside

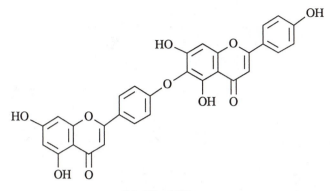

34, Hinokiflavone

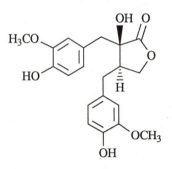

35, (+)-Nortrachelogenin
(Wikstromol)

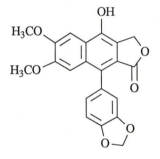

36, Diphyllin

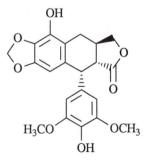

37, α-Peltatin

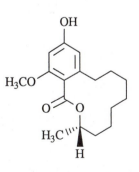

38, Lasiodiplodin

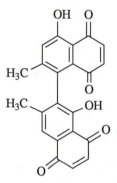

39, Isodiospyrin

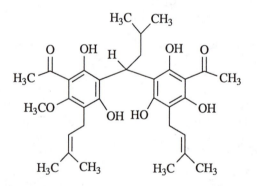

40, Acrovestone

time against KB $(ED_{50} = 0.5 \mu g/ml)$, A-549 $(ED_{50} = 0.98 \mu g/ml)$, P-388 $(ED_{50} = 3.28 \mu g/ml)$, and L-1210 $(ED_{50} = 2.95 \mu g/ml)$ tumor cells (*66*).

Conclusions

As discussed above, plant phenolics play an important role in providing new leads as cytotoxic anti-tumor agents. Some of these new leads and their analogs have already been developed for clinical use as anticancer drugs. Those cytotoxic antitumor agents discovered in the author's laboratory will undoubtedly serve continuously as templates for further synthetic modifications which may lead to more potent and less toxic clinically useful compounds. In addition, these new phenolic natural products and/or their analogs can also be used as tools for studying the biochemical mechanisms involved in the growth and control of tumors.

Acknowledgements. I would like to thank my collaborators who contributed in many ways to the completion of much of this research work. These collaborators are cited in the accompanying references. This investigation was supported by a grant from the National Cancer Institute (CA-17625).

Literature Cited

1. Neuss, N.; Gorman, M.; Johnson, I. S. *Methods in Cancer Res.*; Academic Press: New York, NY, 1967; 633-702, and literature cited therein.
2. Lomax, N. R.; Narayanan, V. L. *Chemical Structures of Interest to the Division of Cancer Treatment*; NCI: 1988; Vol. 6.
3. Barnett, C. J.; Cullinan, G. J.; Gerzon, K.; Hoying, R. C.; Jones, W. E.; Newlon, W. M.; Poore, G. A.; Robinson, R. L.; Sweeney, M. J.; Todd, G. C. *J. Med. Chem.* **1978**, *21*, 88-96, and literature cited therein.
4. Cragg, G.; Suffness, M. *Pharmac. Ther.* **1988**, *37*, 425-461, and literature cited therein.
5. *Etoposide (VP-16) - Current Status and New Developments;* Issell, B. F.; Muggia, F. M.; Carter, S. K., Eds.; Academic Press: Orlando, FL, 1984; 1-355, and literature cited therein.
6. Suffness, M.; Douros, J. In *Methods in Cancer Research;* DeVita, V. T., Jr.; Busch, H., Eds.; New York, NY, 1979; Vol. 16, 73-126.
7. Douros, J.; Suffness, M. *J. Nat. Prod.* **1982**, *45*, 1-14.
8. Driscoll, J. S. *Cancer Treat. Rep.* **1984**, *68*, 63-76.
9. *Anticancer Agents Based on Natural Product Models*; Cassady, J. M.; Douros, J. D., Eds.; Academic Press: New York, NY, 1980.
10. Cassady, J. M.; Chang, C. J.; McLaughlin, J. L. In *Natural Products as Medicinal Agents*; Beal, J. L.; Reinhard, E., Eds.; Hippokrates, Verlag, Stuttgart: 1981; 93-124, and literature cited therein.
11. Aszalos, A. *Antitumor Compounds of Natural Origin*; CRC Press: Boca Raton, FL, 1981.
12. Lee, K. H. In *Adv. in Chinese Med. Materials Res.*; Chang, H. M., Ed.; World Sci. Publ. Co.: Singapore, 1985; 353-367.
13. Pettit, G. R.; Cragg, G. M.; Herald, C. L. *Biosynthetic Products for Cancer Chemotherapy*; Elsevier: New York, NY, 1985; and four earlier volumes under the same title.
14. Lien, E. J.; Li, W. Y. *Anticancer Chinese Drugs*; Oriental Healing Arts Institute: Los Angeles, CA, 1985.
15. Lee, K. H. *Oriental Healing Arts Internat. Bull.* **1986**, *11*, 53-72.
16. Lee, K. H.; Yamagishi, T. *Abs. Chinese Med.* **1987**, *.1*, 606-625.
17. Jardine, I. In *Anticancer Agents Based on Natural Product Models*; Cassady, J.;

Douros, J., Eds.; Academic Press: New York, NY, 1980; 319-351, and literature cited therein.

18. Stahelin, H.; von Wartburg, A. In *Prog. in Drug Res.*; Jucker, E., Ed.; Birkhauser Verlag: Germany, 1989; Vol. 33, pp 169-266.
19. Chen, G. L.; Yang, L.; Rowe, T. C.; Halligan, B. D.; Tewey, K.; Liu, L. *J. Biol. Chem.* **1984**, *259*, 13560-13566.
20. Ross, W.; Rowe, T.; Glisson, B.; Yalowich, J.; Liu, L. *Cancer Res.* **1984**, *44*, 5857-5866.
21. Rowe, T.; Kuppfer, G.; Ross, W. *Biochem. Pharmacol.* **1985**, *34*, 2483-2487.
22. van Maanen, J. M. S.; De Ruiter, C.; Kootstra, P. R.; De Vries, J.; Pinedo, H. M. *Proc. Am. Assoc. Cancer Res.* **1984**, *25*, 384-384.
23. van Maanen, J. M. S.; De Ruiter, C.; Kootstra, P. R.; Broersen, J.; De Vries, J.; Laffeur, M. V. M.; Retel, J.; Kriek, E.; Pinedo, H. M. *Proc. Am. Assoc. Cancer Res.* **1986**, *27*, 308.
24. Haim, N.; Roman, J.; Nemec, J.; Sinha, B. K. *Biochem. Biophys. Res. Commun.* **1986**, *135*, 215-220.
25. van Maanen, J. M. S.; De Ruiter, C.; De Vries, J.; Kootstra, P. R.; Gobars, G.; Pinedo, H. M. *Eur. J. Cancer Chin. Oncol.* **1985**, *21*, 1099-1106.
26. Lee, K. H.; Beers, S. A.; Mori, M.; Wang, Z. Q.; Kuo, Y. H.; Li, L.; Liu, S. Y.; Chang, J. Y.; Han, F. S.; Cheng, Y. C. *J. Med. Chem.* **1990**, *33*, 1364-1368.
27. Wang, Z. Q.; Kuo, Y. H.; Schnur, D.; Bowen, J. P.; Liu, S. Y.; Han, F. S.; Chang, J. Y.; Cheng, Y. C.; Lee, K. H. *J. Med. Chem.* **1990**, *33*, 2660-2666.
28. Chang, J. Y.; Han, F. S.; Liu, S. Y.; Wang, Z. Q.; Lee, K. H.; Cheng, Y. C. *Cancer Res.* **1991**, *51*, 1755-1759.
29. Wall, M. E.; Wani, M. C. In *Anticancer Agents Based on Natural Product Models*; Cassady, J. M.; Douros, J. D., Eds.; Academic Press: New York, NY, 1980; 417-436.
30. Xu, B. *Trends in Pharmacol. Sci.*; 1981; 271-274.
31. Xu, B.; Yang, J. L. *Adv. Chin. Med. Materials Res.*; Chang, H. M.; Yeung, H. W.; Tso, W. W.; Koo, A., Eds.; World Sci. Publ. Co.: Singapore, 1985; 377-389.
32. Hsiang, Y. H.; Hertzberg, R.; Hecht, S.; Liu, L. F. *J. Biol. Chem.* **1985**, *260*, 14873.
33. Hsiang, Y. H.; Liu, L. F.; Wall, M. E.; Wani, M. C.; Nicholas, A. W.; Manikumar, G.; Kirschenbaum, S.; Silber, R.; Potmesil, M. *Cancer Res.* **1989**, *49*, 4385-4389.
34. Johnson, P. K.; McCabe, F. L.; Faucette, L. F.; Hertzberg, R. P.; Kingsbury, W. D.; Boehm, J. C.; Caranfa, M. J.; Holden, K. G. *Proc. Amer. Assoc. Cancer Res.* **1989**, *30*, 623.
35. Wani, M. C.; Nicholas, A. W.; Wall, M. E. *J. Med. Chem.* **1986**, *29*, 2358-2363.
36. Fukuoka, M.; Negoro, S.; Niitani, H.; Taguchi, T. *Proc. Amer. Soc. Clin. Oncol.* **1990**, *9*, 874-884.
37. *Washington Insight*, Vol. 2, Dec. 15, 1989, p 6, and Vol. 3, March 15, 1990, p 6.
38. The Merck Index, 11th Edition; Merck & Co., Inc.: Rahway, NJ, 1989; pp 4435 & 5501, and literature cited therein.
39. *Yao Ping Zhi, Vol. 5, Antitumor Drugs*; Shanghai Institute for Med. & Pharm. Ind., Ed.; Shanghai Sci. & Tech. Publ. Co.: Shanghai, China, 1983.
40. Xiao, P. In *Natural Products as Medicinal Agents*; Beal, J. L.; Reinhard, E., Eds.; Hippokrates Verlag Stuttgart, 1981; 351-394.
41. Lin, Y. M.; Juichi, M.; Wu, R. Y.; Lee, K. H. *Planta Medica* **1985**, *6*, 545-546.
42. Geran, R. I.; Greenberg, N. H.; MacDonald, M. M.; Schumacher, A. M.; Abbott, B. J. *Cancer Chemother. Rep. (Part 3)* **1972**, *3*, 1-5.
43. Shoemaker, R. H.; Monks, A.; Alley, M. C.; Scudiero, D. A.; Fine, D. L.;

McLemore, T. L.; Abbott, B. J.; Paull, K. D.; Mayo, J. G.; Boyd, M. R. *Prediction of Response to Cancer Therapy* **1988**, 265-286.
44. Alley, M. C.; Scudiero, D. A.; Monks, A.; Hursey, M. L.; Czerwinski, M. J.; Fine, D. L.; Abbott, B. J.; Mayo, J. C.; Shoemaker, R. H.; Boyd, M. R. *Cancer Res.* **1988**, *48*, 589-601.
45. Scudiero, D. A.; Shoemaker, R. H.; Paull, K. D.; Monks, A.; Tierney, S.; Nofzigger, T. H.; Currens, M. J.; Seniff, D.; Boyd, M. R. *Cancer Res.* **1988**, *48*, 4827-4833.
46. Boyd, M. R. *Principles of Practice of Oncology* **1989**, *3*, 1-12.
47. Wu, Y. C.; Lu, S. T.; Chang, J. J.; Lee, K. H. *Phytochem.* **1988**, *27*, 1563-1564.
48. Fukamiya, N.; Okano, M.; Aratani, T.; Negoro, K.; McPhail, A. T.; Juichi, M.; Lee, K. H. *J. Nat. Prod.* **1986**, *49*, 428-434.
49. Fukamiya, N.; Okano, M.; Aratani, T.; Negoro, K.; Lin, Y. M.; Lee, K. H. *Planta Medica* **1987**, *53*, 140-143.
50. Wu, Y. C.; Liou, Y. F.; Lu, S. T.; Chen, C. H.; Chang, J. J.; Lee, K. H. *Planta Medica* **1989**, *55*, 163-165.
51. Chang, P.; Lee, K. H.; Shingu, T.; Hirayama, T.; Hall, I. H.; Huang, H. C. *J. Nat. Prod.* **1982**, *45*, 206-210.
52. Chang, P.; Lee, K. H. *Phytochem.* **1984**, *23*, 1733-1736.
53. Chang, P.; Lee, K. H. *J. Nat. Prod.* **1985**, *48*, 948-951.
54. Sayer, J. M.; Yagi, H.; Wood, A. W.; Conney, A. H.; Jerina, D. M. *J. Amer. Chem. Soc.* **1982**, *104*, 5562-5564.
55. Lee, K. H.; Hayashi, N.; Okano, M.; Nozaki, H.; Juichi, M. *J. Nat. Prod.* **1984**, *47*, 550-551.
56. Lee, K. H.; Tagahara, K.; Suzuki, H.; Wu, R. Y.; Haruna, M.; Hall, I. H.; Huang, H. C.; Ito, K.; Iida, T.; Lai, J. S. *J. Nat. Prod.* **1981**, *44*, 530-535.
57. Hall, I. H.; Tagahara, K.; Lee, K. H. *J. Pharm. Sci.* **1982**, *71*, 741-744.
58. Liou, Y. F.; Hall, I. H.; Lee, K. H. *J. Pharm. Sci.* **1982**, *71*, 745-749.
59. Lin, Y. M.; Chen, F. C.; Lee, K. H. *Planta Medica* **1989**, *55*, 166-168.
60. Torrance, S. J.; Hoffmann, J. J.; Cole, J. R. *J. Pharm. Sci.* **1979**, *68*, 664-665.
61. Fukamiya, N.; Lee, K. H. *J. Nat. Prod.* **1986**, *49*, 348-350.
62. Thurston, L. S.; Irie, H.; Tani, S.; Han, F. S.; Liu, Z. C.; Cheng, Y. C.; Lee, K. H. *J. Med. Chem.* **1986**, *29*, 1547-1550.
63. Lee, K. H.; Hayashi, N.; Okano, M.; Hall, I. H.; Wu, R. Y.; McPhail, A. T. *Phytochem.* **1982**, *21*, 1119-1121.
64. Aldridge, D. C.; Galt, S.; Giles, D.; Turner, W. B. *J. Chem. Soc. C.* **1971**, 1623-1627.
65. Yan, X. Z.; Kuo, Y. H.; Lee, T. J.; Shih, T. S.; Chen, C. H.; McPhail, D. R.; McPhail, A. T.; Lee, K. H. *Phytochem.* **1989**, *28*, 1541-1543.
66. Wu, T. S.; Wang, M. L.; Jong, T. T.; McPhail, A. T.; McPhail, D. R.; Lee, K. H. *J. Nat. Prod.* **1989**, *52*, 1284-1289.

RECEIVED November 20, 1991

Chapter 30

Sarcophytol A and Its Analogs

Cancer Preventive Activity

H. Fujiki[1], M. Suganuma[1], K. Takagi[1,5], S. Nishiwaki[1], S. Yoshizawa[1],
S. Okabe[1], J. Yatsunami[1], K. Frenkel[2], W. Troll[2], J. A. Marshall[3],
and M. A. Tius[4]

[1]Cancer Prevention Division, National Cancer Center Research Institute,
Tsukiji 5-1-1, Chuo-ku, Tokyo 104, Japan
[2]Department of Environmental Medicine, New York University Medical
Center, New York, NY 10016
[3]Department of Chemistry, University of South Carolina,
Columbia, SC 29208
[4]Department of Chemistry, University of Hawaii at Manoa, 2545 The Mall,
Honolulu, HI 96822

The aim of our research is screening for new cancer chemopreventive
agents which inhibit tumor promotion in two-stage carcinogenesis
experiments on mouse skin and studying their mechanisms of action.
Sarcophytol A, isolated from the soft coral, Sarcophyton glaucum, inhibits
the tumor promoting activities of okadaic acid as well as 12-O-tetra-
decanoylphorbol-13-acetate (TPA)-type tumor promoters on mouse skin.
Based on these results, we extended the study to its analogs, (+)-α-2, 7,
11-cembratriene-4, 6,-diol (α-CBT), 3, 7, 11-trimethylcyclodeca-3E, 7E,
11E-triene-1-ol (Compound 1) and 2, 8, 12-trimethyldeca-1, 5Z, 7E, 11-
tetraene-4-ol (Compound 4). Compound 1 inhibited tumor promotion of
okadaic acid on mouse skin more strongly than sarcophytol A or α-CBT.
Compound 4 was less effective than sarcophytol A or α–CBT. A diet
containing 0.05% sarcophytol A extended survival of AKR mice by 5
weeks. These mice die due to development of spontaneous thymic
lymphoma. Distribution of ^{3}H-sarcophytol A after a single oral administration was also studied. In addition, we briefly reviewed the inhibitory
effects of sarcophytol A on chemical carcinogenesis in large bowel cancer
and spontanenous tumor development in the mammary gland and liver in
mice. It is worthwhile to investigate Compound 1, sarcophytol A and α–
CBT as promising cancer chemopreventive agents.

Sarcophytol A and Its Analogs

In 1979, Kobayashi and his colleagues found significant amounts of cembrane-type
diterpenes in the lipid extract of the soft coral, Sarcophyton glaucum, and
characterized the structures of the four main compounds, based on the spectral data
and degradative studies by ozonolysis (1). Sarcophytol A, with a molecular weight
of 288, was one of the main compounds (Fig. 1). In 1985, Mizusaki and his

[5]Current address: Gifu University School of Medicine, Gifu 500, Japan

0097–6156/92/0507–0380$06.00/0

colleagues isolated two diastereoisomers of 2, 7, 11-cembratriene-4, 6-diol (α- and β-CBT) from the neutral fractions of cigarette smoke condensate (Fig. 1), that inhibited induction of early antigen of Epstein-Barr virus in lymphoblastoid Raji cells by 12-O-tetradecanoylphorbol-13-acetate (TPA) (Fig. 2) and inhibited tumor promotion of TPA in a two-stage carcinogenesis experiment on mouse skin (2). Sarcophytol A is structurally similar to α-CBT (Fig. 1). In 1989, our group in Tokyo reported that sarcophytol A inhibits tumor promotion induced by teleocidin, one of the TPA-type tumor promoters on mouse skin (3). In 1989, J. A. Marshall and E. D. Robinson succeeded in the total synthesis of (+)-α-CBT from the achiral 17-membered ketone 3 by a sequence featuring asymmetric reduction, diastereoselective Wittig ring contraction and hydroxyl directed epoxidation (4). In 1990, Takayanagi et al. reported the stereo- and enantioselective total synthesis of sarcophytol A (5). Tius et al. had already synthesized the other two analogs of sarcophytol A, Compound 1 and Compound 4[6] (Fig. 1). Compound 1 is 3, 7, 11-trimethylcyclodeca-3E, 7E, 11E-triene-1-ol, which differs from sarcophytol A in lacking the isopropyl group and one alkene group. Compound 4 is 2, 8, 12-trimethyldeca-1, 5Z, 7E, 11-tetraene-4-ol, namely, an acyclic analog. In this experiment, sarcophytol A isolated from soft coral was used, and the other analogs used were chemically synthesized.

Inhibition of Tumor Promotion Induced by Okadaic Acid

Okadaic acid is a polyether compound of a C_{38} fatty acid (Fig. 2), isolated from the black sponge, Halichondria okadai (6), and a new tumor promoter as strong as TPA on female CD-1 mouse skin initiated with 7, 12-dimethylbenz(a)anthracene (DMBA) (7). Okadaic acid acts differently on cells of mouse skin than TPA, namely, it inhibits the activities of protein phosphatases 1 and 2A, resulting in an increase of phosphoproteins, which is called the apparent activation of protein kinases (8) (Fig. 3). Since the okadaic acid pathway is a general mechanism of tumor promotion in various organs, such as mouse skin, rat glandular stomach and rat liver, we think that the okadaic acid pathway is similar to the process of tumor promotion in human cancer development (9). In this experiment, therefore, we studied inhibition of tumor promotion of okadaic acid by sarcophytol A and its analogs. According to the two-stage carcinogenesis experiment, mouse skin was initiated with a single application of 100 μg DMBA followed by repeated applications of 1 μg (1.2 nmol) okadaic acid twice a week. Sarcophytol A and its analogs were applied 15 min before each application of okadaic acid. The amount of sarcophytol A and its analogs applied was 10 times (12 nmol) more than that of okadaic acid in this experiment (Table 1). Figure 4 shows percentages of tumor-bearing mice as well as average numbers of tumors per mouse. The control group treated with DMBA and okadaic acid resulted in 86.7% of mice with tumors and 4.7 average numbers of tumors per mouse in week 20 of tumor promotion. Table 1 summarizes the inhibition of tumor promotion of okadaic acid by sarcophytol A and its analogs. α-CBT inhibited tumor promotion of okadaic acid as effectively as sarcophytol A in week 20. Compound 4, an acyclic analog, was less effective than sarcophytol A, indicating that there is a cyclic structural requirement for inhibitory activity. The inhibitory activity of Compound 1 was exciting. The treatment with Compound 1 reduced the percentage of tumor-bearing mice from 86.7% to 26.7% and the average numbers of tumors per mouse from 4.7 to 0.3 in week 20 (Table 1). The absence of the isopropyl group and one alkene group in Compound 1 enhanced the inhibitory activity and was more effective than sarcophytol A and α-CBT. Compound 1 is a synthetic compound and is not naturally occurring. As for the mechanisms of action of sarcophytol A, we studied how sarcophytol A interacts with the tumor promoting process of okadaic acid. The

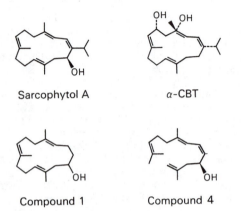

Sarcophytol A   α-CBT

Compound 1   Compound 4

Fig. 1.   Structures of sarcophytol A and its analogs.

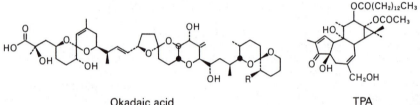

Okadaic acid   TPA

Fig. 2.   Structures of the tumor promoters okadaic acid and TPA.

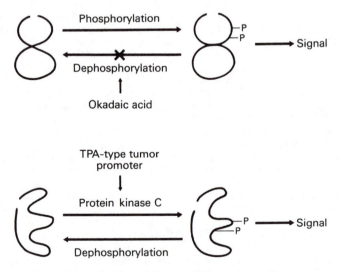

Fig. 3.   Mechanisms of action of the two different types of tumor promoters. Reproduced with permission from ref. 19. Copyright 1990 University of Nagoya Press.

Table 1   Inhibition of Tumor Promotion of Okadaic Acid by
Sarcophytol A and Its Analogs

Inhibitors	Amount per application μg	% of tumor-bearing mice in week 20	Average numbers of tumors per mouse in week 20
Control		86.7	4.7
Compound 4	2.8	73.3	2.7
α-CBT	3.7	46.7	2.0
Sarcophytol A	3.5	46.7	1.5
Compound 1	3.0	26.7	0.3

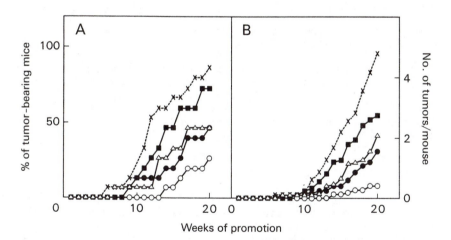

Fig. 4.   Inhibitory effects of sarcophytol A and its analogs on tumor promotion of
okadaic acid. The groups treated with DMBA and okadaic acid (x); DMBA
and okadaic acid plus Compound 4 (■), DMBA and okadaic acid plus α-CBT
(△), DMBA and okadaic acid plus sarcophytol A (●) and DMBA and okadaic
acid plus Compound 1 (○).

specific ^{3}H-okadaic acid binding to the particulate fraction of mouse skin was not inhibited by sarcophytol A at concentrations of up to 1 mM. Sarcophytol A also did not reverse the inhibition by okadaic acid of protein phosphatase 1 and 2A activities, whose enzymes are receptors of okadaic acid (10). Therefore, we think that sarcophytol A inhibits the process of tumor promotion after the binding of okadaic acid to protein phosphatases.

Inhibition of Tumor Promotion of TPA-type Compounds

As we previously reported, sarcophytol A inhibited tumor promotion by teleocidin (3). Sarcophytol A also inhibited tumor promotion by the TPA-types, TPA and aplysiatoxin (data not shown). We previously demonstrated that sarcophytol A inhibited H_2O_2 formation by TPA-activated human polymorphonuclear leukocytes (11). We extended the study to its analogs. Inhibitory potency of Compound 1 was much stronger than sarcophytol A. Therefore, the order of their inhibitory potency against H_2O_2 formation: Compound 1 > sarcophytol A >, probably , α–CBT > Compound 4 (data not shown), is comparable to that of their inhibitory activity of tumor promotion, although TPA rather than okadaic acid was used as a tumor promoter in the experiment. We think Compound 1 might inhibit tumor promotion of the TPA-types more strongly than sarcophytol A.

Inhibitory Effects of Spontaneous Thymic Lymphoma in AKR mice

It is well known that AKR mice die due to compression of the trachea by enlarged thymic lymphoma within one year (12). To test the effects of sarcophytol A on development of spontaneous thymic lymphoma, female AKR mice were purchased from the Jackson Laboratory, Bar Harbor, ME, U.S.A. The control group (29 mice) was given a basal diet. The experimental group (29 mice) was given a diet containing 0.05% sarcophytol A from 7 weeks of age. Figure 5 shows percentages of survival of the two groups. The 50% mortality of the control group was reached at 33 weeks of age, whereas that of the experimental group was at 38 weeks of age. Thus, the survival of AKR mice was extended 5 weeks by sarcophytol A treatment, probably by inhibiting development of thymic lymphoma. It has been reported that a diet supplemented with choline and methionine extended the survival of AKR mice (13).

Distribution of ^{3}H-Sarcophytol A after Oral Administration

A single oral administration of ^{3}H-sarcophytol A (0.255 mCi/ml in sesame oil, specific activity of 12.2 Ci/mmol) was carried out by intubation. Two mice were sacrificed by decapitation 19 hours after housing in metabolic cages, and the tissues and feces were oxidized by combustion in a Packard Sample Oxidizer 306. The radioactivity of ^{3}H$_2$O produced by oxidization derived from ^{3}H-sarcophytol A were measured by a liquid scintillation counter. Table 2 shows the distribution of radioactivity of ^{3}H-sarcophytol A 19 hours after oral administration. The yield was 71%. Radioactivity was mainly found in the feces and urine. The 50% radioactivity extracted from the feces was found to be at the same Rf as that of sarcophytol A on thin-layer chromatography, whereas the other 50% migrated to different Rfs than that of the parent compound. Radioactivity extracted from urine migrated differently than ^{3}H-sarcophytol A on thin-layer chromatography (data not shown). In addition, our autoradiographical study of the

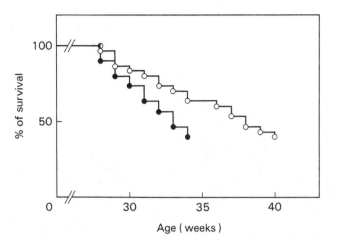

Fig. 5.   Inhibitory effects of sarcophytol A in the diet on development of spontaneous thymic lymphoma in female AKR mice. The groups treated with sarcophytol A (O) and control diet (●).

Table 2   Distribution of Radioactivity 19 Hours
After Oral Administration of
^{3}H-Sarcophytol A in Mice

Total 5.6 x 10^7 dpm

	% of Radioactivity
Feces	53.60
Urine	16.10
Liver	0.60
Stomach	0.30
Skin	0.30
Lung	0.09
Kidney	0.05
Spleen	0.01
Total yield	71.05

distribution of ^{3}H-sarcophytol A supported the evidence that most ^{3}H-sarcophytol A was excreted during the first 24 hours. The rapid excretion observed strengthened our previous evidence that sarcophytol A is a compound with low toxicity potential (3).

Additional Anticarcinogenic Activities

We have previously reported anticarcinogenic activities of sarcophytol A in various organs. It is important to briefly review these anticarcinogenic effects. Dietary sarcophytol A inhibited the development of large bowel cancer induced by N-methyl-N-nitrosourea in female rats (14), spontaneous mammary tumors in female SHN mice (15) and spontaneous liver tumors in male C3H/HeN mice (16). However, sarcophytol A was not effective at inhibiting duodenal carcinogenesis by N-ethyl-N'-nitro-N-nitrosoguanidine in male C57BL/6 mice. (Yamane, unpublished results). Recently, we have demonstrated that the inhibitory effect of sarcophytol A is enhanced by cotreatment with medroxyprogesterone acetate, an angiogenesis inhibitor, on tumor promotion by okadaic acid in mouse skin (17). As Moon and Mehta reported, combination chemoprevention approaches to various organs seems to be promising (18). Compound 1 indicates there is great potential in pursuing combination approaches to cancer chemoprevention.

Acknowledgments

This work was supported in part by Grants-in-Aid for cancer research from the Ministry of Education, Science and Culture, and the Ministry of Health and Welfare for a Comprehensive 10-Year Strategy for Cancer Control, Japan, and by grants from the Foundation for Promotion of Cancer Research, the Princess Takamatsu Cancer Research Fund, the Uehara Memorial Life Science Foundation and the Smoking Research Foundation. We thank Dr. T. Sugimura at National Cancer Center for his encouragement of the work and Dr. H. Takahara at the Smoking Research Foundation for his interest in the work. The authors (K. T., S. N. and J. Y.) thank the Foundation for Promotion of Cancer Research, Japan for support of their work at the National Cancer Center Research Institute, Tokyo. Racemic Compounds 1 and 4 were prepared by Jean M. Cullingham and Xue-quin Gu, University of Hawaii at Manoa.

Literature Cited

1. Kobayashi, M.; Nakagawa, T.; Mitsuhashi, H. *Chem. Pharm. Bull. (Tokyo)* **1979**, 27, 2382.
2. Saito, Y.; Takizawa, H.; Konishi, S.; Yoshida, D.; Mizusaki, S. *Carcinogenesis* **1985**, 6, 1189.
3. Fujiki, H.; Suganuma, M.; Suguri, H.; Yoshizawa, S.; Takagi, K.; Kobayashi, M. *J. Cancer Res. Clin. Oncol.* **1989**, 115, 25.
4. Marshall, J. A.; Robinson, E. D.. *Tetrahedron Lett.* **1989**, 30, 1055.
5. Takayanagi, H.; Kitano, Y.; Morinaka, Y. *Tetrahedron Lett.* **1990**, 31, 3317.
6. Tachibana, Y.; Scheuer, P. J.; Tsukitani, Y.; Kikuchi, H.; Van Engen, D.; Clardy, J.; Gopichand, Y.; Schmitz, F. J. *J. Am. Chem. Soc.* **1981**, 103, 2469.
7. Suganuma, M.; Fujiki, H.; Suguri, H.; Yoshizawa, S.; Hirota, M.; Nakayasu, M.; Ojika, M.; Wakamatsu, K.; Yamada, K.; Sugimura, T. *Proc. Natl. Acad. Sci. U.S.A.* **1988**, 85, 1768.

8. Sassa, T.; Richter, W. W.; Uda, N.; Suganuma, M.; Suguri, H.; Yoshizawa, S.; Hirota, M.; Fujiki, H. *Biochem. Biophys. Res. Commun.* **1989**, 159, 939.
9. Fujiki, H.; Suganuma, M.; Nishiwaki, S.; Yoshizawa, S.; Yatsunami, J.; Matsushima, R.; Furuya, H.; Okabe, S.; Matsunaga, S.; Sugimura, T. In *Relevance of Animal Studies to Evaluate Human Cancer Risk;* D'Amato, R., Slaga, T. J., Farland, W., Henry, C., Eds.; John Wiley & Sons, Inc. New York, N. Y. in press.
10. Nishiwaki, S.; Fujiki, H.; Suganuma, M.; Ojika, M.; Yamada, K.; Sugimura, T. *Biochem. Biophys. Res. Commun.* **1990**, 170, 1359.
11. Frenkel, K.; Zhong, Z.; Rashid, K.; Fujiki, H. In *Anticarcinogenesis and Radiation Protection: Strategies in Protection from Radiation and Cancer;* Nygaard, O. F., Ed.; Plenum Press, New York, in press.
12. Gross, L. In *Mouse Leukemia in Oncogenic Viruses*, Vol 1; Pergamon Press, New York, **1983**, p. 305.
13. Wainfan, E.; Dizik, M.; Kelkenny, M.; O'Callaghan, J. P. *Carcinogenesis* **1990**, 11, 361.
14. Narisawa, T.; Takahashi, M.; Niwa, M.; Fukaura, Y.; Fujiki, H. *Cancer Res.* **1989**, 49, 3287.
15. Fujiki, H.; Suganuma, M.; Suguri, H.; Takagi, K.; Yoshizawa, S.; Ootsuyama, A.; Tanooka, H.; Okuda, T.; Kobayashi, M.; Sugimura, T. In *Antimutagenesis and Anticarcinogenesis Mechanisms II*; Kuroda, K., Shankel, D. M., Waters, M. D., Eds.; Plenum Press, New York, London, **1990**, p. 205.
16. Yamauchi, O.; Omori, M.; Ninomiya, M.; Okuno, M.; Moriwaki, H.; Suganuma, M.; Fujiki, H.; Muto, Y. *Jpn. J. Cancer Res.* **1991**, 82, 1234.
17. Suganuma, M.; Yoshizawa, S.; Yatsunami, J.; Nishiwaki, S.; Furuya, H.; Okabe, S.; Nishiwaki-Matsushima, R.; Frenkel, K.; Troll, W.; Verma, A. K.; Fujiki, H. In *Antimutagenesis and Anticarcinogenesis Mechanisms III;* Bronzetti, G., De Flora, S., Shankel, D. M., Waters, M. D., Eds.; Plenum Press, New York, London, in press.
18. Moon, R. C.; Mehta, R. G. In *Antimutagenesis and Anticarcinogenesis Mechanisms II;* Kuroda, K., Shankel, D. M., Waters, M. D., Eds. ; Plenum Press, New York, London, **1990**, p. 213.
19. Fujika, H. et al. In *Epidemiology and Prevention of Cancer;* Sasaki, R. and Aoki, K., Eds. University of Nagoya Press: Nagoya, Japan; 1990, pp36–41.

RECEIVED January 13, 1992

Author Index

Affiliation Index

Subject Index

Production: Peggy D. Smith
Indexing: Deborah H. Steiner
Acquisition: Barbara C. Tansill
Cover design: Pat Cunningham

Printed and bound by Maple Press, York, PA